Communications for Control in Cyber Physical Systems

Communications for Control in Cyber Physical Systems

Theory, Design and Applications in Smart Grids

Husheng Li

AMSTERDAM • BOSTON • HEIDELBERG • LONDON
NEW YORK • OXFORD • PARIS • SAN DIEGO
SAN FRANCISCO • SINGAPORE • SYDNEY • TOKYO
Morgan Kaufmann is an imprint of Elsevier

Morgan Kaufmann is an imprint of Elsevier
50 Hampshire Street, 5th Floor, Cambridge, MA 02139, USA

Library of Congress Cataloging-in-Publication Data
A catalog record for this book is available from the Library of Congress

British Library Cataloguing-in-Publication Data
A catalogue record for this book is available from the British Library

ISBN: 978-0-12-801950-4

For information on all Morgan Kaufmann publications
visit our website at https://www.elsevier.com/

Publisher: Todd Green
Acquisition Editor: Brian Romer
Editorial Project Manager: Amy Invernizzi
Production Project Manager: Mohana Natarajan
Cover Designer: Vicky Pearson Esser

To my dear wife, Min Duan, and my lovely children, Siyi Li and Mohan Li.

Contents

Biography

Husheng Li received BS and MS degrees in electronic engineering from Tsinghua University, Beijing, China, in 1998 and 2000, respectively, and a PhD degree in electrical engineering from Princeton University, Princeton, NJ, in 2005. From 2005 to 2007, he worked as a senior engineer at Qualcomm Inc., San Diego, CA. In 2007, he joined the EECS Department of the University of Tennessee, Knoxville, TN, as an assistant professor. He became an associate professor in 2013. His research is mainly focused on wireless communications, cyber physical systems, networking, statistical signal processing, smart grids, and game theory. He has received the Best Paper Awards from EURASIP Journal of Wireless Communications and Networks 2005, EURASIP Journal of Advances in Signal Processing 2015, IEEE ICC 2011, and IEEE SmartGridComm 2012 and the Best Demo Award of IEEE Globecom 2010.

Preface

It is prohibitively difficult to introduce all the aspects of a cyber physical system (CPS), since it consists of many technologies such as computing, control and communications, and concerns both theoretical elements such as information theory in the context of control and practical issues such as software or hardware design. In this book, I focus only on the interplay of communications and control in CPSs, in particular how to design a communication system for the purpose of control, from the viewpoint of a communications engineer or scientist. Of course this does not mean that this book will benefit only the communications community. Experts on controls or computing, theorists or practical engineers may find interesting points in this book since it is a multidisciplinary subject.

This book consists of two parts. In the first part, which is composed of the first four chapters, the basics about CPS, communications, and control are briefly introduced. By reading through these four chapters, working knowledge can be acquired for readers unfamiliar with both communications and control. Then, the second part (the remaining four chapters), which is the main part of this book, explains various aspects of communication design for controlling CPSs, namely the capacity requirements, network topology design, networking, and physical layer design. These subjects are discussed specifically in the context of smart grids, using illustrated examples.

The writing of this book and many research results of mine incorporated in this book were supported by the National Science Foundation (NSF) under grants CNS-1239366, ECCS-1407679, CNS-1525418, and CNS-1543830. I sincerely thank the NSF for this generous support. I also sincerely thank my colleagues, Prof. Kevin Tomsovic and Prof. Seddik Djouadi, who collaborated with me on this NSF project and taught me the basics of power systems and automatic control. Last but not least, I would like to thank my family: without their full support, I could not have written this book.

Husheng Li

CHAPTER

Introduction to cyber physical systems

1

1.1 INTRODUCTION

In the last decade, there have been numerous studies on cyber physical systems (CPSs), illustrated in Fig. 1.1. In CPSs, there exist physical, computing, and communication elements. To control the physical dynamics, sensors and controllers are connected via a communication network over which measurements on the system state can be conveyed. Many systems can be categorized as CPSs, such as smart grids, unmanned aerial vehicles, and robotic networks. The study on CPSs can provide a comprehensive and interdisciplinary framework for analyzing and designing these practical systems.

As a special example of CPS, smart grids may consist of a wide area monitoring system (WAMS), which measures the synchronized metrics such as voltage, currents, phase and frequency, and a set of controllers, e.g., the flexible AC transmission system [1]. In a microgrid, multiple distributed energy resources (DERs), such as wind turbines, are connected to a main power network; hence the power electronics interfaces of these DERs, which convert DC to AC power, should be controlled using the voltage measurements at the points of common coupling (PCCs; i.e., the connection point between the DER and the main power grid), which requires a communication link for the measurement feedback [2]. Networked control in smart grids can also be found in many other operations such as power consumption marketing (e.g., how to set a dynamic price for the electric power) and protective relays (which are triggered to isolate possible faulting devices when faults occur). In these CPSs, the communication infrastructure is of key importance since it plays a role analogous to that of the nervous system in the human body. When the communication system malfunctions, severe damage can be incurred to the dynamical system. Most existing studies on smart grids focus on isolated topics. Hence it is necessary to apply the framework of CPSs for holistic analysis and design.

Communications for Control in Cyber Physical Systems. http://dx.doi.org/10.1016/B978-0-12-801950-4.00001-9

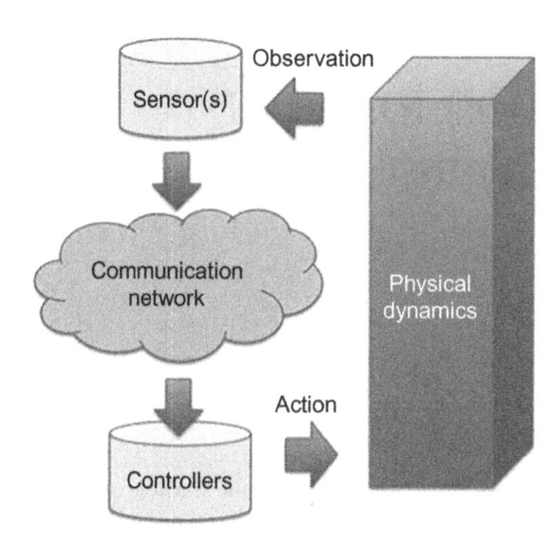

FIG. 1.1

Illustration of a CPS.

1.2 ELEMENTS OF A CPS

A typical CPS is illustrated in Fig. 1.1. It consists of the following components [3]:

- Physical dynamics: The "P" aspect of CPS is characterized by the physical dynamics (also called plant in the community of controls and systems), which could be the frequency fluctuations in power grids, or positions and velocities of robots, or vehicle densities along a highway. The physical dynamics change with time, and the evolution law is determined by the physical system itself and the control actions.
- Sensors: Sensors are used to sense the physical dynamics. The observations may be direct, i.e., the system state of physical dynamics, or indirect observations. There may be multiple sensors in a CPS. In smart grids, a sensor might be a synchronized phasor that samples the electric dynamics and computes the corresponding frequencies and phases.
- Controllers: A controller receives reports from sensors; it then computes and carries out the control action. Sometimes, the computation and actuation of the control action may be in different locations, thus needing communications. In this paper, we assume that the controller can carry out computation and actuation simultaneously. In smart grids, a frequency controller can compute the governor valve (which regulates the flow of steam pushing the mover) position and actuate it, based on the reports of frequencies in the power grid.
- Communication networks: If the sensors and controllers are not co-located, a communication network is needed for the sensors to send their reports to the controllers. The communication network can be either wired or wireless.

Dynamics of CPS

In a typical CPS such as power grids [4] and robotic networks [5], the physical dynamics can be described by a differential equation:

$$\frac{d\mathbf{x}}{dt} = \tilde{f}(\mathbf{x}(t), \mathbf{u}(t), \mathbf{n}(t)), \quad (1.1)$$

where $\mathbf{x}$ is the system state, $\mathbf{u}$ is the control action, $\mathbf{n}$ is the random noise, and $\tilde{f}$ is the physical law governing the evolution of system dynamics. If a discrete-time domain is considered, a difference equation can be used to describe the system in a similar manner:

$$\mathbf{x}(t+1) = f(\mathbf{x}(t), \mathbf{u}(t), \mathbf{n}(t)). \quad (1.2)$$

Such a differential equation description of dynamical systems has been used since the original use of Lagrangian mechanics and Hamiltonian mechanics, and has thus been studied intensively [6].

Note that there are other descriptions of physical dynamics, such as:

- Discrete-time systems such as Petri nets and stochastic timed automata, which are summarized in Ref. [7].
- Differential-algebraic equations, which consider both continuous-time and discrete events [8].
- Symbolic dynamical systems, which study the dynamics using graphs [9].

Since dynamical systems are a very broad area of study, it is prohibitively difficult to provide an exhaustive introduction to dynamical systems. A comprehensive introduction can be found in Ref. [10].

Linear time-invariant systems

In many situations, e.g., when the system is time invariant and oscillates around the equilibrium point, the system dynamics can be simplified to linear time-invariant (LTI) dynamics either in the continuous-time domain:

$$\frac{d\mathbf{x}(t)}{dt} = \tilde{\mathbf{A}}\mathbf{x}(t) + \tilde{\mathbf{B}}\mathbf{u}(t) + \mathbf{n}(t), \quad (1.3)$$

or in the discrete-time domain:

$$\mathbf{x}(t+1) = \mathbf{A}\mathbf{x}(t) + \mathbf{B}\mathbf{u}(t) + \mathbf{n}(t), \quad (1.4)$$

where $\tilde{\mathbf{A}}$, $\tilde{\mathbf{B}}$, $\mathbf{A}$, and $\mathbf{B}$ are time-invariant parameters of the system. The parameters of the continuous- and discrete-time systems are related by

$$\begin{cases} \mathbf{A} = e^{\tilde{\mathbf{A}}T_s}, \\ \mathbf{B} = \int_0^{T_s} e^{\tilde{\mathbf{A}}\tau} d\tau \tilde{\mathbf{B}}, \end{cases} \quad (1.5)$$

where T_s is the sampling period of the discrete-time system. Note that LTI systems are widely used in many practical scenarios due to their simplicity. Even if a system is nonlinear, it can still be approximated by a linear system near its operation point.

Observation model

The controller(s) in a CPS needs observations on the system dynamics to estimate and control the system state. In the generic case, the observation is a function of the system state and random perturbations. For simplicity, we consider a linear observation model in this chapter:

$$\mathbf{y}(t) = \mathbf{C}\mathbf{x}(t) + \mathbf{w}(t), \tag{1.6}$$

where $\mathbf{y}$ is the observation, $\mathbf{w}$ is random measurement noise, and $\mathbf{C}$ is the measurement matrix. Different sensors may observe different dimensions of $\mathbf{y}$.

Feedback control

The observation $\mathbf{y}$ is sent from the sensor(s) to the controller(s) via the communication infrastructure in order to control the system dynamics (e.g., stabilize it). We denote by $\hat{\mathbf{y}}$ the observations received at the controller, which may have experienced quantization error, delay or packet loss. An estimate of the system state, denoted by $\hat{\mathbf{x}}$, is made at the controller(s), which is a function of $\hat{\mathbf{y}}$. For simplicity we assume that a linear feedback control is used:

$$\mathbf{u} = -\mathbf{K}\hat{\mathbf{x}}, \tag{1.7}$$

where $\mathbf{K}$ is the feedback gain matrix. There are many approaches for the synthesis of $\mathbf{K}$, e.g., the linear quadratic regulator (LQR) in which the cost function is a quadratic function of the system state and control actions. We will use LQR in our subsequent discussion due to its elegant framework and its capability to handle stochastic dynamics. The same principle can be applied in other control strategies such as model predictive control (MPC).

1.2.1 COMMUNICATIONS FOR CPSs: THEORETICAL STUDIES

There have been several theoretical studies on the communications in CPSs from the viewpoint of networked control systems.

Communication requirements

A comprehensive introduction to the communication requirements in CPS, mainly in terms of communication channel capacities between the sensors and controllers, can be found in Ref. [11]. For the case of deterministic dynamical systems with an unknown initial state, it has been found that the requirement of the communication rate of sensors is dependent on the topological entropy [12] of the system dynamics [11]. The topological entropy is a metric measuring the uncertainty in the generation rate of a given dynamical system. Information theoretic arguments are also introduced to study the control through communication channels, such as the anytime capacity of a channel [13] and communication for stochastic linear control [14]. Another approach to analyze the communication requirements is to apply the second law of thermodynamics and use the reduction of entropy in the physical dynamics as a lower bound of the communication requirement [15].

Impact of communications on control

Given the communication network connecting the sensors and controllers in a CPS, the corresponding impacts of communication imperfections have been studied in the realm of networked control systems. The major impacts of communications include:

- Communication delay: Communication delay introduces memory to the system dynamics. Hence, extra variables need to be added to the system dynamics. When the communication delay is a constant, Ref. [16] provided the equivalent system dynamics, as well as the criterion for system stability. For the case of random delays, the corresponding system stability has been studied in Ref. [11].
- Packet drop: For random packet drops, the system can be considered as a Markovian jumping process whose equivalence can be found in Ref. [17]. Then, the stability conditions for Markovian jumping processes can be applied to analyze the system dynamics with random packet drops.
- Quantization: Quantization introduces errors into the measurement reports conveyed to the controllers. Hence the analysis can be based on small perturbations. Some related analysis, which analytically shows the impact of quantization on the control of a system, can be found in Ref. [18].

When more than one communication imperfection is present, the analysis will be more complicated, although still mathematically tractable [19]. For example, when there exist both significant packet drops and communication delay, the Markov jumping process can be applied to the equivalent system incorporating the communication delay, which makes the equivalent system complicated and possibly unsolvable.

Communication design

Seminal works have been carried out for communication system design in CPSs [20–22]. In Ref. [20], the MAC layer design is studied for distributed control systems, mainly based on numerical simulations. It is extended to the wireless communication in CPS within the framework of the linear quadratic Gaussian (LQG), in which the system dynamics are linear, the cost function is a quadratic one of the system state and control actions, and the random noise is assumed to be Gaussian [21]. In Ref. [22], the resource allocation in communication systems for CPS is formulated as an optimization problem with the awareness of the control system. In recent years, there have been more studies on this topic. For example, in Ref. [23], the IEEE 802.11 standard is applied and tuned (e.g., adjusting the network operation parameters such as the retransmission limit) for networked control systems. In Ref. [24], the routing problem (i.e., finding the path(s) from information source(s) such as sensors to information sink(s) such as controllers) is studied for the purpose of system state estimation. In Ref. [25], the communication topology is designed for distributed control, where the communication delay is omitted, which is reasonable when the delay is very small. In Ref. [26], various uncertainties, such as the uncertainties of delays and packet drops, in wireless networking endangering real-time estimation and control have been addressed.

1.2.2 COMMUNICATIONS FOR CPS: INDUSTRIAL SYSTEMS

Communication systems for industrial control systems are typical examples of those for CPS. Compared with commercial data communication systems such as cellular systems, the communication networks for industrial control systems have extra requirements such as predictable throughput, extremely low down time, operation in hostile environments, scalability and maintenance by other than communication specialists [27]. Below we introduce two types of popular communication systems for industrial controls. A comprehensive introduction can be found in Ref. [28].

- Foundation Fieldbus [29]: Fieldbus, a family of industrial control communication systems, has been widely used in industrial controls, which has been standardized as IEC61158. The Foundation Fieldbus is Type 1 of fieldbus. The largest deployment of the Foundation Fieldbus is China Nanhai and SECCO, each having about 15,000 fieldbus devices. It contains four layers, namely the Physical, Datalink, Application, and User layers. Different from many modern communication systems, the physical layer of fieldbus uses current modulation, in which each device senses the voltage drops at the terminating resistors. The Datalink layer manages the token passing for access control. Fieldbus is real-time and reliable, which is important for control systems. However, its data rate is low, around 30 kbps. Although it is very useful in many industrial controls with slow dynamics, it cannot be used in tasks having fast dynamics and requiring high data speed.
- Supervisory control and data acquisition (SCADA) [30]: The SCADA system is used to monitor and control the dynamics of industrial infrastructures such as power grids. It consists of a human-machine interface, supervisory computer system, remote terminal units (RTUs) connecting to sensors, and communication infrastructure. The communication network is either wired or wireless. For large area networks, SONET/SDH optical communications can be used. A disadvantage of SCADA is its slow speed, which may take minutes to collect data from the power grid [31]. Hence it cannot be used in fast data communications, such as data from phasor measurement units (PMUs) that measure the instantaneous phase and frequencies of different locations of the power grid, which are important sensors in power grids and will be explained later, and real-time control.

1.3 WHAT IS INCLUDED AND WHAT IS MISSING

In this section, we provide an outline of the book and also point out what has not been included.

1.3.1 CONTENT

This book mainly consists of the following two parts:

- Basics of CPS: We give brief introductions to the basics of communications and controls in Chapters 2 and 3, thus providing working knowledge for the subsequent specific discussions. In Chapter 4, we introduce some typical CPSs, such as smart grids and aerial vehicles, which serve as working examples for the subsequent theoretical studies.
- Communication system design for CPS: Chapters 5–8 cover various topics on the design of communication systems in CPSs. It follows the order of system design. We begin with the theoretical analysis on the communication capacity requirements for the purpose of control in Chapter 5. Once the communication capacity requirement is well understood, we then consider how to design the communication network topology in Chapter 6, based on the requirement of throughput. Given a fixed communication network topology, we study the networking protocol in CPS in Chapter 7, ranging from the MAC layer to the network layer. Finally we study the physical layer design for communications in CPS, such as modulation, source coding, and channel coding. The key feature of these studies is the incorporation of the characteristics of control and physical dynamics in the design procedure of communications, thus resulting in a physical dynamics aware communication system.

1.3.2 MISSING TOPICS

This book is far from an exhaustive introduction to CPSs. The following aspects of CPSs are not included in this book, due to the scope of the authors' research and limitations of space:

- Computing in CPS: It is well recognized that computing, control, and communications are interleaved and integrated in CPS. The computing for control, and communication for computing, and the overall interactions of these three components are important for many typical CPSs. There have been numerous studies on communications for computing (e.g., the theory of communication complexity), and also some studies on the computing aspects of control and decision making (e.g., the computational complexity of Markov decision processes). However, there have been no significant studies on the corresponding interactions in the context of CPS.
- Sensor networks: This is an important topic since a CPS needs to use sensors to measure physical dynamics. There are many research issues in sensor networks, e.g., the power consumption schedule due to the limited battery power of many sensors, various networking issues (such as scheduling and routing) for communications, security issues (such as data privacy), and in-network computation, which saves communication resources. Since there have been many monographs on sensor networks, we do not provide a systematic introduction of this topic in this book.

- Security issues: Information security such as data privacy and message authentication. There have been some studies on this aspect of CPSs; e.g., data injection attack in smart grids, Global Positioning System (GPS) spoofing attacks in PMUs of power systems, and resilient control for generic CPSs. Many existing techniques in the area of cryptography, such as oblivious transfer and differential privacy, can be applied; however, these generic approaches have not been thoroughly studied in the context of CPSs.

CHAPTER

Basics of communications

2

2.1 INTRODUCTION

In this section, we introduce the basics of communications, which will serve as a foundation for the topics covered in the subsequent chapters. A reader acquainted with the area of communications can skip this chapter, or take a quick glance to review the main concepts in communications. For readers interested in more detailed explanations of communications, a guidance to further reading will be given at the end of this chapter.

This chapter is organized as follows:

1. Since a communication system essentially conveys information from source to destination, we first introduce the measure of information (i.e., how much information is carried by a message), which falls in the area of information theory.
2. Since the information must be transmitted through a certain channel, we then explain the models of typical communication channels and the corresponding channel capacity (i.e., the maximum amount of information that can be carried by the channel for reliable communications).
3. The first step of digital communications is to convert the original information source to bit sequences. Hence we will explain the basic concepts of source coding.
4. These bits are then channel coded in order to combat noise by using redundancies, and then modulated from coded bits to physical signals (such as sinusoidal signals). The schemes of typical modulation and coding will be detailed.
5. Once the point-to-point communications are understood, we turn to communication networks and explain the functionalities of different layers.
6. Finally, we provide an introduction to several typical communication systems, either wired or wireless.

Communications for Control in Cyber Physical Systems. http://dx.doi.org/10.1016/B978-0-12-801950-4.00002-0

2.2 INFORMATION MEASURES

In this section, we introduce how to measure information.

2.2.1 SHANNON ENTROPY

Information and communications are triggered by uncertainties of information source (which can be represented by a random variable or random process). When the random variable is realized and has a deterministic value, the more uncertain the random variable is, the more information we can obtain.

For a random variable X with finite alphabet $\mathcal{X}$ and probability distribution P, its Shannon entropy is defined as

$$H(X) = -\sum_{x \in \mathcal{X}} P(x) \log P(x), \tag{2.1}$$

where the unit is bit (or nat) when the logarithm base is 2 (or e). In the special case of uniform distribution, the entropy of X is given by $\log |\mathcal{X}|$, where $|\mathcal{X}|$ is the cardinality of $\mathcal{X}$. For example, if X has eight possible values, then we obtain 3 bits when X is determined.

Similarly, for two random variables X and Y, we can also define the joint information as

$$H(X;Y) = -\sum_{x \in \mathcal{X}, y \in \mathcal{Y}} P(x,y) \log P(x,y) \tag{2.2}$$

and the conditional entropy as

$$H(X|Y) = -\sum_{x \in \mathcal{X}, y \in \mathcal{Y}} P(x,y) \log P(x|y). \tag{2.3}$$

It is easy to verify the following chain rule:

$$H(X,Y) = H(X|Y) + H(Y). \tag{2.4}$$

2.2.2 MUTUAL INFORMATION

For two random variables X and Y, the mutual information is to measure the amount of randomness they share. It is defined as

$$I(X;Y) = H(X) - H(X|Y) = H(Y) - H(Y|X). \tag{2.5}$$

When $I(X;Y) = H(X)$, X and Y completely determine each other; on the other hand, if $I(X;Y) = 0$, X and Y are mutually independent. As we will see, the mutual information is of key importance in deriving the capacity of communication channels.

2.3 COMMUNICATION CHANNELS

A communication channel is the medium through which information is sent from the source to the destination, which is illustrated in Fig. 2.1. It is an abstraction of practical communication systems and can be modeled using mathematics. Once the mathematical model is set up for the communication channel, we can estimate the capacity of the channel.

2.3.1 TYPICAL CHANNELS

Below we introduce some typical communication channels and their corresponding characteristics:

- Wireless channels: The information is carried by electromagnetic waves in the open space. There are many typical wireless communication systems such as 4G cellular systems and WiFi (IEEE 802.11) systems, which enable mobile communications. However, wireless channels are often unreliable and have relatively small channel capacities.
- Optical fibers: The information can also be sent through optical fibers in wired communication networks. This communication channel is highly reliable and can support very high communication speeds. It is often used as the backbone network in a large area communication network such as the Internet.
- Underwater channels: Underwater acoustic waves can convey information with very low reliability and very low communication rates, which can be used in underwater communications such as submarine or underwater sensor communications.

2.3.2 MATHEMATICAL MODEL OF CHANNELS

A communication channel can be modeled using conditional probabilities. Given an input x, the channel output y is random due to random noise, and is characterized by the corresponding conditional probability $P(y|x)$. Several typical models of communication channels are given below, as illustrated in Fig. 2.2:

- Binary symmetric channel: Both the input and output are binary with alphabet $\{0, 1\}$. The input is swapped at the output with probability $1 - p$. Such a model

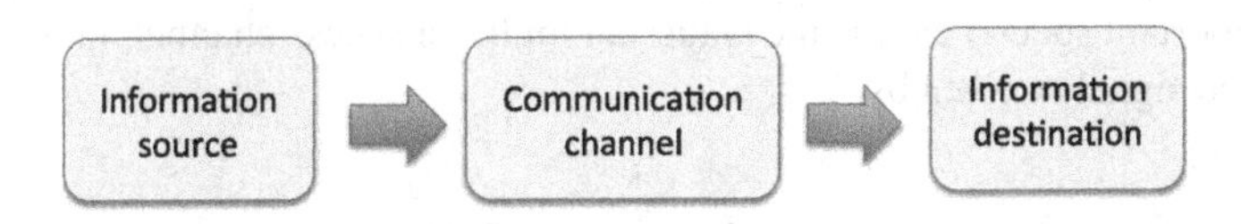

FIG. 2.1

Model of communication systems.

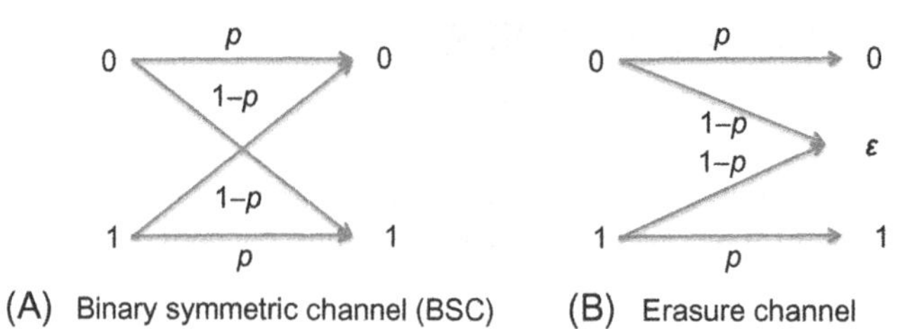

FIG. 2.2

Examples of channel models.

can be used to describe the errors in demodulation and decoding, where $1 - p$ is the error probability of each information bit.

- Erasure channel: The input is binary. A new element ϵ, which denotes the loss of information, is added to the output, besides bits 0 and 1. Such a model can depict communication systems with possible packet losses, such as the Internet.
- Gaussian channel: Both the input and output of the channel are real numbers. The output is a superposition of the input and Gaussian noise, i.e.,

$$y(t) = x(t) + n(t), \tag{2.6}$$

where t is the time index. This can be used to model a physical layer signal with Gaussian noise. Denote by P_s and σ_n^2 the powers of the signal and the noise. The signal-to-noise ratio (SNR) is defined as

$$\text{SNR} = \frac{P_s}{\sigma_n^2}. \tag{2.7}$$

As will be seen, the SNR is of key importance in determining the performance of Gaussian channels.

- Gaussian band-limited channel: This channel is continuous in time. The additive noise has power spectral density N_0 and the signal is allowed to transmit within a frequency band with bandwidth W.
- Multiple access channel: When there are multiple (say K) transmitters and a single receiver, the transmitted signals will have an effect on each other. For the discrete case, the channel is characterized by the conditional probability given the input symbols $X_1, \ldots, X_K$, namely

$$P(Y|X_1, \ldots, X_K), \qquad \forall Y \in \mathcal{Y}, X_1, \ldots, X_K \in \mathcal{X}. \tag{2.8}$$

An important special case is the Gaussian multiple access channel, in which the received signal is given by

$$y = \sum_{k=1}^{K} x_k + n, \tag{2.9}$$

where n is Gaussian noise.

2.3.3 CHANNEL CAPACITY

In his groundbreaking paper [32], Shannon proposed the concept of channel capacity C (bits/channel use) for communication channels; i.e., if the transmission rate R (also bits/channel use) is less than C, then there exists an encoding and decoding scheme to make the transmission error arbitrarily small; otherwise, the data cannot be reliably transmitted.

For the simplest case of discrete memoryless channels, the channel capacity is given by

$$C = \max_{P_X} I(X; Y), \tag{2.10}$$

where P_X is the distribution of the input.

The capacities of the channels introduced in the previous subsection are given below:

- The binary symmetric channel has a channel capacity of $1 - H(p)$, where $H(p) = -p \log p - (1-p)\log(1-p)$ is the Shannon entropy of a binary distribution with probabilities p and $1-p$.
- The erasure channel has a channel capacity p, where p is the probability that the transmitted bit is not erased.
- The Gaussian channel has the following channel capacity:

$$C = \frac{1}{2}\log\left(1 + \frac{P_s}{\sigma_n^2}\right), \tag{2.11}$$

where P_s is the maximum average power, if the constraint is the average power (instead of the peak power).
- The band-limited Gaussian channel has the following channel capacity:

$$C = \frac{W}{2}\log\left(1 + \frac{P_s}{WN_0}\right), \tag{2.12}$$

if the constraint is on the average power.
- Multiple access channel: When there are K transmitters, the communication capability is described by the capacity region Ω, within which each point $(R_1, \ldots, R_K)$ is achievable. For the Gaussian multiple access channel, we have

$$\Omega = \left\{ (R_1, \ldots, R_K) \middle| \sum_{i \in I} R_i \leq \frac{1}{2}\log\left(1 + \frac{\sum_{k \in I} P_k}{\sigma_n^2}\right), I \subset \{1, \ldots, K\} \right\}. \tag{2.13}$$

Although the concept of channel capacity has been widely used in the practical design of communication systems, the following points need to be noted in the context of communications in a CPS.

- The channel capacity defined in traditional information theory is based on asymptotic analysis and infinite codeword length. However, in the context of

CPS, due to the requirement of real-time for controlling the dynamics, the infinite codeword length, which also implies infinite communication delay, cannot be tolerated.

- The achievability of channel capacity is usually based on the argument of random coding, which is of theoretical importance but is not practical. Although for simple binary channels the low-density parity-check (LDPC) codes have almost attained the channel capacity, the practical capacity of coding schemes has not been obtained for many typical channel models.

2.4 SOURCE CODING

Source coding means converting from the original information, which could be either discrete or continuous, to a series of information bits. It can be categorized into lossless source coding, in which no information is lost during the source coding, and lossy source coding, in which a portion of the information is lost. For simplicity, we assume that the information symbols are discrete-time and i.i.d.

2.4.1 LOSSLESS SOURCE CODING

The information source of lossless source coding, which is represented by a random variable X, must be finite. Denote by $\mathcal{X} = \{x_1, \ldots, x_n\}$ the alphabet of the information source symbols. It is assumed that the distribution is $\{P(X = x_j)\}_j$. One simple approach is to use $\lceil \log_2 n \rceil$ bits to represent these n symbols. However, we can do better and use variable length code, in which the information symbols may be represented by different numbers of bits. To achieve the lower average code length, we should assign fewer bits to more frequent symbols. It has been shown that the minimum average number of bits is given by $H(P)$ (i.e., the entropy of X) and can be achieved by properly assigning the bits to the symbols with known probabilities.

The above discussion assumes known distributions of the information symbols. In practice, it is possible that the distribution is unknown and we only have a sequence of bits. In the situation of unknown statistics, a technique of universal source coding, such as the Lempel-Ziv coding scheme, can be applied. A surprising conclusion is that, even though the statistical information is lacking, the universal coding can still achieve the optimal compression rate asymptotically (as the length of sequence tends to infinity), given that the information source is stationary and ergodic.

2.4.2 LOSSY SOURCE CODING

When the information source is continuous in value, it is impossible to encode the source information symbols using finitely many bits. Hence some information has to be lost when digital communications are used. We say that the source coding is lossy in this situation. Note that we can also apply lossy source coding for discrete information sources, if some information loss is tolerable.

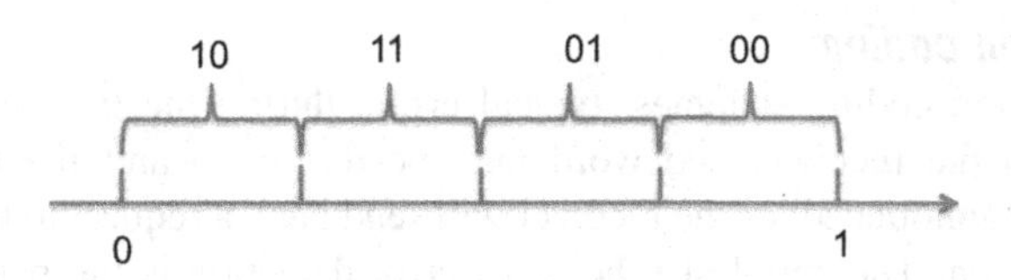

FIG. 2.3

Example of a scalar quantizer.

Theoretically, lossy source coding can be studied using rate-distortion theory [33]. First a distortion function is defined as $d(X, \hat{X})$, where X is the information symbol and $\hat{X}$ is the recovered one. For example, when X is real valued, the distortion can be defined as the mean square error; i.e., $d(X, \hat{X}) = E[|X - \hat{X}|^2]$. Then a rate-distortion function $R(D)$ is defined for each given average distortion D, which means the minimum source coding rate ensuring that the average distortion is no larger than D. It is well known that $R(D)$ can be obtained from the optimization of mutual information between X and $\hat{X}$.

In practice, lossy source coding is carried out by a quantizer. Essentially a quantizer divides the scope of the information source symbol X into many regions, each of which corresponds to a bit string. Fig. 2.3 shows a simple scalar quantizer, in which the range of X is the interval $[0, 1]$ and it is divided into four intervals. When X falls in one of the intervals, the quantizer outputs the corresponding two bits. When X is a vector, a vector quantizer is needed, which is much more complicated. A powerful design algorithm for vector quantizers is Lloyd's algorithm [34].

2.5 MODULATION AND CODING

When receiving a series of information bits, the transmitter will convert them to physical symbols ready for transmission. Usually two steps are needed for this conversion. First the information bits are encoded by adding redundancies in order to combat the communication channel unreliability and the potential errors arising from this. Then the coded bits are converted to physical symbols that are ready to be passed to the hardware for transmission.

2.5.1 CHANNEL CODING

There are two types of channel coding used to protect the transmission reliability from the channel uncertainty. In error detection coding, mistakes during the transmission may be detected. In error correction coding, the errors may be corrected without retransmitting the message.

Error detection coding

In error detection coding schemes, typical errors during the transmission can be detected. Then the received codeword may be discarded and the failure of the transmission is announced; or the receiver will send back a request to the transmitter for retransmission. The simplest scheme of error detection is the parity check bit, namely the sum of the information bits is appended to the codeword. For example, when the original information bits are 1100111, the transmitted bit sequence is 11001111, where the last bit 1 is the parity check bit. If an error occurs in one of the information bits and the receiver receives 11001110, then the error can be detected since the parity check bit is unequal to the sum of the information bits. Although it is also possible that the parity check bit, or two information bits, is wrong, the corresponding probability is much smaller.

Error correction coding

In the error correction coding scheme, typical errors in transmission can be corrected at the receiver without retransmission. The capability of error correction is assured by adding redundancies to the information bits. Take the simplest repetition code, for example. When we transmit bit 1, we can transmit it three times, thus sending out bit sequence 111. The decoder uses the majority rule; i.e., the decision is the most frequent bit in the received sequence. Hence if there is one transmission error and the decoder receives bits 011, the decoder can claim that the information bit is actually 1 and thus correct the error. The transmission rate, usually denoted by R, is defined as the ratio between the number of information bits and the number of coded bits. The above example of repetition code has a transmission rate 1/3. The smaller R is, the more redundancies are added and the more reliable the transmission is; however, a smaller R also uses more communication bandwidth since the transmitter needs to send out more coded bits in the same time. A tradeoff is needed between the reliability and resource consumption.

A typical error correction coding scheme used in practice is the linear block code, in which the coded bits $\mathbf{c}$ are given by

$$\mathbf{c} = \mathbf{bG}, \tag{2.14}$$

where $\mathbf{b}$ represents the information bits (as a row vector) and $\mathbf{G}$ is called the generation matrix. The receiver receives a senseword $\mathbf{s} = \mathbf{c} + \mathbf{e}$, where $\mathbf{e}$ is the error vector. The decoder applies a matrix $\mathbf{H}$, which satisfies $\mathbf{G}^T\mathbf{H} = 0$, to calculate $\mathbf{d} = \mathbf{sH} = \mathbf{eH}$, which is called the syndrome. The error vector $\mathbf{e}$ can be found by looking up a table indexed by the syndrome.

A more powerful error correction code is the convolutional code, which does not have a fixed codeword length and is thus different from the linear block code. In the convolutional code, memory is introduced by streaming the input information bits through registers. Take the structure in Fig. 2.4, for example. There are two registers; thus the memory length is 2. For each input information bit $b(t)$, the two coded bits, denoted by $c_1(t)$ and $c_2(t)$, are generated by $b(t)$ and the previous two information bits $b(t-1)$ and $b(t-2)$, which are given by

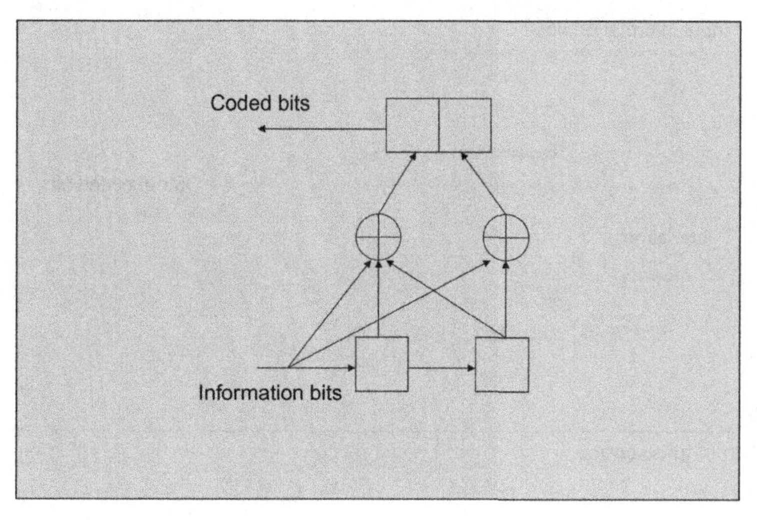

FIG. 2.4

Illustration of convolutional code.

$$\begin{cases} c_1(t) = b(t) + b(t-1) + b(t-2), \\ c_2(t) = b(t) + b(t-2). \end{cases} \quad (2.15)$$

Then the two registers are occupied by $b(t-1)$ and $b(t)$. Since one input information bit triggers the generation of two coded bits, the transmission rate is $\frac{1}{2}$. By changing the number of simultaneous output coded bits and the number of branches of the input bits, we can achieve any rational number for transmission rates. As we will see in later chapters, the structure of the convolutional code is actually similar to that of linear dynamical systems, thus facilitating the proposed turbo channel decoding and system state estimation in CPSs.

A breakthrough in the area of error correction codes is the turbo code, which was proposed by Berrou et al. in 1993 [35] and achieves transmission rates very close to the Shannon limit of channel capacity (Fig. 2.5). For simplicity, we use the original form of turbo codes given in Ref. [35]. The encoder essentially uses a parallel convolutional code. Two convolutional encoders are used. The raw information bits, and a randomly permeated version of the bits (carried out by an interleaver), are fed into the two convolutional codes. Then the output of the turbo encoder consists of the original information bits and the output bits of the two convolutional encoders.

The key novelty in the turbo code is in its decoder, which is illustrated in Fig. 2.6. The received encoded bits at the decoder consist of three parts: m (the original information bits with possible errors), X_1 (the output of convolutional encoder 1), and X_2 (the output of convolutional encoder 2). m and X_1 are fed into a standard decoder, which generates the soft decoding output (namely the probability distribution of each bit being 0 or 1), instead of hard decisions. The soft output is fed into decoder 2 as prior information. Decoder 2 combines m, X_2 and the prior information from

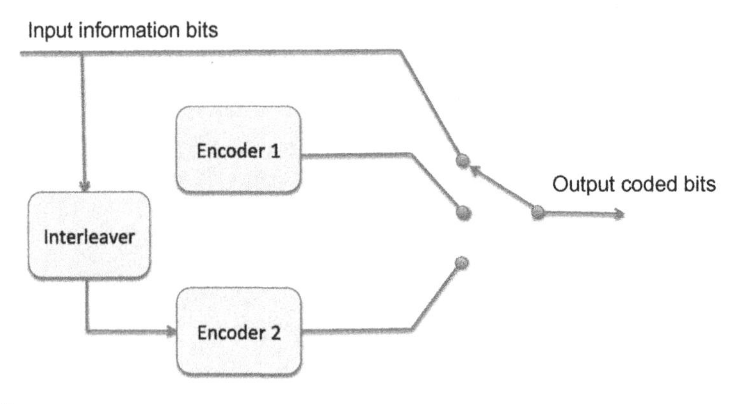

FIG. 2.5

Illustration of turbo code.

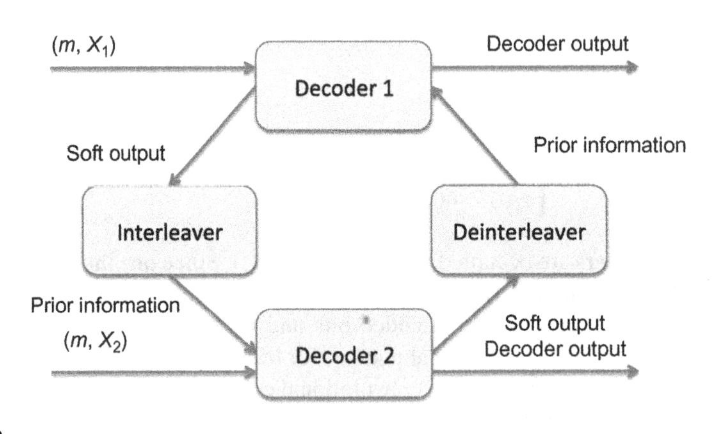

FIG. 2.6

Illustration of the turbo decoder.

decoder 1, and also generates soft outputs. Then the outputs of decoder 2 are fed into decoder 1 as prior information. Such a procedure is repeated iteratively (thus named "turbo") until convergence is achieved. The performance of decoding is improved through this iterative procedure. Such a "turbo" principle of information processing has been applied in many other tasks such as multiuser detection and equalization. As we will see, it can also be used for system state estimation in a CPS.

2.5.2 MODULATION

When the information bits are encoded into codewords (note that in some situations the channel coding procedure is not necessary), the procedure of modulation maps the abstract information to physical signals. In many typical communication systems, the information is conveyed by sinusoidal carriers, namely

$$s(t) = A\cos(2\pi ft + \phi) = Re[Ae^{j2\pi ft + \phi}], \tag{2.16}$$

where A, f, and ϕ are the amplitude, frequency, and phase of the sinusoidal signal, respectively. Hence the information can be carried in the amplitude, frequency or phase, thus resulting in the following types of modulation schemes:

- Pulse amplitude modulation (PAM): There are M possible signal amplitudes, which represent $\log_2 M$ bits. The simplest case is $M = 2$, namely the transmitter either transmits or not ($A = 0$), which is called on-off keying (OOK).
- Frequency shift keying (FSK): The transmitter can choose one of M frequencies for the sinusoidal carrier signal. Hence each switch of the frequency represents $\log_2 M$ bits.
- Phase shift keying (PSK): The transmitter switches among M possible phases for transmitting $\log_2 M$ bits.

The above three typical modulation schemes can be combined; e.g., quadrature amplitude modulation (QAM) is a combination of PAM and PSK. To explain the mechanism of QAM, we consider the following signal:

$$\begin{aligned} s(t) &= A\cos\phi\cos(2\pi ft) - A\sin\phi\sin(2\pi ft) \\ &= A_1\cos(2\pi ft) + A_2\sin(2\pi ft), \end{aligned} \tag{2.17}$$

which is decomposed to the quadrature and in-phase terms (A_1 and A_2 are the corresponding amplitudes). We can consider the carriers $\cos(2\pi ft)$ and $\sin(2\pi ft)$ as two orthogonal signals, each of which carries one branch of independent information. Hence (A_1, A_2) can be considered as a point in the signal plane (called a constellation), which is used to convey information. When there are M possible points in the constellation, we call the modulation scheme M-QAM. Fig. 2.7 shows illustrations of 16QAM and 64QAM.

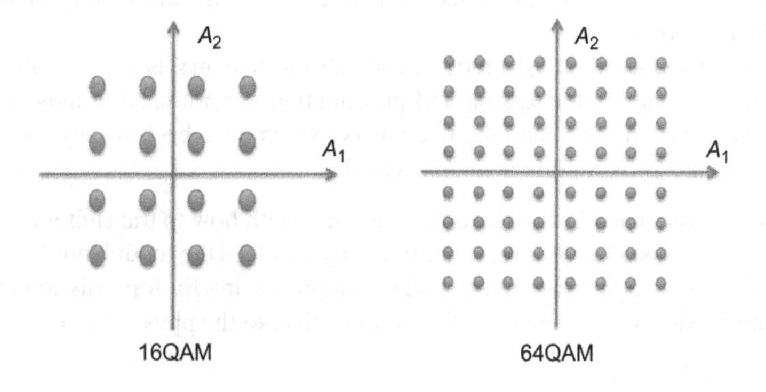

FIG. 2.7

Illustration of QAM.

When selecting a suitable modulation scheme, there is a tradeoff between the transmission rate and transmission reliability. Usually, the symbol rate (i.e., the number of modulation signals transmitted in one second) is fixed. Hence if a denser constellation is used, more coded bits can be transmitted in the same time, thus improving the data transmission rate. However, the points in a dense modulation are close to each other, and thus can be confused by noise and results in transmission errors. As we will see in subsequent chapters, this tradeoff is important for the modulation selection in CPSs.

2.6 NETWORKING

We now discuss communication networks with an emphasis on the layered structure.

2.6.1 GRAPH REPRESENTATION

A communication network can be represented by a graph, in which each vertex represents a communication node and each edge represents a communication link. Hence graph theory (e.g., traditional graph theory such as connectivity and coloring, algebraic graph theory and random graph theory) is of great importance in the research on communication networks.

2.6.2 LAYERED STRUCTURE

To facilitate the design of communication networks, the functionalities of networks can be organized into a stack of layers. Each layer takes charge of different tasks and communicates with adjacent layers. Certain protocols are also designed for the interfaces between two adjacent layers. The most popular definition of network layers is the Open Systems Interconnection (OSI) reference model developed by the International Standards Organization (ISO). Another typical model is the TCP/IP reference model. Both are illustrated in Fig. 2.8. The details of the models are explained as follows.

We first introduce the OSI model, in which the network is divided into seven layers. Since the layers for session and presentation are not used in most network designs, we focus on the remaining five layers, which have been widely used in the design and analysis of communication networks.

- Physical layer: The physical layer is concerned with how to transmit information bits from a transmitter to a receiver. It mainly involves the modulation/demodulation, coding/decoding,[1] and signal processing for transmission and reception. Most of our previous discussions relate to the physical layer.

[1] In the original definition of the OSI model, coding and decoding are issues in the data link layer; however, in practice, they are usually considered as physical layer issues.

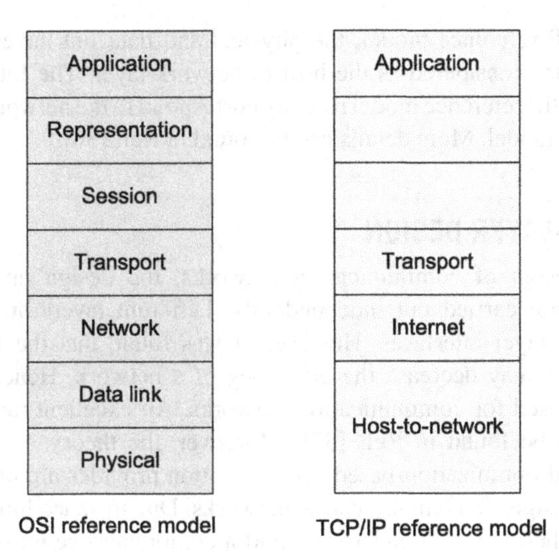

FIG. 2.8

Illustration of layers in OSI and TCP/IP.

- Data link layer: This layer takes charge of tasks such as frame acknowledgment (ACK), flow regulation, and channel sharing. The latter, called medium access control (MAC), is the most important for wireless networks due to the broadcast nature of wireless transmissions, and is therefore usually considered as an independent layer. Essentially, the MAC layer addresses resource allocation, e.g., how to allocate different communication channels to different users, and scheduling, e.g., when there is competition among users, which user should obtain the priority to transmit.
- Network layer: This layer determines how to find a route in the network from the source to the destination. For example, we need to design an addressing mechanism for routing. Moreover, when the addresses of the source and destination are known, we need to design algorithms for the network to find a path with the minimum cost (e.g., the number of hops) to the destination. When a path is broken by an emergency, the network layer needs to find a new path for the data flow.
- Transport layer: This layer receives data from the application layer, splits it into smaller units if needed, and then passes it to the network layer. The main task of the transport layer is congestion control, i.e., how to control the source rate according to the congestion situation in the network.
- Application layer: This provides various protocols for different applications. For example, the HyperText Transfer Protocol (HTTP) is used for websites.

Note that the physical, data link, and network layers are concerned with the intermediate nodes in the network, while the transport and application layers involve only the two ends of data flow, namely the source and destination.

In the TCP/IP reference model, the physical and data link layers are not well specified. They are considered as the host-to-network layer. The Internet and TCP layers in the TCP/IP reference model roughly correspond to the network and transport layers in the OSI model. More details can be found in Ref. [36].

2.6.3 CROSS-LAYER DESIGN

In traditional design of communication networks, the design and operation of different layers are carried out independently. Different layers are coupled only through the inter-layer interfaces. However, it was found that the isolated design of different layers may decrease the efficiency of a network. Hence a cross-layer design was proposed for communication networks. An excellent tutorial on cross-layer design can be found in Ref. [37]. Moreover, the theory of network utility maximization and optimization-based decomposition provides a unified framework for cross-layer designs in communication networks. Due to space limitations, we do not discuss this theory here. Readers can find a comprehensive introduction to this topic in Ref. [38].

A motivating example for cross-layer design is the opportunistic scheduling in cellular systems [39]. Consider a base station serving multiple users. Different users may have different channel gains. The base station needs to schedule the transmission of multiple users. Recall that scheduling is an MAC layer issue while channel gain is a physical layer quantity. In traditional layered design, the scheduling algorithm does not take the channel gains into account. However, it has been shown that, to maximize the sum capacity, it is optimal to schedule only the user having the largest channel gain. Hence we see that, if the sum capacity is the performance metric, it is more desirable to incorporate physical layer issues into the scheduling algorithm in the MAC layer, thus resulting in a cross-layer design.

In the subsequent discussions, we will focus more on the MAC layer, network layer, and transport layer, which are the main focuses of studies on networking. Since modulation and coding have been introduced, we skip the discussion of the physical layer.

2.6.4 MAC LAYER

A major task of the MAC layer is to schedule the transmissions of different users. We first introduce multiple access schemes in which multiple transmitters access the same receiver, which is typical in cellular systems. Then we briefly introduce scheduling in generic networks.

Multiple access schemes

Consider a multiuser system in which multiple transmitters access the same receiver. The following multiple access schemes are used in typical wireless multiuser communication systems, illustrated in Fig. 2.9:

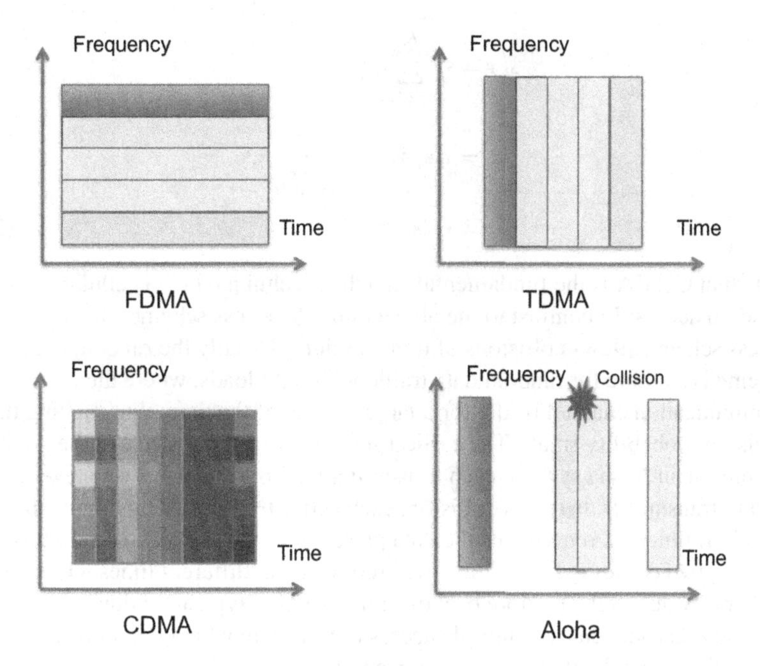

FIG. 2.9

Illustration of multiple access schemes, in which bars with the same colors represent the data packets of one user.

- Frequency division multiple access (FDMA): Different transmitters transmit over different frequency channels, thus avoiding possible collisions. In first-generation cellular systems, FDMA is used to separate different users. In 4G cellular systems, the separation in the frequency domain is achieved by discrete Fourier transform and is called orthogonal frequency division multiple access (OFDMA).
- Time division multiple access (TDMA): The transmitters use different time slots for transmission; this is used in the second generation of cellular systems.
- Code division multiple access (CDMA): The transmitters transmit at the same time and use the same frequency band, thus resulting in collisions on the air. However, each transmitter has a spreading code, say $\mathbf{s}_k$ for user k, which can be considered as an N-dimensional vector. Hence the received signal is given by (noise is omitted here)

$$\mathbf{r} = \sum_{i=1}^{K} x_i \mathbf{s}_i, \tag{2.18}$$

where x_k is the information symbol of user k. When the spreading codes of the users are orthogonal to each other, i.e., $\mathbf{s}_i^T \mathbf{s}_j = 0$ when $i \neq j$, the receiver can detect x_k by correlating the received signal with the spreading codes of user k:

$$
\begin{aligned}
\mathbf{s}_k^T\mathbf{r} &= \mathbf{s}_k^T \sum_{i=1}^{K} x_i \mathbf{s}_i \\
&= x_k \mathbf{s}_k^T \mathbf{s}_k + \mathbf{s}_k^T \sum_{i=1, i\neq k}^{K} x_i \mathbf{s}_i \\
&= x_k \|\mathbf{s}_k\|^2.
\end{aligned} \tag{2.19}
$$

Note that CDMA is the fundamental signaling technique in 3G cellular systems.

- Random access: In contrast to the above multiple access schemes, the random access scheme allows collisions of transmissions. Usually the random access scheme is suitable for random data traffic with light loads, where the communication channel is idle for a large portion of the time, thus making the collision probability small. The earliest scheme of random access is the Aloha system. In an Aloha system, each transmitter begins to transmit whenever it has data to transmit. If there is a collision, each active transmitter randomly chooses a backoff time and retransmits the data packet after the backoff time. In this manner, two colliding transmitters will retransmit at different times with a large probability due to the random backoff time. Another typical random access scheme is carrier sensing multiple access (CSMA), in which the transmitter senses the channel before transmission and transmits only when it finds that the channel is idle.

Scheduling

Scheduling in the MAC layer is to allocate communication resources to different transmitters. For example, in 4G OFDMA systems, different subcarriers in the frequency spectrum are allocated to different users. In TDMA-based communication networks, the scheduler assigns different time slots to different users, thus avoiding collisions. Here we provide a brief introduction to TDMA-based networks with generic topology.

We consider a wireless communication network with N users. We denote by $n \sim m$ if users n and m are neighbors in the network. We assume that there are a total of F data flows in the network. We denote by S_f and D_f the source node and destination node of flow f, respectively. We assume that the number of packets arriving at the source node of data flow f satisfies a Poisson distribution with expectation a_f. The routing paths of the F data flows can be represented by an $F \times N$ matrix $\mathbf{R}$ in which $R_{fn} = 1$ if data flow f passes through user n and $R_{fn} = 0$ otherwise. We denote by $\mathcal{I}_n$ the set of flows passing through secondary user n. When a packet cannot be transmitted immediately, it is placed in a queue and different data flows at the same node have different queues. The queue length (i.e., the number of packets in the queue) is denoted by Q_{nf} for node n and flow f. Since wireless communications are used for each link, there could be collisions of data packets. Usually the relationship of interference among the network nodes is represented by a graph, in which each node represents a communication link in the network and each edge means that the two corresponding communication links cause interference to each other if transmitting simultaneously.

The goal of the scheduler is to schedule the transmissions of the nodes according to the status of queues and stabilize the queuing dynamics. The groundbreaking paper by Tassiulas and Emphremides [40] identifies the region of stability, i.e., the set of arrival rates $\mathbf{a} = \{a_1, \ldots, a_F\}$ that can be stabilized, and the corresponding scheduling algorithm. It is shown that the closure of the stability region is given by

$$\bar{C} = \{\mathbf{a} | \exists \text{ feasible } \mathbf{f} \text{ and } \mathbf{c} \text{ s.t. } \mathbf{M}^{-1}\mathbf{f} \leq \mathbf{c}\}, \tag{2.20}$$

where the vector **f** consists of the serving rate of each link, **c** is the vector consisting of the capacities of the links (as functions of the scheduling policy), and **M** is a diagonal matrix whose diagonal elements are the probabilities of service completion of each link.

The following scheduling policy is shown to universally stabilize the feasible arrival rates, in the following steps (consider time slot t):

- Step 1. A weight $D_i(t)$ is assigned to link i in the following manner. For each flow j and link i, consider

$$D_{ij}(t) = \begin{cases} (Q_{q(i)j}(t-1) - Q_{h(i)j}(t-1))m_i, & \text{if } i \text{ is not the destination}, \\ Q_{q(i)j}(t-1)m_j, & \text{if } i \text{ is the destination}, \end{cases} \tag{2.21}$$

where $q(i)$ is the receiver node of link i, $h(i)$ is the transmitter node of link i, and m_i is the success probably of link i. Then we set

$$D_i(t) = \max_j D_{ij}(t). \tag{2.22}$$

- Step 2. Find a maximum weighted vector **c**:

$$\hat{\mathbf{c}} = \arg \max_{\text{feasible } \mathbf{c}} \mathbf{D}^T(t)\mathbf{c}, \tag{2.23}$$

where the elements in **c** indicate the active status of each link (1 if active and 0 otherwise) and **c** must satisfy the constraint of no collisions in the interference graph.
- Step 3. If link i is activated, the data flow that maximizes $\{D_{ij}\}$ is allowed to transmit.

From the above procedure, we can see that the scheduling algorithm is more inclined to activate links with large differences in the two corresponding queue lengths. Hence if a node has large queue lengths for the packets, it is more likely to be scheduled. However, although the above scheduling algorithm can stabilize the queuing dynamics, it has the following severe challenges:

- Each communication link needs to send its weight to the scheduler; hence the corresponding two nodes need to exchange their information. These bring significant overheads to the network.

- The optimization in Eq. (2.23) is an integer programming problem and is usually NP-hard. For large-scale networks, it is infeasible to obtain the optimal solution.

Due to the above challenges in practice, there have been many studies on reducing the communication overheads (e.g., locally broadcasting the weights) or the computational cost (e.g., using the greedy algorithm). There have been no systematic studies on data traffic scheduling in the context of CPSs. Although it is reasonable to apply the existing scheduling algorithms to the communications for control, it may not be efficient since the goal of the existing algorithms is to guarantee the conveyance of all data packets, instead of controlling the physical dynamics.

2.6.5 NETWORK LAYER

As has been explained, the main task of the network layer is to find the route for each data flow. Usually it is desirable to find the shortest path for each data flow, where the term "shortest" usually means the minimum number of hops, which may be weighted, between the source node and the destination node.

A typical routing algorithm for finding the shortest path is distance vector routing, which is also called the Bellman-Ford algorithm or the Ford-Fulkerson algorithm. In the routing algorithm, each node maintains a routing table. In the table, each item is indexed by each possible destination node, and records the next-hop node and the total distance to the destination of the shortest path. These nodes exchange information in the routing tables and update their routing tables until the shortest paths are found. The updating procedure is illustrated in Procedure 1.

The key step in Procedure 1 is the updating rule of the routing table. Each node compares the minimum distance in its current routing table, and the minimum distances stored in other neighbors (plus the distances to the neighbors), and chooses the minimal one.

PROCEDURE 1 PROCEDURE FOR DISTANCE VECTOR ROUTING IN AN *N*-NODE NETWORK

1: **for** Each node n, $n = 1, \ldots, N$ **do**
2: Initialization: Set $D_{nn} = 0$ and $D_{nm} = \infty$, for $m = 1, \ldots, N$, $m \neq n$. Here D_{nm} means the minimum distance between nodes n and m.
3: **end for**
4: **for** The routing tables have not converged **do**
5: **for** Each node n, $n = 1, \ldots, N$ **do**
6: Exchange the routing table with neighbors.
7: Update the routing table: $D_{nm} = \min_{k \sim n} w_{nk} + D_{km}$, where $k \sim n$ means that node k is adjacent to node n, and w_{nk} is the weight for the hop between nodes n and k. Keep record of the optimal next hop node.
8: **end for**
9: **end for**

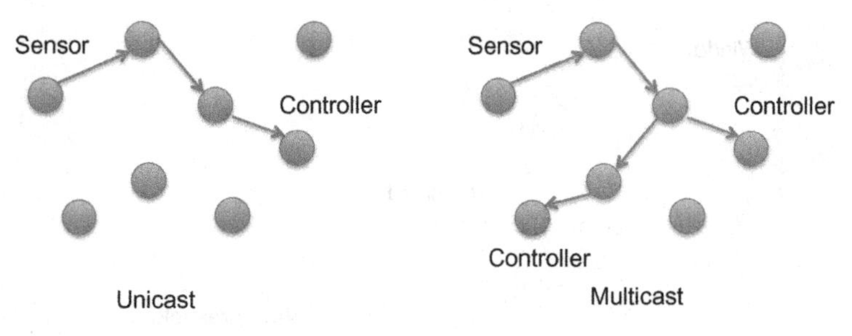

FIG. 2.10

Illustrations of unicasting and multicasting in a CPS.

There are also many other concerns in the design of the routing table; e.g., the possible breakdown of communication links and the corresponding response of the routing algorithm. The characteristics of the network (e.g., wired or wireless) should also be incorporated into the routing algorithm, thus resulting in many algorithms such as hierarchical routing and OSPF routing algorithms.

Note that the above routing algorithms are for unicast routing; i.e., there is only one destination. In many situations, multicast routing is needed; i.e., a packet needs to be sent to more than one destination node. For example, in a CPS, if there are multiple controllers, a sensor may need to send its measurements to multiple destinations. Usually a tree is formed for the multicast, as illustrated in Fig. 2.10.

2.6.6 TRANSPORT LAYER

One of the major tasks of the transport layer is congestion control. In the Transmission Control Protocol (TCP), which is widely used on the Internet, a congestion window is used to control the data traffic volume. The sender can send out all the packets within the congestion window. At the beginning of the connection, the window size is set to the maximum segment size allowed in the connection. Then upon receiving the ACK from the receiver, which means that the corresponding packets have been correctly received, the transmitter doubles its congestion window. When the congestion window is larger than a certain threshold, the increase of the window size becomes linear. If an ACK has not been received before the expiration of the timer, the transmitter assumes that congestion has occurred in the network (thus causing the loss of a packet) and then decreases the congestion window size to the minimum one and also reduces the threshold to a half. This procedure is illustrated in Fig. 2.11.

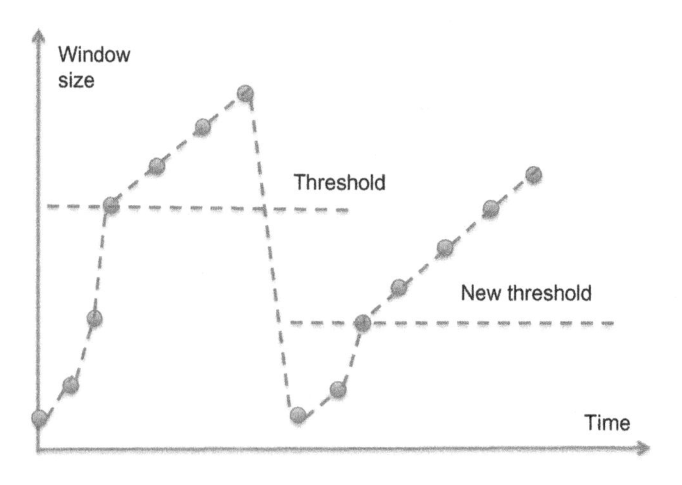

FIG. 2.11

Illustration of congestion control.

2.7 TYPICAL COMMUNICATION SYSTEMS

In this section we list some typical communication systems. They are all potential candidates for communication networks in a CPS.

2.7.1 WIRED NETWORKS

In wired networks, information is sent through a guided medium such as optical fibers. Usually, wired networks have higher throughput and much more reliability. The disadvantages of wired networks include the lack of flexibility and mobility.

According to the geometrical scope, wired networks can be categorized as follows:

- Local area networks (LANs): An LAN may cover a single house, or a campus, whose coverage may be within a few kilometers. The communication agents are connected by cables such as optical fibers. Typical LAN systems include Ethernet (IEEE 802.3) and Token Ring (IEEE 802.5).
- Metropolitan area networks (MANs): The corresponding coverage is a city. A typical system is the cable television network, which can also provide an Internet service.
- Wide area networks (WANs): The coverage of a WAN could be a country or even a continent. The Internet can be considered as a WAN. The hosts (machines) are connected by subnets, which consist of transmission lines and switching elements (routers). Most modern WANs use packet switched subnets, in which the data packets are transmitted in a store-and-forward manner.

2.7.2 WIRELESS NETWORKS

Wireless communication networks have been widely used in both civil and military communications. Many of them can also be adapted to controlling a CPS. Several of the most popular wireless networks are listed below:

- Cellular networks: In cellular networks, the space is divided into cells. Within each cell, a base station is in charge of the communications in it. The users in the cell either transmit to the base station (namely uplink) or receive data from the base station (namely downlink). The base stations are connected with high-speed wired backbone networks. The most recently deployed cellular network is the 4G one, which is called the Long Term Evolution (LTE) network and is mainly based on the signaling of OFDMA. A major challenge in cellular networks is power control within and across cells.
- WiFi networks: In contrast to cellular networks, WiFi networks, which are based on the IEEE 802.11 protocol, cover much smaller areas such as a house or a floor in a building. In contrast to cellular networks, whose deployment is carefully designed, WiFi networks are much more ad hoc and flexible. The cost of the access point (AP) in WiFi, which connects wireless users to the Internet and plays the role of the base station in cellular networks, is much less than that of a base station, thus making careful planning unnecessary.

With reference to the different coverages, wireless networks are usually categorized into the following types, whose coverages and data rates are illustrated in Fig. 2.12:

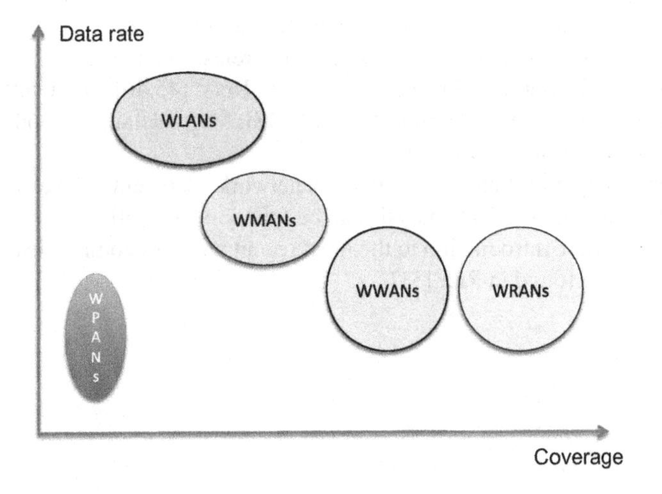

FIG. 2.12

Illustration of congestion control.

- Wireless personal area networks (WPANs): The coverage of WPANs is usually within the reach of a person. Typical systems include Bluetooth and Zigbee.
- Wireless local area networks (WLANs): The coverage of WLANs is usually small, say tens of meters. WiFi is a typical WLAN.
- Wireless mesh networks: A mesh network usually covers a larger area than WLANs. It can be built using WLANs. For example, WiFi can be used to construct a mesh network.
- Wireless metropolitan networks (WMANs): A WMAN may cover a few kilometers. A typical WMAN is WiMAX, which is based on IEEE802.16.
- Wireless wide area networks (WWANs): A WWAN may cover a city and its suburbs. Cellular networks may be used to cover such a large area.
- Wireless regional area networks (WRANs): A WRAN may cover a large distance such as 100 kilometers. A typical standard for WRANs is IEEE 802.22, which may be operated in the TV band.

2.8 CONCLUSIONS

This chapter provides a brief introduction to modern communication theory and systems. Due to space limitations, we cannot provide a comprehensive introduction to all aspects of communications. Readers who are interested in more details are referred to the following literature:

- Excellent comprehensive introductions to communication theory can be found in Refs. [41,42]. Note that these two books focus on wireless communications. A textbook on wired communications, particularly optical communications, is Ref. [43].
- More details on information theory can be found in Ref. [33].
- Ref. [44] provides numerous discussions on wireless channels.
- Introductions to source coding can be found in Refs. [45,46], while introductions to channel coding can be found in Refs. [47,48]. Many details on modulation schemes can be found in Ref. [49].
- There are many books on communication networks. One entry-level textbook is Ref. [36]. More theoretical analysis can be found in Ref. [50].
- A comprehensive introduction to the most recent wireless communication systems can be found in Ref. [51].

CHAPTER

Basics of control 3

3.1 INTRODUCTION

In this chapter, we provide a brief introduction to control systems. We will cover the following aspects of control systems:

- Models of controlled dynamical systems: We will introduce differential equation-based modeling, particularly linear dynamics, of continuous-time systems, as well as models of discrete or hybrid systems. Models of measurements will also be introduced.
- The observability and controllability of linear systems will be briefly discussed.
- As a special case of concrete control strategies, we will explain the key components of optimal control for linear systems with quadratic cost functions.

3.2 MODELING OF CONTROLLED DYNAMICAL SYSTEMS

In this section, we will introduce the modeling of continuous, discrete, and hybrid dynamical systems.

3.2.1 CONTINUOUS-TIME CASE

We first introduce models of dynamics in continuous time, including those with generic dynamics, linear dynamics, and controlled dynamics with feedback.

Dynamics evolution

In a typical dynamical environment such as power grids [4] and robotic networks [5], the physical dynamics can be described by an ordinary differential equation (ODE):

$$\frac{d\mathbf{x}}{dt} = \tilde{f}(\mathbf{x}(t), \mathbf{u}(t), \mathbf{n}(t)), \tag{3.1}$$

Communications for Control in Cyber Physical Systems. http://dx.doi.org/10.1016/B978-0-12-801950-4.00003-2

where $\mathbf{x}$ is the system state, $\mathbf{u}$ is the control action, $\mathbf{n}$ is the random noise, and $\tilde{f}$ is the physical law governing the evolution of system dynamics.

Note that there are many alternative descriptions of physical dynamics, such as:

- Differential-algebraic equations, which consider both continuous-time and discrete events [8].
- Symbolic dynamical systems, which study dynamics using graphs [9].

Since dynamical systems are a very broad area of study, it is prohibitively difficult to provide an exhaustive introduction to dynamical systems. A comprehensive introduction can be found in Ref. [10].

Linear time-invariant systems

In many situations, e.g., when the system is time invariant and oscillates around the equilibrium point, the system dynamics can be simplified to or approximated by linear time-invariant (LTI) dynamics [52]:

$$\frac{d\mathbf{x}(t)}{dt} = \tilde{\mathbf{A}}\mathbf{x}(t) + \tilde{\mathbf{B}}\mathbf{u}(t) + \mathbf{n}(t). \tag{3.2}$$

Observation model

The controller(s) in CPS needs observations on the system dynamics to estimate and control the system state. In the generic case, the observation is a function of the system state and random perturbation:

$$\mathbf{y}(t) = g(\mathbf{x}(t), \mathbf{w}(t)), \tag{3.3}$$

where g is the observation mechanism and $\mathbf{w}(t)$ is the noise.

For simplicity, we consider a linear observation model [52] in this chapter:

$$\mathbf{y}(t) = \mathbf{C}\mathbf{x}(t) + \mathbf{w}(t), \tag{3.4}$$

where $\mathbf{y}$ is the observation, $\mathbf{w}$ is random measurement noise, and $\mathbf{C}$ is the measurement matrix. Different sensors may observe different dimensions of $\mathbf{y}$.

Feedback control

The observation $\mathbf{y}$ is fed back to the controller in order to control the system dynamics (e.g., stabilize it) [53]. We denote by $\hat{\mathbf{y}}$ the observations received at the controller, which may have experienced quantization error, delay or packet loss. An estimation of the system state, denoted by $\hat{\mathbf{x}}$, is made at the controller(s), which is a function of $\hat{\mathbf{y}}$. For simplicity, we assume that linear feedback control [52] is used:

$$\mathbf{u} = -\mathbf{K}\hat{\mathbf{x}}, \tag{3.5}$$

where $\mathbf{K}$ is the feedback gain matrix. There are many approaches for the synthesis of $\mathbf{K}$, e.g., the linear quadratic regulator (LQR) in which the cost function is a quadratic function of the system state and control actions [54]. We will use the LQR in our subsequent discussion due to its elegant framework and its capability to handle stochastic dynamics. The same principle can be applied in other control strategies such as model predictive control (MPC).

3.2.2 DISCRETE-TIME CASE

The dynamics can also be discrete time; e.g., we consider only samples at discrete times. If the discrete-time domain is considered, a difference equation can be used to describe the system in a similar manner:

$$\mathbf{x}(t+1) = f(\mathbf{x}(t), \mathbf{u}(t), \mathbf{n}(t)). \tag{3.6}$$

For linear dynamical systems, the discrete-time version is given by

$$\mathbf{x}(t+1) = \mathbf{A}\mathbf{x}(t) + \mathbf{B}\mathbf{u}(t) + \mathbf{n}(t), \tag{3.7}$$

where **A**, and **B** are time-invariant parameters of the system.

The parameters of the continuous- and discrete-time systems are related by

$$\begin{cases} \mathbf{A} = e^{\tilde{\mathbf{A}} T_s}, \\ \mathbf{B} = \int_0^{T_s} e^{\tilde{\mathbf{A}}\tau} d\tau \tilde{\mathbf{B}}, \end{cases} \tag{3.8}$$

where T_s is the sampling period of the discrete-time system. Note that LTI systems are widely used in many practical scenarios due to their simplicity [52]. Even if a system is nonlinear, it can still be approximated by a linear system near its operation point.

3.2.3 DISCRETE OR HYBRID DYNAMICAL SYSTEMS

We notice that in the above models the system states have continuous values. In practical systems, the system state may be discretely valued. For example, in queuing systems [55] the system state is the vector of the queue lengths at different nodes in the queuing network, which is obviously discrete. A typical model for discrete-valued dynamical systems is the Markov chain, in which there are finitely or infinitely many states $\{s_i\}$. The stochastic dynamics is represented by the transition probabilities $\{P(x_t = s_i | x_{t-1} = s_j)\}$.

It is possible for a system to have both discretely and continuously valued states. As we will see, a CPS consisting of a digital communication network and continuous physical dynamics has both types of states. Such systems are called hybrids. More details will be introduced in Chapter 5.

3.3 OBSERVABILITY AND CONTROLLABILITY

For a controlled dynamical system, the key characteristics include its observability and controllability. The former is more related to communications (or the cyber aspect) in the context of a CPS, since the observations are conveyed by the communication network. Meanwhile, the controllability is more dependent on the capability of the controller, namely the physical aspect of the CPS. For simplicity, we assume the linear dynamics in Eqs. (1.4) and (3.7), and further assume that there is no random noise.

3.3.1 OBSERVABILITY

To determine the system state $\mathbf{x}$, we can use the observations $\{\mathbf{y}(\tau)\}$. If, for any trajectory of the dynamics, $\mathbf{x}(t)$ can be determined in finite time, we say that the system is observable. It is well known that, for linear dynamics and linear observations, if

$$\text{rank}\left(\begin{pmatrix} \mathbf{C} \\ \mathbf{CA} \\ \mathbf{CA}^2 \\ \vdots \\ \mathbf{CA}^{n-1} \end{pmatrix}\right) = n, \tag{3.9}$$

where n is the dimension of $\mathbf{x}$, then the system is observable.

It is easy to observe that the observation matrix $\mathbf{C}$ is dependent on the communication network topology. Therefore the observability can be a criterion for the communication network topology design.

3.3.2 CONTROLLABILITY

The concept of controllability describes the capability of the controller. We say that the system is controllable if the controller can move any initial state to any desired final state in a finite time. Regardless of the continuous- or discrete-time dynamics, the condition for controllability is given by

$$\text{rank}((\mathbf{B}, \mathbf{AB}, \mathbf{A}^2\mathbf{B}, \ldots, \mathbf{A}^{n-1}\mathbf{B})) = n, \tag{3.10}$$

where n is the dimension of the system state.

3.4 OPTIMAL CONTROL

The control of a dynamical system can be cast as an optimization problem, where a cost function is defined and the feedback control law is optimized subject to the evolution law of dynamics. In this book, we provide only an introduction to linear systems.

3.4.1 LQR CONTROL

We first assume that the dynamics is linear and deterministic. For simplicity, we explain the algorithm for only the discrete-time models. We assume that the cost function is given by

$$J = \sum_{t=0}^{T} (\mathbf{x}^T(t)\mathbf{Q}\mathbf{x}(t) + \mathbf{u}^T(t)\mathbf{R}\mathbf{u}(t)), \tag{3.11}$$

where $\mathbf{Q}$ and $\mathbf{R}$ are both nonnegative definite matrices. Obviously the first term $\mathbf{x}^T(t)\mathbf{Q}\mathbf{x}(t)$ means that we desire to drive the system state $\mathbf{x}$ to a vector with lesser norm; meanwhile, the second term $\mathbf{u}^T(t)\mathbf{R}\mathbf{u}(t)$ means that we want to use control actions with less magnitude. Hence the cost function forces us to find the optimal tradeoff between the magnitudes of system state and control action. Since the cost function is quadratic and the dynamics is linear, the corresponding control is called linear quadratic regulation (LQR).

It is well known that the optimal control is given by

$$\mathbf{u}(t) = -\mathbf{K}(t)\hat{\mathbf{x}}(t|t), \tag{3.12}$$

where $\mathbf{K}$ is the feedback gain matrix and $\hat{\mathbf{x}}(t|t)$ is the estimation of $\mathbf{x}(t)$ given all the observations until time t. According to the standard conclusion of linear quadratic Gaussian (LQG) control, we have

$$\mathbf{K}(t) = (\mathbf{R} + \mathbf{B}^T\mathbf{P}(t)\mathbf{B})^{-1}\mathbf{B}^T\mathbf{P}(t)\mathbf{A}, \tag{3.13}$$

and $\mathbf{P}(t)$ satisfies

$$\mathbf{P}(t-1) = \mathbf{Q} + \mathbf{A}^T\mathbf{P}(t)\mathbf{A} - \mathbf{P}^*(t+1), \tag{3.14}$$

where

$$\mathbf{P}^*(t+1) = \mathbf{A}^T\mathbf{P}(t)\mathbf{B}(\mathbf{R} + \mathbf{B}^T\mathbf{P}(t)\mathbf{B})^{-1}\mathbf{B}^T\mathbf{P}(t)\mathbf{A} \tag{3.15}$$

and

$$\mathbf{P}(T) = \mathbf{Q}. \tag{3.16}$$

3.4.2 LQG CONTROL

When the noise is taken into consideration, we usually assume that the noise is Gaussian, thus resulting in LQG control. Since the dynamics becomes stochastic, the cost function is defined as the following expectation:

$$J = E\left[\sum_{t=0}^{T}(\mathbf{x}^T(t)\mathbf{Q}\mathbf{x}(t) + \mathbf{u}^T(t)\mathbf{R}\mathbf{u}(t))\right], \tag{3.17}$$

where the expectation is over the randomness of the noise. It is well known that optimal LQG control can be constructed from LQR.

3.5 CONCLUSIONS

In this chapter, we have briefly explained the basic aspects of control systems. Note that all these theories are based on the assumption that the feedback for control is perfect (i.e., there is no need for communications, since the sensor and controller are located in the same place, which is usually assumed in traditional control systems). As we will see, when communications with imperfect channels are needed, the characteristics of communications, such as delay, packet loss, and quantization error, will have a considerable effect on the physical dynamics and thus the control strategies.

CHAPTER

Typical cyber physical systems 4

4.1 INTRODUCTION

In this chapter, we introduce several typical cyber physical systems (CPSs), such as power networks and robotic networks. For each typical CPS, we focus on the following aspects:

- Physical dynamics: We will introduce the background of the CPS and formulate the mathematical models, in particular the differential equations, of the physical dynamics.
- Communication networks: We will introduce the practical protocols and mathematical models for networked control.

4.2 POWER NETWORKS

Power networks are among the most long-standing CPSs. In power networks, the physical dynamics is the dynamics of voltages, currents, frequencies, and phases. The communication system helps to better estimate and control the current system state. In the subsequent subsections, we will explain models of physical dynamics and the corresponding communication systems.

4.2.1 PHYSICAL DYNAMICS

For simplicity, we consider only the dynamics of generators and ignore the details of the loads and transmission lines. We consider the power grid as a graph G in which each node represents a geographical site and each edge denotes a transmission line. For simplicity, we assume that each site has one synchronous generator and all generators are the same. If sites i and j are adjacent, we denote this by $i \sim j$. It is challenging to incorporate the difference of generators into the model; however, it is complicated and beyond the scope of this section.

Communications for Control in Cyber Physical Systems. http://dx.doi.org/10.1016/B978-0-12-801950-4.00004-4

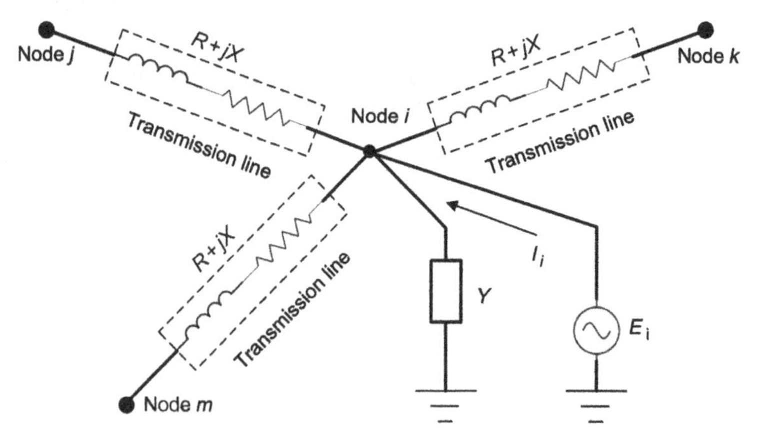

FIG. 4.1

Illustration of one node in the power grid.

Similarly to Ref. [56], we assume that each generator supplies a time-varying current, as well as time-varying power, using a constant voltage. The state of a single generator is represented by rotor rotation angle δ, which is described by the following *swing equation* [4]:

$$M\ddot{\delta} + D\dot{\delta} = P_{\mathrm{m}} - P_{\mathrm{e}}, \tag{4.1}$$

where P_{m} is the mechanical power, P_{e} is the electrical power, M is the rotor inertia constant, and D is the mechanical damping constant.

Fix an arbitrary node i shown in Fig. 4.1, where $Z = R + jX$ is the impedance of the transmission line, E_i is the voltage of the generator, and Y is the shunt admittance. Then for any node $k \sim i$, the current flowing through the transmission line between these two nodes is given by

$$I_{i,k} = \frac{E_i - E_k}{Z}. \tag{4.2}$$

Then the current flowing from the voltage source to the node, denoted by I_i, is given by

$$\begin{aligned} I_i &= \sum_{k \sim i} I_{i,k} + YE_i \\ &= \frac{1}{Z} \sum_{k \sim i} (E_i - E_k) + YE_i, \end{aligned} \tag{4.3}$$

due to Kirchoff's current law. Since we assume that the voltage of the generator is constant in magnitude, we have $E_i = Ve^{j\delta_i}$. Meanwhile, the electric power is given by

$$
\begin{aligned}
P_e^i &= Re[E_i I_i^*] \\
&= \frac{V^2}{|Z|^2}\left[R\left(N_i - \sum_{k\sim i}\cos(\delta_i - \delta_k)\right)\right. \\
&\left. -X\sum_{k\sim i}\sin(\delta_i - \delta_k)\right] + V^2 Re[Y],
\end{aligned}
\tag{4.4}
$$

which is obtained by substituting Eq. (4.3) into the expression of P_e^i.

Substituting Eq. (4.4) into the swing Eq. (4.1), we obtain the differential equation describing the dynamics of a power grid, which is given by

$$
\begin{aligned}
M\ddot{\delta}_i + D\dot{\delta}_i = P_m^i - \frac{V^2}{|Z|^2}&\left[R\left(N_i - \sum_{k\sim i}\cos(\delta_i - \delta_k)\right)\right. \\
&\left. -X\sum_{k\sim i}\sin(\delta_i - \delta_k)\right] + V^2 Re[Y],
\end{aligned}
\tag{4.5}
$$

where N_i is the number of neighbors of node i. We notice that the dynamics of different nodes are coupled to the difference of rotation angles.

4.2.2 PROTECTION

In power networks, protection devices are used to handle faults and also to prevent damage. For example, if the protection system detects a fault in a transmission line, the corresponding protective relays at both ends will be triggered to isolate the transmission line. The protection actions need to be very fast in order to prevent the propagation of the damage incurred by the fault. There are numerous types of protections such as overcurrent relays protecting the devices from large currents, and distance protection distinguishing the faults occurring in different parts of the power network. Note that both overcurrent relay and distance protection do not need communications, since local measurements (e.g., the local current) are used to make the decision.

In a sharp contest, communications are needed for differential protection, whose mechanism is illustrated in Fig. 4.2 and explained as follows. Suppose that two relays are installed at the two ends of a power transmission line. Both relays have communication transceivers; they can communicate with each other. They measure the currents flowing through the two ends, denoted by I_1 and I_2, simultaneously. We define the directions of the currents as flowing into the power line. When there is no fault in the transmission line, we have

$$
I_1 + I_2 = 0, \tag{4.6}
$$

according to Kirchhoff's current law.

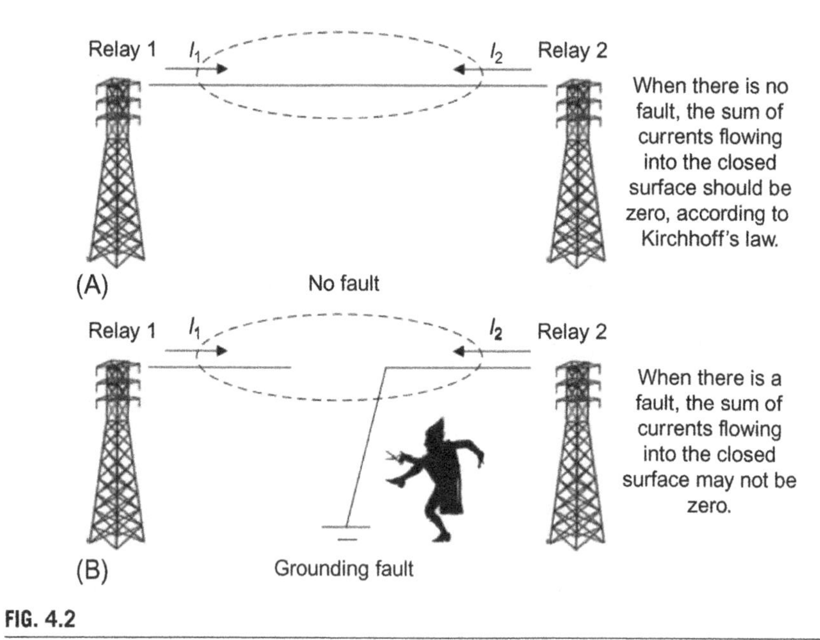

FIG. 4.2

Illustration of differential protection.

When a fault occurs (e.g., the transmission line breaks and falls to the ground), there is usually a significant current flowing into the ground since the ground usually has a low impedance. It is possible that the ground has a high impedance. In this situation, the leaking current is very weak and is very difficult to detect. For simplicity, we assume that the ground has a low impedance; thus Eq. (4.6) is broken due to the leaking current. Therefore when the two relays find that $I_1 + I_2$ is significantly different from zero, they trigger the breakers to take action such that the fault is isolated. If the two relays are not located at the same place (e.g., at the two ends of a transmission line), telecommunication is needed across the two relays. In a more generalized scenario [57,58], the measurements are propagated through a communication network (no longer point-to-point communications) in order to provide backup protections. The corresponding communication network for differential protection has been optimized to maximize the overall reliability in Ref. [58].

Besides differential protection, communications may be needed in other protection schemes. It is well known that phasor measurement units (PMUs) are very useful for protection. For example, the reports from PMUs can be used to detect and locate transmission line faults [59]. They can also be used in adaptive out-of-step control and distance relaying of multiterminal transmission lines [1]. Since the action of protection should be very fast (thus very short delays), communication could be a bottleneck of PMU-based protection schemes.

4.2.3 SMART METERING

Communications are needed in advanced metering infrastructure (AMI) [60], which is a critical mechanism in smart grids [61,62]. The AMI is needed by the demand response mechanism in future power grids. Demand response is defined as "changes in electric usage by end-use customers from their normal consumption patterns in response to changes in the price of electricity over time, or to incentive payments designed to induce lower electricity use at times of high wholesale market prices or when system reliability is jeopardized" [63]. It has been intensively studied, for example the demand response for areas with large populations [64] and the charging of electric vehicles [65].

In the AMI system, each power consumer (e.g., factory, household, workshop, building, etc.) is equipped with smart meters capable of two-way communications. A smart meter can monitor the activities of the power consumer. It reports the most recent power consumption to a control center; meanwhile it receives a power price from the power market. These reports can be used to achieve the balance of power demand and supply, and avoid power consumption peaks (e.g., the dinner time or summer noons), by setting the correct power price for demand response and adjusting the power generations [66,67]. Although the communication requirements, such as throughput and delay, are quite loose for AMI (e.g., a report with at most tens of bytes per few minutes) compared with the capability of modern data communications, the aggregated data traffic can be large in densely populated areas (consider an area with thousands of smart meters). Therefore an event-triggered communication mechanism has been proposed in Ref. [68] such that the throughput requirement is substantially relaxed. Among many modern communication technologies, wireless communication has great potential for AMI since it is inexpensive and can be deployed quickly.

4.2.4 COMMUNICATION SYSTEMS IN INDUSTRIAL CONTROL AND SMART GRIDS

Compared with modern data communication systems such as 4G cellular systems, WiFi and Internet, the communication networks in industrial control systems have special requirements, e.g., predictable throughput, very high reliability, robustness in hostile environments, scalability, and the maintenance capability [27]. Here we explain two typical types of communication systems for industrial controls. Readers can consult Ref. [28] for a comprehensive introduction.

- Foundation Fieldbus [29]: Fieldbus is a family of industrial control communication systems. It has been widely used in numerous industrial control applications. It corresponds to IEC61158. Foundation Fieldbus is a Type 1 Fieldbus. In sharp contrast to modern communication systems, the physical layer of Fieldbus is based on current modulation, in which each device senses the voltage drops at the terminating resistors, thus still being analog. The access control is managed by the Datalink layer using tokens. Fieldbus is real-time and reliable, which is important for control systems. However, its data rate is around 30 kbps, thus being too low for applications with fast dynamics.

- Supervisory Control and Data Acquisition (SCADA) [30]: SCADA has been used to monitor or control the plant dynamics of industrial infrastructures such as power grids. It consists of a human-machine interface, supervisory computer system, remote terminal units (RTUs) connected to sensors, and communication infrastructure. Either wired or wireless communications are used. For large area networks, SONET/SDH optical communication is an option. A major disadvantage of SCADA is the slow speed, which may take minutes to collect data from the power grid [31]. Hence it cannot be used in fast data communications.

4.3 ROBOTIC NETWORKS

A robotic network is a typical CPS, in which multiple robots interact, using either communications or sensing, with each other in order to accomplish certain tasks. Hence the physical dynamic is the motion of the robots, while the cyber part consists of the communications, sensing, and computing of each robot. In this section, we provide a brief introduction to basic models of robotic networks. This section will serve as a working example for which the subsequent discussions on communications and controls in CPS can be applied. Note that we follow the framework in Refs. [5,69] to introduce basic models of robotic networks.

4.3.1 PHYSICAL DYNAMICS

The physical dynamics mainly describes the locations and motions of the robots. They can be modeled in either a deterministic or probabilistic manner.

Deterministic model

We assume that each robot moves in a continuous space in continuous time. The state of each robot consists of the following four components:

- Robot state (location and velocity vector): $x \in R^d$.
- Control action: $u \in R^m$.
- Initial state: x_0.
- Dynamics law: f, which is given by

$$\dot{x}(t) = f(x(t), u(t)). \tag{4.7}$$

The space of the state is denoted by X^i for robot i. A simple example is the planar vehicle model, in which each robot has a constant speed and the control is on the direction. The corresponding dynamics of motion is described by

$$\begin{cases} \dot{x}_1 = v\cos\theta, \\ \dot{x}_2 = v\sin\theta, \\ \dot{w} = w, \end{cases} \tag{4.8}$$

where $x = (x_1, x_2, \theta)$ and $u = w$.

Another example is first-order robots with range-limited communications. Denote by $p^{[i]}(t)$ the location of robot i, whose evolution is given by

$$\dot{p}^{[i](t)} = u^{[i]}(t), \tag{4.9}$$

where $u^{[i]}(t) \in [-u_{\max}, u_{\max}]$ is the control action. It can be approximated by the following discrete-time dynamics:

$$p^{[i]}(l+1) = p^{[i]}(l) + u^{[i]}(t), \quad i = 1, \ldots, n, \tag{4.10}$$

where $p^{[i]}(l)$ is the location of robot i at time slot l.

Then a robotic network consists of multiple robots described as above. In particular, the communication network of the robotic network can be represented by a graph, in which each vertex is a robot while each edge is a communication link. Whether two robots can establish a communication link is mainly determined by their distance.

Probabilistic model

The motion of a robot can also be modeled in a probabilistic manner. Denote by $x(t)$ and $u(t)$ the state and control of a robot, similarly to the deterministic model. Then the dynamics can be modeled by a Markov chain, whose transition probability is given by

$$p(x(t)|u(t), x(t-1)). \tag{4.11}$$

In probabilistic models, the modeling of the robot motion is of key importance. Two models are possible for robot motion:

- Velocity motion model: In this model, a robot can be controlled by manipulating its rotational and translational velocities (denoted by $v(t)$ and $w(t)$). Hence the control action is given by

$$\mathbf{u}(t) = (v(t), w(t)). \tag{4.12}$$

  To evaluate the transition probability $p(x(t)|u(t), x(t-1))$, one first assumes that the control $\mathbf{u}(t)$ is exact and calculates the expressions for the location after a small time of motion. Then we relax the assumption that the control is exact and assume that the actual control action is contaminated by random perturbations, which is given by

$$(\hat{v}, \hat{w}) = (v, w) + (\epsilon_v, \epsilon_w), \tag{4.13}$$

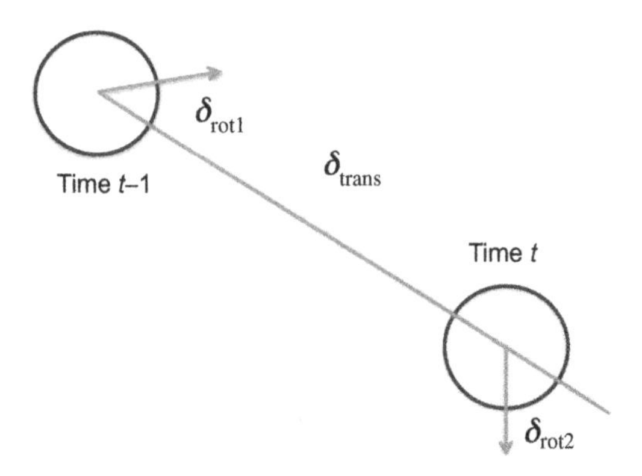

FIG. 4.3

Illustration of the three parameters of relative motion.

where (v, w) is the expected control and (ϵ_v, ϵ_w) is the random perturbation, which is assumed to be centered at zero and has a bounded variance. Typical distributions of (ϵ_v, ϵ_w) include Gaussian distribution and triangular distribution. From these distributions of random perturbation, $p(x(t)|u(t), x(t-1))$ can be evaluated. The detailed algorithm can be found in Ref. [69].

- Odometry motion model: In this model, odometry measurements are used. Usually odometry is more accurate than velocity. At time t, the odometry of a robot estimates the relative location change; i.e., it reports the advance from $x(t-1)$ to $x(t)$. The location change has three components, namely δ_{rot1}, δ_{trans}, and δ_{rot2}, as illustrated in Fig. 4.3. In the odometry motion model, it is assumed that these three components experience independent random perturbations. Given the distributions of these random perturbations, the conditional probability $p(x(t)|u(t), x(t-1))$ can be obtained, whose details can be found in Ref. [69].

4.3.2 COMMUNICATIONS AND CONTROL

The procedure of communication and control of a robot is illustrated in Fig. 4.4. In each cycle, a robot exchanges information with neighbors through its communication module. Then it updates its own processor state and calculates the control action, which will be carried out subsequently. Hence the communication and control law of each robot (say robot $i \in I$, where I is the index set of robots) consists of the following elements:

- Communication alphabet A, which consists of the symbols for communications.
- Processor state sets W^i, which describe the local state of computing.
- Allowable initial values W_0^i, which denote the initial state of computing.
- Message generation law: $msg^i : X^i \times W^i \times I \rightarrow A$, which means the message generated according to the physical state, computing state, and robot index.

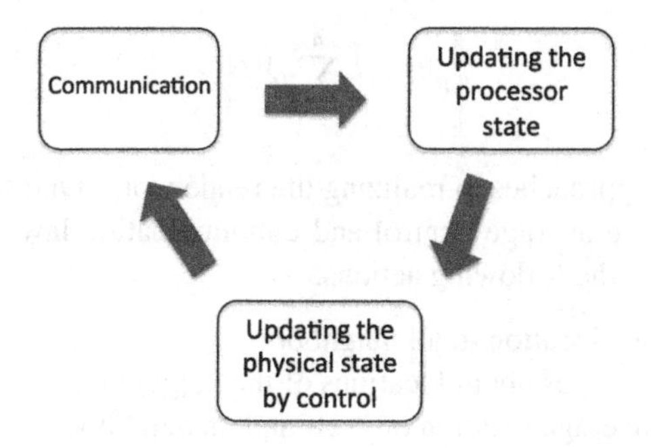

FIG. 4.4

Communication and control in robots.

- State transition law: $stl^i : X^i \times W^i \times A^{|I|} \to W^i$, which makes the robot transit to a new state of computing according to the current physical state, computing state and all messages.
- Control law: $ctl^i : X^i \times X^i \times W^i \times A^n \to U^i$, which calculates the control action at time $t \in [l, l+1)$, where l is the time instant of the previous update of the computing state; hence the control action $u(t)$ is a function of $x(l)$, $x(t)$, current local computing state, and all received messages.

Several simplifications can be made to the above rather general model:

- The communication and control law can be static, which does not change with time and thus means that W^i is a singleton.
- The communication and control law can be data sampled, if the control action $u(t)$ is independent of $x(t)$ and is dependent on only $x(l)$ (the physical state of the last sample time).

4.3.3 COORDINATION OF ROBOTS

The robots in a robotic network need coordination in order to accomplish their tasks. In the subsequent discussion, we introduce two major types of coordination tasks, namely rendezvous and connectivity maintenance.

Rendezvous

One of the major tasks of robotic network is the rendezvous problem, i.e., achieving agreement of the locations of the robots. The following two models of rendezvous can be used for the analysis and design of the corresponding algorithm:

- Exact rendezvous: The task succeeds only when the physical states of all robots coincide; i.e., $x^{[i]} = x^{[j]}$, for all $i, j = 1, \ldots, n$.
- ϵ-Rendezvous: The task succeeds only when each robot is within an ϵ-neighborhood of the average location; i.e.,

$$\left\| p^{[i]} - \frac{1}{n}\sum_{k=1}^{n} p^{[k]} \right\| \leq \epsilon. \tag{4.14}$$

There are many approaches to realizing the rendezvous. One simple approach to communications is the average control and communication law. In each time slot, each robot carries out the following actions:

1. It transmits its own location to all neighbors.
2. It receives the messages about locations of the neighbors.
3. It calculates the average location of itself and its neighbors.
4. It drives itself to that average location.

Another approach, called the circumcenter control and communication law, carries out the following actions for each robot in each time slot:

1. It exchanges position information.
2. It computes the circumcenter of these positions (including itself).
3. It moves to the circumcenter.

Connectivity maintenance

The other task of coordination is the maintenance of network connectivity. First we define the r-disk graph of geographical nodes, where two nodes are connected when their distance is no more than r. When the r-graph is connected (i.e., for any pair of nodes in the graph we can always find a path between them), then the connectivity maintenance problem is to enable the robots to form a connected r-disk graph.

We first consider the maintenance of the connectivity of two robots. Suppose that two robots i and j have positions $p^{[i]}$ and $p^{[j]}$ with $\|p^{[i]} - p^{[j]}\| < r$. Then we define the connectivity constraint set as

$$\chi(p^{[i]}, p^{[j]}) = B\left(\frac{p^{[i]} + p^{[j]}}{2}, \frac{r}{2}\right), \tag{4.15}$$

where $B(c, a)$ denotes a ball with center c and radius a.

Assume that robots i and j are within distance r at time t. Then it can be shown that, if the control actions $u^i(t)$ and $u^j(t)$ satisfy

$$u^i(t) \in \chi\left(p^{[i]}(t), p^{[j]}(t)\right) - p^{[i]}(t) \tag{4.16}$$

and

$$u^j(t) \in \chi\left(p^{[i]}(t), p^{[j]}(t)\right) - p^{[j]}(t), \tag{4.17}$$

as illustrated in Fig. 4.5, then the following two conclusions hold:

- Robots i and j will still be in the connectivity constraint set $\chi\left(p^{[i]}, p^{[j]}\right)$.
- Their positions are still within distance r.

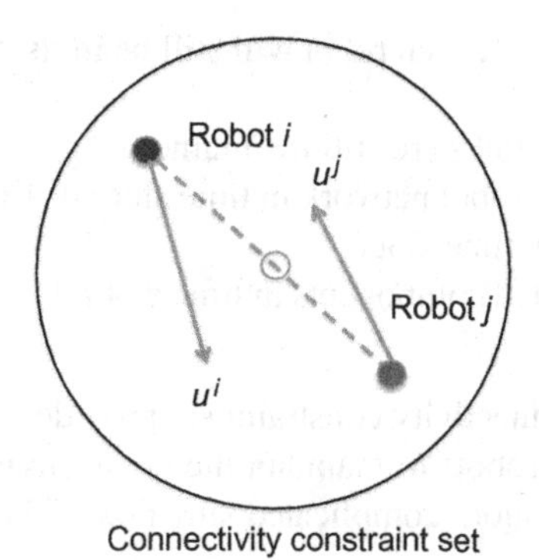

FIG. 4.5

Illustration of the connectivity constraint set for two robots.

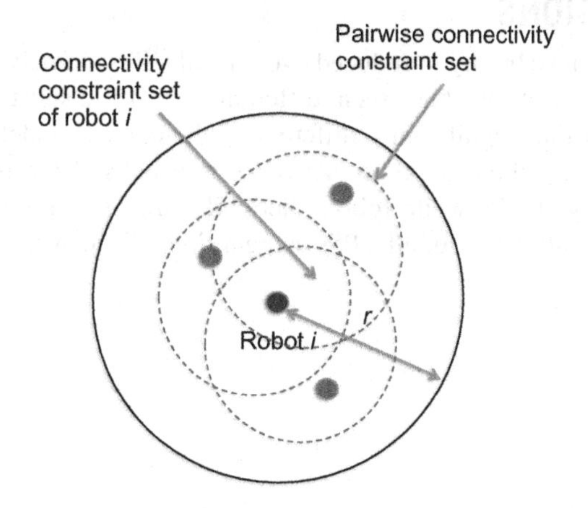

FIG. 4.6

Illustration of the connectivity constraint set for robot *i*.

Then we can extend the two-robot case to the generic case with arbitrarily many robots. When the locations of the robots are $p^{[1]}, p^{[2]}, \ldots, p^{[n]}$, the corresponding connectivity constraint set is defined as (as illustrated in Fig. 4.6)

$$\chi_i\left(p^{[1]}, \ldots, p^{[n]}\right) = \left\{x \in \chi_{p^{[i]}, p^{[j]}} | j \neq i, \text{s.t.}, \left\|p^{[i]} - p^{[j]}\right\| \leq r\right\}, \tag{4.18}$$

which means the joint set of the connectivity constraint set of robot i with all other robots that are within distance r of robot i. Then it is shown in Ref. [5] that, if the robot network keeps the control actions at time l to satisfy the following condition:

$$u^{[i]}(l) \in \chi_i\left(p^{[1]}(l), \ldots, p^{[n]}(l)\right) - p^{[i]}, \tag{4.19}$$

then the following goals can be achieved:

- At the next time slot $l + 1$, each robot will still be in its connectivity constraint set of time slot l.
- All the communication links are still maintained.
- The connectivity of the robot network in time slot $l + 1$ is maintained, if the network is connected in time slot l.
- The number of connected components in time slot $l + 1$ is not larger than that of time slot l.

The concept of the connectivity constraint set provides a sufficient condition for the control actions of the robots to maintain the connectivity of the network. It can be relaxed and applied in more complicated situations. The details can be found in Chapter 4 of Ref. [5].

4.4 CONCLUSIONS

In this chapter, we have briefly introduced two typical CPSs, namely power networks and robotic networks. As we have seen, different CPSs have substantially different characteristics and thus significantly different mathematical models. For example, the mathematical modeling of power networks is usually deterministic, since its operation is usually stable, while robotic networks are more random. Hence it is impossible to cover all details of all CPSs using a single framework.

CHAPTER

Communication capacity requirements

5

5.1 INTRODUCTION

In this chapter, we study the communication capacity requirements (usually in terms of bits) for controlling the physical dynamics in cyber physical systems (CPSs). This is the first step of communication network design in CPSs, since we need to understand how much communication is needed before we design the details of the communication network.

5.1.1 METHODOLOGIES AND CONTEXTS

There is no unified framework to study communication requirements, since this is still an open question. Moreover, the problem may be formulated in different ways for different contexts. In this chapter, we will consider the following methodologies and contexts:

- Deterministic physical dynamics with uncertain initial states, which is appropriate to model the physical dynamics of a sudden random perturbation but deterministic subsequent evolution: topological entropy is used to describe the communication requirements. The introduction mainly follows the work of Ref. [11].
- State estimation of stochastic systems, which is appropriate to model a system subject to random perturbations and a control policy of separated estimation and control: We study the communication requirements for reliably estimating the system state in stochastic systems. The introduction mainly follows the work of Ref. [14].
- Stability control for stochastic systems: It has been found that the anytime capacity measures the capability of communication channels for feedback control to stabilize the physical dynamics [13]. We provide a brief introduction to the concept of anytime capacity and the corresponding applications.
- Shannon entropy-based approach: The above approaches can provide the communication requirements for system stability. However, it is useful to further study the communication requirements subject to different levels of control precision (e.g., the disorderliness of the system state). We adopt the concept of

Communications for Control in Cyber Physical Systems. http://dx.doi.org/10.1016/B978-0-12-801950-4.00005-6

Table 5.1 Summary of Methodologies and Setups of the Studies in Different Sections

Section	Dynamics Types	Communication Type	Key Concept	Goal	Source
5.2	Deterministic	Sensor to controller	Topological entropy	Stability	[11]
5.3	Stochastic	Sensor to controller	Rate distortion	Estimation precision	[14]
5.4	Stochastic	Sensor to controller	Anytime capacity	Stability	[13]
5.5	Stochastic	Sensor to controller	Shannon entropy	Control precision	[15]
5.6	Deterministic	Controller to controller	Communication complexity	Achieving desired state	[70]
5.7	Thermodynamics	Sensor to controller	Fluctuation theorem	Entropy generation	[71]

entropy and the intuition from the second law of thermodynamics [15] to study the communication requirements if a certain entropy of the physical dynamics needs to be achieved.

- Communication complexity: Here we study the case of two agents wanting to achieve the same desired system state and consider it as a distributed computing problem. We follow the study in Ref. [70] to apply the theory of communication complexity to obtain bounds for the communication requirements.
- Thermodynamics argument: We consider a physical system governed by thermodynamics laws, and use the theory of nonequilibrium statistical mechanics to study the impact of communications on entropy generation. This mainly originates from Ref. [71].

The contents of the different sections are summarized in Table 5.1.

5.1.2 BASIC MODELS

We denote by $\mathbf{x}$ the N-dimensional system state. For the generic case, the system state of physical dynamics evolves in the following manner:

$$\begin{cases} \dot{\mathbf{x}}(t) = f(\mathbf{x}(t), \mathbf{u}(t), \mathbf{w}(t)), \\ \mathbf{y}(t) = g(\mathbf{x}(t), \mathbf{n}(t)), \end{cases} \tag{5.1}$$

for the continuous-time case, where the functions f and g represent the evolution of physical dynamics and the observation mechanism, and

$$\begin{cases} \dot{\mathbf{x}}(t+1) = f(\mathbf{x}(t), \mathbf{u}(t), \mathbf{w}(t)), \\ \mathbf{y}(t) = g(\mathbf{x}(t), \mathbf{n}(t)), \end{cases} \tag{5.2}$$

for the discrete-time case.

A simpler but very useful model is that of linear dynamics, where the physical dynamics evolve as follows. For the continuous-time case we have

$$\begin{cases} \dot{\mathbf{x}}(t) = \mathbf{A}\mathbf{x}(t) + \mathbf{B}\mathbf{u}(t) + \mathbf{w}(t), \\ \mathbf{y}(t) = \mathbf{C}\mathbf{x}(t) + \mathbf{n}(t), \end{cases} \tag{5.3}$$

while for the discrete-time case, we have

$$\begin{cases} \mathbf{x}(t+1) = \mathbf{A}\mathbf{x}(t) + \mathbf{B}\mathbf{u}(t) + \mathbf{w}(t), \\ \mathbf{y}(t) = \mathbf{C}\mathbf{x}(t) + \mathbf{n}(t). \end{cases} \tag{5.4}$$

5.2 DETERMINISTIC SYSTEM: STABILITY

We begin with the case of a deterministic dynamical system, in which there is no random perturbation in the system state evolution and the observation mechanism. For simplicity, we consider only the discrete-time system. Then the system dynamics can be described as

$$\begin{cases} \mathbf{x}(t+1) = f(\mathbf{x}(t), \mathbf{u}(t)), \\ \mathbf{y}(t) = g(\mathbf{x}(t)). \end{cases} \tag{5.5}$$

If everything is deterministic, then there is no need for communications since we can calculate the control action in advance. In this section, we assume that the initial system state $\mathbf{x}(0)$ is unknown to us and the distribution of $\mathbf{x}$ is unknown. Then communications are needed to convey the information about $\mathbf{x}(0)$ from the observation and thus estimate the current system state $\mathbf{x}(t)$. The uncertainty incurred by the initial state $\mathbf{x}(0)$ will be changed by the system dynamics and thus will create a need for communications. If the uncertainty in the system state is quickly removed (e.g., the system state converges to a deterministic value in open-loop control), then there is no need for communications in order to stabilize the system dynamics; at this point, we say that the system is simple. If the uncertainty in the system dynamics is amplified, then we refer to the system as complex. The complexity of the physical dynamics is measured by the topological entropy, which will be explained subsequently. As we will see, the topological entropy creates a fundamental limit for the communication capacity requirement. Consider the illustrations in Fig. 5.1, which show the state trajectory in two-dimensional phase space; we observe that the linear case is much simpler than the Lorentz transform.

5.2.1 TOPOLOGICAL ENTROPY

In this subsection, we define topological entropy. There are three equivalent definitions, which will be introduced separately. For all these definitions, we consider a topological dynamical system, which is represented by the triplet (X, T, S). X is a metric space (namely, a distance is defined for any pair of points), in which the metric is denoted by d. We assume that X is compact. We define a transformation T which

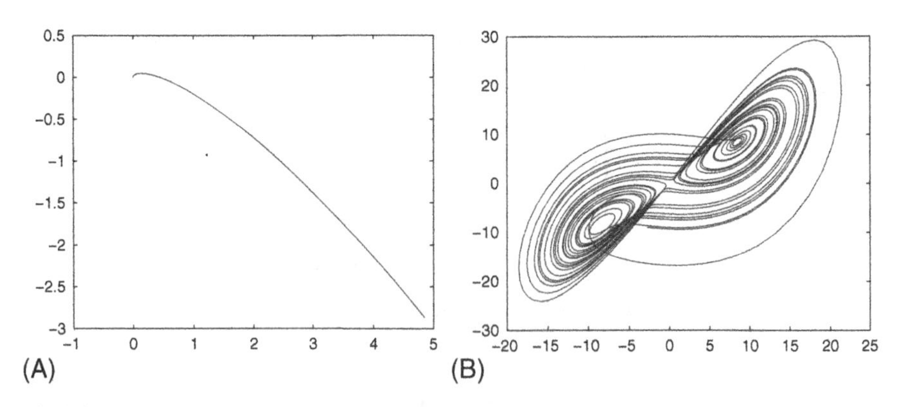

FIG. 5.1

Comparison of simple and complex dynamics. (A) Linear transformation. (B) Lorentz transformation.

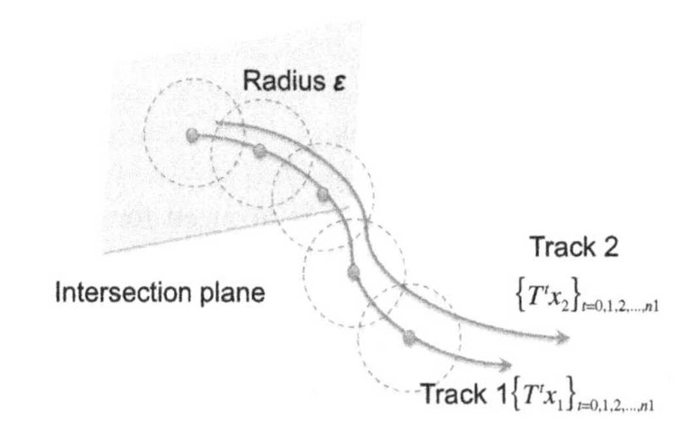

FIG. 5.2

Illustration of two close tracks.

maps from X to X. Moreover, we assume that T is continuous; i.e., T^{-1} maps from open sets to open sets.

Spanning orbit-based definition

We define the nth order ball of point $x \in X$, denoted by $B^n(x, \epsilon)$, as the ball around x with radius ϵ with respect to metric d^n defined as

$$d^n(x, y) = \max\{d(T^i x, T^i y), i = 0, \ldots, n-1\}. \tag{5.6}$$

Fig. 5.2 shows that a track beginning from x_2 is within $B^n(x_1, \epsilon)$.

We say a set F is (n, ϵ)-spanning if it intersects every (n, ϵ)-ball (i.e., intersecting $B^n(x, \epsilon)$ for every x). Then we define

$$r(n, \epsilon) = \min\{\#F : F \text{ is } (n, \epsilon)\text{-spanning}\}. \tag{5.7}$$

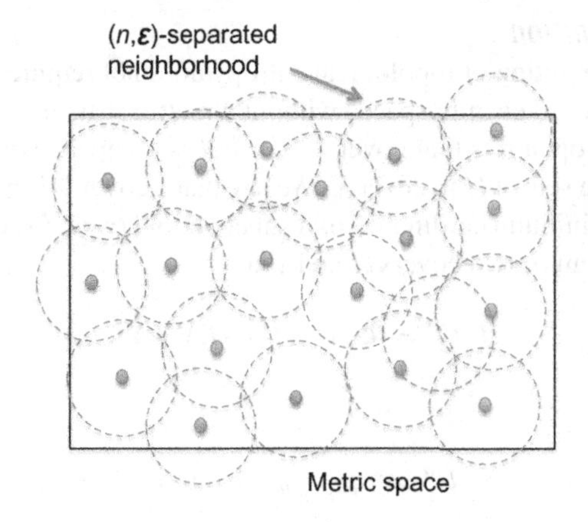

FIG. 5.3

Illustration of an (n, ϵ)-separated set. Note that the circle representing the neighborhood is just an illustration, not a real geometric image.

Then we define the spanning orbit-based topological entropy as follows, where the subscript s denotes "spanning":

Definition 1. The topological entropy is then defined as

$$\begin{cases} H_s(T,n,\epsilon) = \log r(n,\epsilon), \\ h_s(T,\epsilon) = \lim_{n\to\infty} \frac{1}{n} H_s(T,\epsilon), \\ h_s(T) = \lim_{\epsilon\to 0} h_s(T,\epsilon). \end{cases} \tag{5.8}$$

Separated orbit-based definition

We say a subset F of X is (n,ϵ)-separated if $d^n(x_i,x_j) > \epsilon$ for all $x_i, x_j \in F$ and $x_i \neq x_j$. Since X is compact, all (n,ϵ)-separated sets have finite elements, which is illustrated in Fig. 5.3. The maximal cardinality of all (n,ϵ)-separated sets is denoted by

$$s(n,\epsilon) = \max\{|F| : F \text{ is } (n,\epsilon)\text{-separated}\}. \tag{5.9}$$

We denote by $s(n,\epsilon)$ the maximum cardinality of a product (n,ϵ)-separated set, namely

$$s(n,\epsilon) = \min\{\#G(F) : F \in \Omega_{n,\epsilon}\}. \tag{5.10}$$

Then we define the metric-based product topological entropy as follows:

Definition 2. We define the topological entropy as

$$\begin{cases} H_m(n,\epsilon) = \log s(n,\epsilon), \\ h_m(T,\epsilon) = \limsup_{n\to\infty} \frac{1}{n} H_m(n,\epsilon), \\ h_m(T) = \lim_{\epsilon\to 0} h_m(T,\epsilon). \end{cases} \tag{5.11}$$

Cover-based definition

The cover-based definition of topological entropy does not require the definition of a metric; hence it is valid even in spaces without a metric structure. We define a cover of X as a family of open sets that cover X. Since X is compact, we can always find a finite subset of open sets of $\mathcal{U}$ to cover X. We say that a cover $\mathcal{V}$ is a subcover of cover $\mathcal{U}$ if $\mathcal{V} \subset \mathcal{U}$. The minimum cardinality of a subcover of cover $\mathcal{U}$ is denoted by $N(\mathcal{U})$.

We define the join of two covers $\mathcal{U}$ and $\mathcal{V}$ as

$$\mathcal{U} \vee \mathcal{V} = \{U \cap V : U \in \mathcal{U}, V \in \mathcal{V}\}. \tag{5.12}$$

We further define

$$\mathcal{U}^n = \vee_{n=0,\ldots,n-1} T^{-n}(\mathcal{U}). \tag{5.13}$$

Based on the above definitions, we can now define the cover-based topological entropy:

Definition 3. The topological entropy is defined as

$$\begin{cases} H_c(\mathcal{U}) = \log N(\mathcal{U}), \\ h_c(T,\mathcal{U}) = \lim_n \frac{1}{n} H_c(\mathcal{U}^n)), \\ h_c(T) = \lim_{\mathcal{U}} \uparrow h_c(T,\mathcal{U}). \end{cases} \tag{5.14}$$

Equivalence

The above three definitions of topological entropy look quite different, focusing on different aspects of the dynamical system. Interestingly, they are equivalent in metric spaces:

Theorem 1. *In metric spaces, we have*

$$h_s(T) = h_m(T) = h_c(T). \tag{5.15}$$

The proof is highly nontrivial. The details can be found in Chapter 6 of Ref. [12]. Due to their equivalence, we use only the notation $h_s(T)$ in the subsequent discussion.

5.2.2 COMMUNICATION CAPACITY REQUIREMENTS

As we have seen, the topological entropy measures the complexity of the transform in the dynamical system. Intuitively, a more complex dynamical system requires more communications for estimating and controlling the dynamics, since the controller needs more information to obtain the precise status. In this subsection, we will prove that the topological entropy provides a tight bound for the communication requirements in deterministic dynamical systems having uncertain initial states. The argument follows Chapter 2 of Ref. [11].

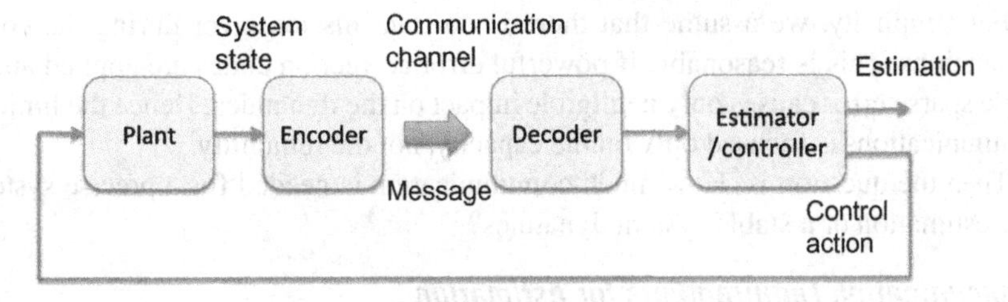

FIG. 5.4

Communication system model for topological entropy.

System model

The model of the CPS is given in Fig. 5.4. We consider the following system dynamics:

$$\mathbf{x}(t+1) = f(\mathbf{x}, \mathbf{u}(t)), \tag{5.16}$$

which is a simpler version of Eq. (5.1). The initial value $\mathbf{x}(0)$ is unknown. An encoder observes the system state and sends out a message $h(jT)$ every T time slots, where j is the index of the message. The total number of messages is denoted by L. Hence the average transmission rate is given by $\frac{\log_2 L}{T}$. If the capacity of the communication channel is R, then we have

$$\frac{\log_2 L}{T} \leq R. \tag{5.17}$$

The transmitted message is essentially a function of the observation history, i.e.,

$$h(jT) = F_j(\mathbf{x}(1:jT)), \tag{5.18}$$

where the subscript j in F_j means that the encoding mechanism can be time varying.

Upon receiving a message from the encoder, the controller decodes it and carries out the following possible actions:

- System state estimation: The controller estimates the system states between times $(j-1)T+1$ and jT (i.e., the system state in the previous period of the message), namely $\hat{\mathbf{x}}((j-1)T+1), \ldots, \hat{\mathbf{x}}(jT)$. The outcome of the estimator is given by

$$\hat{\mathbf{x}}((j-1)T+1:jT) = G_j(h(T:jT)), \tag{5.19}$$

  where the subscript j in G_j means that the estimation algorithm can be time varying.
- System state control: The controller computes the control action in the next T time slots (i.e., before receiving the next message), namely $\mathbf{u}(jT+1), \ldots, \mathbf{u}((j+1)T)$. The outcome of the controller is given by

$$\mathbf{u}((j-1)T+1:jT) = U_j(h(T:jT)), \tag{5.20}$$

  where the subscript j in G_j means that the control algorithm can be time varying.

For simplicity, we assume that there is no transmission error during the communications. This is reasonable if powerful error correction codes are applied and a single sparse error causes only negligible impact on the dynamics. Hence the limit of communications is focused only on the capacity, not the reliability.

Then the question is: How much communication is needed for a precise system state estimation or a stable system dynamics?

Communication requirements for estimation

We first focus on the system state estimation. We define the observability of the system as follows:

Definition 4. The system in Eq. (5.16) is observable if, for all $\epsilon > 0$, there exists a message period $T \in \mathcal{N}$ and an encoding-decoding mechanism such that[1]

$$\|\mathbf{x}(t) - \hat{\mathbf{x}}(t)\|_\infty < \epsilon, \quad \forall t = 1, 2, 3, \ldots \tag{5.21}$$

For the observability of the system, we have the following conclusion (Theorem 2.3.6 in Ref. [11]):

Theorem 2. *Consider a compact space X of a system state and transformation T in Eq. (5.16). Then we have:*

- *If $R < h_s(T)$, the system is not observable.*
- *If $R > h_s(T)$, the system is observable.*

Here we provide an intuitive explanation of the encoding and decoding mechanism to achieve observability, as well as the reason for the unobservability when the transmission rate is too low. A rigorous proof can be found in Chapter 2 of Ref. [11].

- $R < h_s(T)$: We assume that a coder-decoder pair can make the system observable. Then for n time slots and any $\epsilon > 0$, we can find $2^{\lceil nR \rceil}$ spanning points in time $[0, n-1]$ such that they form an (n, ϵ)-separated orbits, due to the definition of observability. Then the topological entropy of the system will be no larger than $\lim_{n\to\infty} \frac{1}{n} \log_2 2^{nR} < h_s(T)$, which contradicts the assumption that the topological entropy is $h_s(T)$.
- $R > h_s(T)$: Due to the definition of $h_s(T)$, for any $n > 1$ and $\epsilon > 0$, we can always find an (n, ϵ)-spanning set with cardinality no larger than 2^{nR} in the state space. Then we can simply transmit the index of the corresponding point in the spanning set (thus taking a transmission rate no larger than R) and achieve an error less than ϵ. This encoding procedure is illustrated in Fig. 5.5.

Communication requirements for linear system control

The communication requirements for control are more complicated. We focus on only the special case of linear systems without noise and with direct observation of the system state:

[1] For a vector $\mathbf{x}$, $\|\mathbf{x}\|_\infty = \max_j |x_j|$.

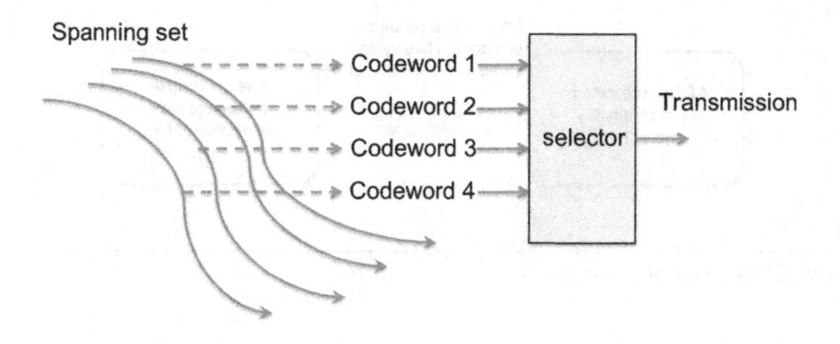

FIG. 5.5

Illustration of the encoding process.

$$\begin{cases} \mathbf{x}(t+1) = \mathbf{A}\mathbf{x}(t) + \mathbf{B}\mathbf{u}(t), \\ \mathbf{y}(t) = \mathbf{x}(t). \end{cases} \tag{5.22}$$

The control action is for the purpose of system stabilization, which is defined as follows:

Definition 5. The system in Eq. (5.22) is stabilizable if there exist a communication period $T \geq 1$ and a 3-tuple of encoder, decoder and controller such that the system is stable in the following sense: for any $\epsilon > 0$ there is an integer t_0 such that

$$\|\mathbf{x}(t)\|_\infty < \epsilon, \quad \forall t \geq t_0, \tag{5.23}$$

for any traces of the dynamics and any initial condition $\mathbf{x}(0)$.

Intuitively, that the system is stabilizable means that we can arbitrarily control the system state given sufficient time. The following two assumptions are subsequently made:

- The initial state $\mathbf{x}(0)$ lies in a compact set $\mathcal{X}_1$ which has the origin as an interior point; i.e., there exists a $\delta > 0$ such that $\|\mathbf{x}\|_\infty < \delta$ implies $\mathbf{x} \in \mathcal{X}_1$.
- The system $(\mathbf{A}, \mathbf{B})$ is stabilizable given a perfect system state feedback.

Then the following theorem provides conditions for the stabilizabilities of the linear system:

Theorem 3. *Suppose that the above two assumptions hold. We denote by $h_s(\mathbf{A})$ the topological entropy of system $\mathbf{x}(t+1) = \mathbf{A}\mathbf{x}(t)$. Then the following two statements hold:*

- *When $R < h_s(\mathbf{A})$, the system is not stabilizable.*
- *When $R > h_s(\mathbf{A})$, the system is stabilizable.*

The first conclusion ($R < h_s(\mathbf{A})$) in the above theorem can be obtained from contradiction, as illustrated in Fig. 5.6. A detailed proof can be found in Section 2.7 of Ref. [11]. Here we provide a sketch of the proof. We assume that the linear system is still stabilizable when $R < h_s(\mathbf{A})$. Since $R < h_s(\mathbf{A})$, we can find a constant $\epsilon > 0$ such that

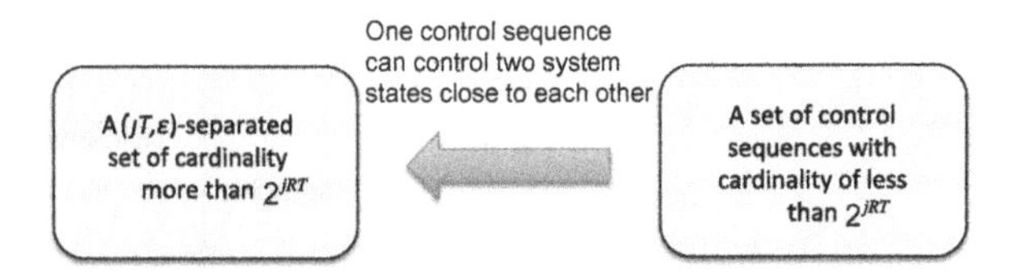

FIG. 5.6

Illustration of the proof of Theorem 3.

$$\limsup_{k \to \infty} \frac{1}{k} \log_2 s(k, \epsilon) > R, \tag{5.24}$$

where $s(k, \epsilon)$ is the metric-based topological entropy. This implies that there is a (jT, ϵ)-separated set S of cardinality N such that

$$\frac{\log_2 N}{jT} > R, \tag{5.25}$$

and for any two different points $\mathbf{x}_1$ and $\mathbf{x}_2$ in S satisfying

$$\|\mathbf{x}_1(t) - \mathbf{x}_2(t)\|_\infty \geq \epsilon. \tag{5.26}$$

On the other hand, we set $\epsilon_0 = \frac{1}{2}\epsilon$. Since the system is stabilizable, we can always find control actions for each element $\mathbf{x}$ in S such that $\|\mathbf{x}(t)\|_\infty \leq \epsilon_0$. There are no more than 2^{jRT} possible control action sequences. Hence we can always find a sequence of controls, $\mathbf{u}^*(0), \ldots, \mathbf{u}^*(jT)$, in these sequences such that we can find $\mathbf{x}_1$ and $\mathbf{x}_2$ that are both well controlled by the sequence $\{\mathbf{u}^*\}$; i.e., both $\mathbf{x}_1'(t) = \mathbf{x}_1(t) + \mathbf{x}_u(t)$ and $\mathbf{x}_2'(t) = \mathbf{x}_2(t) + \mathbf{x}_u(t)$ satisfy Eq. (5.23) with respect to ϵ_0. Then it is easy to verify

$$\|\mathbf{x}_1(t) - \mathbf{x}_2(t)\| \geq \epsilon. \tag{5.27}$$

This contradicts the assumption of the (jT, ϵ)-separated set S. Hence the cardinality of the message set cannot be less than $2^{h_s(\mathbf{A})}$.

The second conclusion ($R > h_s(\mathbf{A})$) in the theorem can be derived from the conclusions in the subsequent treatment of the optimal control of linear systems.

Communication requirements for optimal control

In the previous discussion, the feedback is to stabilize the system, regardless of the corresponding cost (e.g., a very large control action $\mathbf{u}$). In the optimal control of linear systems, the costs of both system state magnitude and control action power are considered; i.e., the system cost with initial state $\mathbf{x}(0)$ is given by

$$J(\mathbf{x}(0)) = \sum_{t=1}^{\infty} \mathbf{x}^T(t)\mathbf{Q}\mathbf{x}(t) + \mathbf{u}^T(t)\mathbf{R}\mathbf{u}(t), \tag{5.28}$$

where both $\mathbf{Q}$ and $\mathbf{R}$ are positive definite matrices.

When the feedback is perfect (i.e., there is no constraint on the communication rate), it is well known that the optimal feedback control is given in [72]

$$\mathbf{u}(t) = -\mathbf{K}\mathbf{x}(t), \tag{5.29}$$

where

$$\mathbf{K} = (\mathbf{R} + \mathbf{B}^T\mathbf{P}\mathbf{B})^{-1}\mathbf{B}^T\mathbf{P}\mathbf{A}, \tag{5.30}$$

and the matrix $\mathbf{P}$ satisfies the following equation:

$$\mathbf{A}(\mathbf{P} - \mathbf{P}\mathbf{B}(\mathbf{R} + \mathbf{B}^T\mathbf{P}\mathbf{B})^{-1}\mathbf{B}^T\mathbf{P})\mathbf{A} + \mathbf{Q} - \mathbf{P} = 0. \tag{5.31}$$

The optimal cost function is given by

$$J_{\text{opt}}(\mathbf{x}(0)) = \mathbf{x}^T(0)\mathbf{P}\mathbf{x}(0). \tag{5.32}$$

Now we consider the constraint on the communication rate; i.e., R is bounded. We define the solvability of the optimal control as follows [11]:

Definition 6. The optimal control of the linear system in Eqs. (5.29)–(5.32) is solvable with respect to the communication model introduced above, if for any $\epsilon > 0$, there exist a $T \geq 1$ and a 3-tuple of encoder, decoder, and controller such that

- The closed-loop controlled system is stable in the following sense: For any $\epsilon > 0$, there exists a $t_0 \geq 1$ such that Eq. (5.23) holds.
- The cost satisfies

$$J(\mathbf{x}(0)) \leq J_{\text{opt}}(\mathbf{x}(0)) + \epsilon, \tag{5.33}$$

where $J_{\text{opt}}(\mathbf{x}(0))$ is given in Eq. (5.32).

The conclusion on the communication requirements for the solvable optimal control is given in the following theorem:

Theorem 4. *Suppose that the assumptions in Theorem 3 hold and the pair* $(\mathbf{A}, \sqrt{\mathbf{Q}})$ *has no unobservable nodes on the unit circle. Then the following statements are correct:*

- *If* $R < h_s(\mathbf{A})$*, then the optimal control is not solvable.*
- *If* $R > h_s(\mathbf{A})$*, then the optimal control is solvable.*

The conclusion for $R < h_s(\mathbf{A})$ is a straightforward conclusion of Theorem 3, since the system is not even stabilizable when $R < h_s(\mathbf{A})$. The proof for the case of $R > h_s(\mathbf{A})$ is much more involved (note that the corresponding conclusion in Theorem 3 is merely a special case). A rigorous proof can be found in Section 2.7 of Ref. [11]. Here we provide a sketch of the proof. First we choose a reasonable period T. Then we consider the system:

$$\mathbf{x}_a(t+1) = \mathbf{A}\mathbf{x}_a(t). \tag{5.34}$$

We can find a set Ω of initial point $\{\mathbf{x}_n^*(0)\}$ with cardinality $2^{(T+1)\alpha}$, where $h_s(\mathbf{A}) < \alpha < R$, with the following property: For any initial state $\mathbf{x}(0)$, we can always find a point $\mathbf{x}_n^*(0)$ in Ω such that the trace in Eq. (5.34) triggered by $\mathbf{x}_n^*(0)$ is very close to that triggered by $\mathbf{x}(0)$ within the time duration $[0, T]$. For each point $\{\mathbf{x}_n^*(0)\}$ in Ω, we can compute the corresponding control actions within time period $[0, T]$, namely $\mathbf{u}^*(0), \ldots, \mathbf{u}^*(T-1)$.

Then we define the following encoder and controller pair for time interval $[0, T-1]$:

- Encoder: Given the initial state $\mathbf{x}(0)$, we choose the corresponding point in Ω, namely $\{\mathbf{x}_n^*(0)\}$, such that the traces are sufficiently close to each other in $[0, T-1]$:

$$h(1) = \text{Enc}(\mathbf{x}(0)) = n. \tag{5.35}$$

- Controller: In the time interval $[0, T-1]$, the control action is set as the optimal control action for $\{\mathbf{x}_n^*(0)\}$, namely

$$(\mathbf{u}^T(0), \ldots, \mathbf{u}^T(T-1))^T = ((\mathbf{u}_n^*)^T(0), \ldots, (\mathbf{u}_n^*)^T(T-1))^T. \tag{5.36}$$

For the jth time period, namely $[(j-1)T, jT-1]$, we can use the same encoding and control mechanism by scaling the system state and control actions, namely

$$h(j) = \text{Enc}\left(\frac{1}{b^j}\mathbf{x}(0)\right) \tag{5.37}$$

and

$$(\mathbf{u}^T(jT), \ldots, \mathbf{u}^T((j+1)T-1))^T = b^j((\mathbf{u}_n^*)^T(0), \ldots, (\mathbf{u}_n^*)^T(T-1))^T. \tag{5.38}$$

The whole procedure is illustrated in Fig. 5.7.

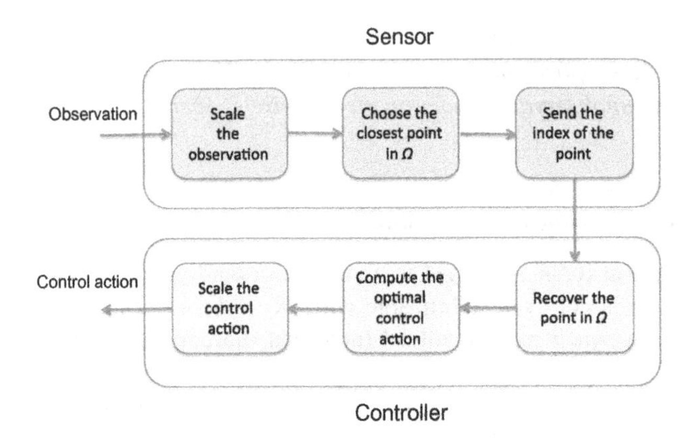

FIG. 5.7

Illustration of the encoding, decoding, and control action procedure.

5.2.3 CALCULATION OF TOPOLOGICAL ENTROPY

The above discussions show the importance of the topological entropy on the communication requirements for controlling the physical dynamics in a CPS. Hence it is very important to evaluate the topological entropy of a given dynamical system.

Linear systems

The topological entropy of linear systems is fully explained in the following theorem (the rigorous proof is given in Theorem 2.4.2 of Ref. [11]):

Theorem 5. *The topological entropy of linear system* $\mathbf{x}(t+1) = \mathbf{A}\mathbf{x}(t)$*, where* $\mathbf{A}$ *is an* $N \times N$ *square matrix, is given by*

$$h_s(\mathbf{A}) = \sum_{n=1}^{N} \log_2(\max\{1, |\lambda_n|\}), \tag{5.39}$$

where $\{\lambda_n\}_{n=1,\ldots,N}$ *are the eigenvalues of* $\mathbf{A}$.

An immediate conclusion is that, when all the eigenvalues of $\mathbf{A}$ are within the unit circle, the corresponding topological entropy is zero. Hence the corresponding communication requirement is zero; i.e., no communication is needed for observing or stabilizing the physical systems. This is obvious since the system state will converge to zero spontaneously.

Generic systems

For the generic case of dynamical systems, there have been no systematic approaches to compute the topological entropy. The existing approaches can be categorized into the following two classes:

- Consider some special types of dynamics, such as one-dimensional piecewise monotone mappings [73].
- Discretize the mapping and then using the approaches of calculating the topological entropy of symbol dynamics.

In this book, we provide a brief introduction to the second approach by following the argument in Ref. [74]. We consider a compact topological space M and partition it into a set of subsets $Q = \{A_1, \ldots, A_q\}$. Consider a continuous mapping T on M and a discrete time period $\{0, \ldots, t-1\}$. We define

$$W_t(T, Q) = \{(a_0, \ldots, a_{t-1}) | \exists x \in M \text{ and } 0 \le i \le t-1, \text{ s.t., } T^i x \in A_{a_i}\}, \tag{5.40}$$

namely the set of all possible t-strings, each of which represents a series of regions in Q that a trajectory of T passes. An example is given in Fig. 5.8, where the trajectory generates a 6-string $(1, 4, 2, 6, 7, 3)$.

Then we define the following quantity:

$$h^*(T, Q) = \lim_{t \to \infty} \frac{|\log W_t(T, Q)|}{t}. \tag{5.41}$$

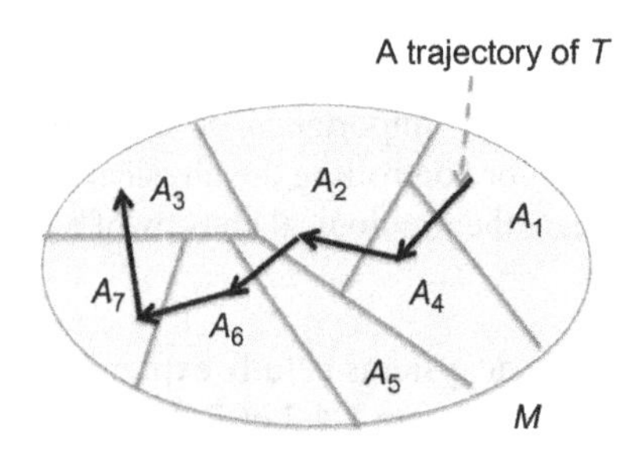

FIG. 5.8

Illustration of the partition and trajectory.

We further define the concept of generating partition:

Definition 7. We say that a partition Q of the space M is generating if

$$\vee_{i=0}^{\infty} T^{-i} = \mathcal{B}, \quad \text{or } \vee_{i=0}^{\infty} T^{i} = \mathcal{B}, \tag{5.42}$$

where $\mathcal{B}$ is the Borel σ-algebra of M.

Then we obtain the following conclusion that provides bounds for the topological entropy of T:

Theorem 6. *Denote by $h(T)$ the topological entropy of transform T. Then it satisfies*

$$h(T) \leq \lim \inf_{\text{diam}(Q)\to 0} h^*(T, Q). \tag{5.43}$$

Moreover, if Q is generating, we have

$$h(T) \leq h^*(T, Q). \tag{5.44}$$

The proof of the theorem can be found in Ref. [74].

Based on this theorem, an algorithm for computing an upper bound of the topological entropy was proposed in Ref. [74], consisting of the following four steps:

- Step 1: Select a coarse partition of M, which is denoted by A.
- Step 2: Select a partition $B = \{B_1, \ldots, B_K\}$ that is much finer than A. Each element in A is the union of a set of elements in B. Then we define a topological Markov chain. The transition matrix is given by

$$\mathbf{B}_{ij} = \begin{cases} 1, & \text{if } B_i \cap T^{-1}B_j \neq \phi, \\ 0, & \text{otherwise,} \end{cases} \tag{5.45}$$

which means that $\mathbf{B}_{ij}$ equals 1 if B_i and B_j are two successive regions that a trajectory of T visits. A sequence $(b_0, \ldots, b_{N-1})$ is called a B-word if

$\mathbf{B}_{b_i,b_{i+1}} = 1$. Each B-word corresponds to an A-word $(a_0, \ldots, a_{N-1})$, where $B_{b_i} \subset A_{a_i}$. Then we define the set of A-words as

$$W_N(B,A) = \{\mathbf{a})|\exists B\text{-word } \mathbf{b} \text{ generating } \mathbf{a}\}. \tag{5.46}$$

It is easy to verify

$$W_N(T,A) \subset W_N(B,A), \tag{5.47}$$

which implies

$$\begin{aligned} h(B,A) &= \lim_{N\to} \frac{\log|W_N(B,A)|}{N} \\ &\geq \lim_{N\to} \frac{\log|W_N(T,A)|}{N} = h^*(T,A). \end{aligned} \tag{5.48}$$

It has been shown that $h(B,A)$ converges to $h(T,A)$ if the diameter of B tends to zero.

- Step 3. The set of all B-words forms a sofic shift, whose detailed definition can be found in Ref. [74]. First a directed labeled graph can be constructed, in which each node corresponds to one element in B and an edge exists between nodes i and j if $\mathbf{B}_{ij} = 1$. Moreover, if $B_i \in \alpha \in A$, we label the edge ij (if it exists) as α. The following example from Ref. [74] illustrates the procedure:
  Example 1. Consider a one-dimensional mapping T from [0, 1] to [0, 1], which is given by

$$Tx = \begin{cases} 2x, & x < 1/2, \\ 1.5(x - 0.5), & x \geq 1/2. \end{cases} \tag{5.49}$$

  The partitions A and B are selected as

$$\begin{cases} A = \{A_1, A_2\} = \{[0, 0.5), [0.5, 1)\}, \\ B = \{B_1, B_2, B_3, B_4\} = \{[0, 0.25), [0.25, 0.5), [0.5, 0.75), [0.75, 1)\}. \end{cases} \tag{5.50}$$

  It is easy to verify that the matrix $\mathbf{B}$ is given by

$$\mathbf{B} = \begin{pmatrix} 1 & 1 & 0 & 0 \\ 0 & 0 & 10 & 1 \\ 1 & 1 & 0 & 0 \\ 0 & 1 & 1 & 0 \end{pmatrix}. \tag{5.51}$$

  The graph generated by the matrix $\mathbf{B}$ is given in Fig. 5.9. A path in the graph can represent a string.
- Step 4. We compute the entropy of the sofic shift, which is shown to be equal to the topological entropy of the corresponding Markov chain. To that end, we need to find a new graph $G'(B,A)$ satisfying
  - $W_N(G'(B,A)) = W_N(G(A,B))$;
  - All outgoing edges of the same node have different labels.

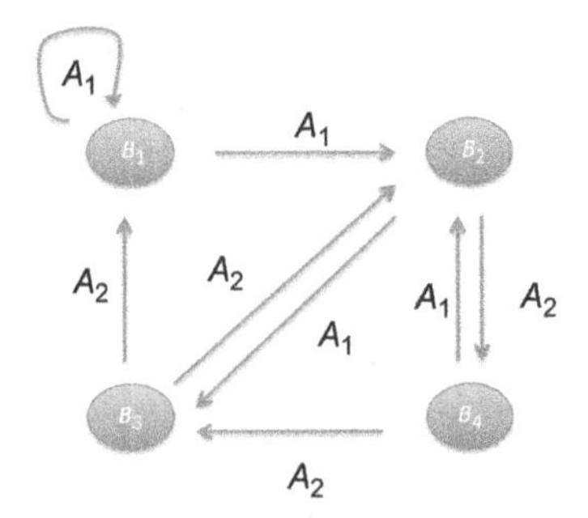

FIG. 5.9

The graph generated in the example.

The following algorithm from Ref. [74] has been proposed to find such a subgraph of G', which is called a reduced right-resolving presentation of G:

1. Begin from an empty graph R. Add a single node in $G(B, A)$ to R.
2. If there is at least one hyper node without outgoing hyper edges, choose one of them and label it by H. Otherwise, go to Step 6.
3. Since the hyper node H consists of multiple nodes in $G(B, A)$, define H' as the set of all nodes in $G(B, A)$ reached by the outgoing edges of the nodes in H that have the same label (say, A_i). Define a new hyper node H' and a hyper edge pointing from H to H'. Label it as A_i.
4. Repeat Step 3 for all possible labels in A.
5. Return to Step 2.
6. If a hyper node has no incoming hyper edges, remove it.
7. If a hyper node is removed, return to Step 6. Otherwise, stop and output R.

Take the example in Fig. 5.9, for instance. If we begin from node B_1, then we label B_1 as H_1. Since all the outgoing edges of B_1 are labeled as A_1, in Step 3 we form a hyper node H_2 consisting of B_1 and B_2 (since the two outgoing edges reach B_1 and B_2). Since H_2 does not have any outgoing edges, we form a hyper node H_3 consisting of B_1, B_2, B_3 and $B4$. Then we consider H_3 whose outgoing edges have two labels A_1 and A_2. For A_1, we form a hyper node H_4 consisting of B_1, B_2, and B_3; for A_2, the outgoing edge returns to H_4. This results in the graph shown in Fig. 5.10. Finally, we remove the hyper node H_1.
From the graph R, we can also obtain its adjacency matrix. The example in Fig. 5.10 has the matrix $\mathbf{R}$ given by

$$\mathbf{B} = \begin{pmatrix} 0 & 0 & 1 \\ 1 & 0 & 1 \\ 0 & 1 & 1 \end{pmatrix}. \tag{5.52}$$

We can obtain the eigenvalues of $\mathbf{R}$, which are related to the topological entropy that we want to compute, in the following theorem [74]:

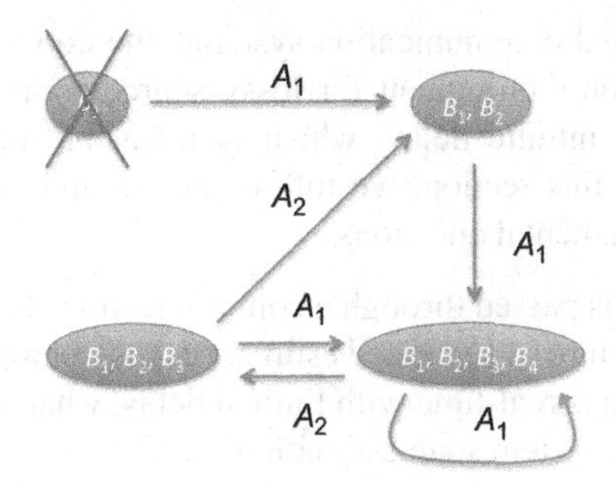

FIG. 5.10

The graph R generated in the example.

Theorem 7. *Denote by $\lambda_{\max}(\mathbf{R})$ the maximum eigenvalue of $\mathbf{R}$. If T is transitive, we have*

$$h^*(T,A) \leq h(B,A) = \log \lambda_{\max}(\mathbf{R}). \tag{5.53}$$

Note that although the above procedure provides a systematic approach for the concrete calculation of topological entropy, it does not completely solve the problem for the following reasons: (1) It is still not clear how to verify that a partition is generating. (2) It is still not clear how to choose the radius of the partitions. Hence we can only choose as small a radius as possible, within the capability of computing.

5.3 STOCHASTIC SYSTEMS: ESTIMATION

Many control systems use the structure of separated estimation and control, as illustrated in Fig. 5.11, in which an estimator provides an estimation $\hat{\mathbf{x}}$ of the system state, and then a controller considers the estimate $\hat{\mathbf{x}}$ as the true system state and computes the corresponding control action. However, such a separated structure may not be optimal, although it is more feasible in practice. On the other hand, source coding is needed for the communication of observations $\mathbf{y}$ in order to estimate the system state $\mathbf{x}$, which is usually lossy when the system state is

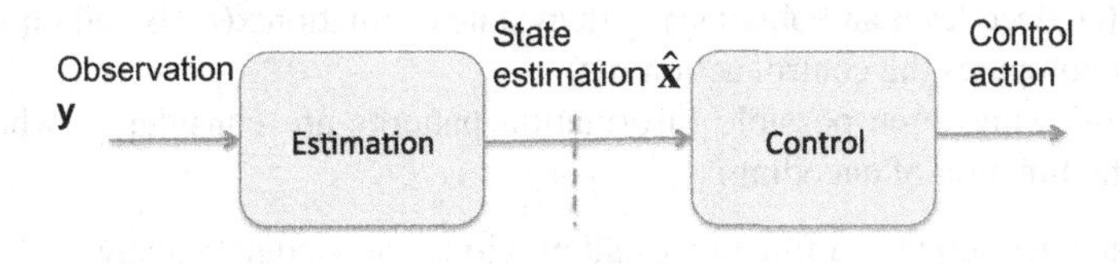

FIG. 5.11

Illustration of separated estimation and control.

continuous. In traditional communication systems, the corresponding information-theoretic communication requirement for lossy source coding is based on infinite block length and thus infinite delay, which is infeasible in the context of real-time control. Hence in this section, we follow the research in Ref. [14] and study the following two fundamental questions:

- When the feedback is passed through a communication channel with limited capacity, is the structure of separated estimation and control still optimal?
- When the estimation is real-time with limited delay, what are the communication requirements for the system state estimation?

5.3.1 SYSTEM MODEL

We first introduce the system model, which is adopted in Ref. [14]. The overall system has the same structure as that in Fig. 5.4.

We assume that the system state dynamics is described by the linear system dynamics in Eq. (5.4). The noise $\{\mathbf{w}(t)\}_t$ is a sequence of i.i.d. Gaussian distributed random variables. The initial state $\mathbf{x}(0)$ is assumed to be Gaussian with zero expectation and covariance matrix $\Sigma_x(0)$. For simplicity, we assume that the system state can be observed directly by the encoder at the sensor.

The communication channel is assumed to be stochastic, characterized by stochastic kernels $P(dy|x)$, where y and x are the output and input of the channel. For simplicity we assume that the channel is time-invariant and memoryless. The following two special channels are discussed in Ref. [14]:

- Noisy digital channel with rate R: The input and the output of the channel are the same. There are a total of 2^R messages.
- Memoryless vector Gaussian channel: Both the input and output alphabets are d-dimensional real spaces. The output of the channel is given by

$$\mathbf{y}(t) = \mathbf{x}(t) + \mathbf{v}(t), \tag{5.54}$$

  where $\mathbf{v}(t)$ is a series of i.i.d. Gaussian distributed random vectors with zero expectation and covariance matrix Σ_v. The input is average power constrained; i.e., $E[\|\mathbf{x}\|^2] \leq P_0$.

At each time slot t, the sensor obtains an observation $\mathbf{x}(t)$, generates and transmits a symbol $x(t)$ through the communication channel. Then the controller receives a symbol $y(t)$, decodes it and obtains a system state estimation $\hat{\mathbf{x}}(t)$. Based on $\hat{\mathbf{x}}(t)$, the controller computes the control action $\mathbf{u}(t)$.

The following three possible information patterns are considered, where Enc denotes the function of encoding:

- The encoder output is a function of all previous knowledge, namely

$$x(t) = \text{Enc}(\mathbf{x}(0:t), x(0:t-1), y(0:t-1), \hat{\mathbf{x}}(0:t-1), \mathbf{u}(0:t-1)). \tag{5.55}$$

- The encoder does not know the channel and decoder outputs, namely

$$x(t) = \text{Enc}(\mathbf{x}(0:t), x(0:t-1), \mathbf{u}(0:t-1)). \tag{5.56}$$

- The encoder uses only the current system state to encode the message:

$$x(t) = \text{Enc}(\mathbf{x}(t)). \tag{5.57}$$

The decoder uses a stochastic mechanism for decoding, namely generating the system state estimation according to the following conditional distribution:

$$P(d\hat{\mathbf{x}}(t)|\mathbf{y}(0:t), \hat{\mathbf{x}}(0:t-1), \mathbf{u}(0:t-1)); \tag{5.58}$$

namely, it is dependent on the previous received communication symbols, the previous system state estimations and control actions.

The controller is a stochastic one, which is generated according to the condition probability $P(d\mathbf{u}(t)|\hat{\mathbf{x}}(t))$, which is dependent only on the current system state estimation.

5.3.2 SEPARATION

Now we study whether the communication channel will affect the optimality of the separated structure of system state estimation and control. In traditional linear quadratic Gaussian (LQG) control with perfect feedback, the certainty equivalence guarantees the optimality of the following control law:

$$\mathbf{u}(t) = -\mathbf{K}\hat{\mathbf{x}}(t) \tag{5.59}$$

and

$$\hat{\mathbf{x}}(t) = E[\mathbf{x}(t)|\mathbf{y}(0:t-1), \mathbf{u}(0:t-1)], \tag{5.60}$$

which represent separated control and system estimation. We define the estimation error as

$$\delta\mathbf{x}(t) = \mathbf{x}(t) - E[\mathbf{x}(t)|\mathbf{y}(0:t-1), \mathbf{u}(0:t-1)]. \tag{5.61}$$

We say that the control has no dual effect if (here $\bar{y}(t)$ denotes the channel output when there is no feedback control)

$$E[\delta\mathbf{x}(t)\delta\mathbf{x}^T(t)|\mathbf{y}(0:t), \mathbf{u}(0:t-1)] = E[\delta\mathbf{x}(t)\delta\mathbf{x}^T(t)|\bar{\mathbf{y}}(0:t)], \tag{5.62}$$

which means that the error covariance matrix is independent of the control action taken by the controller. An intuitive explanation of the no dual effect is given in Ref. [14]: the control action can be used to both control the system state evolution and prob the system in order to estimate the system state; the no dual effect means that the control action is fully used to control the system state (since the system state estimation error is irrelevant to the control action). The following theorem shows that the no dual effect is equivalent to the certainty equivalence [14]:

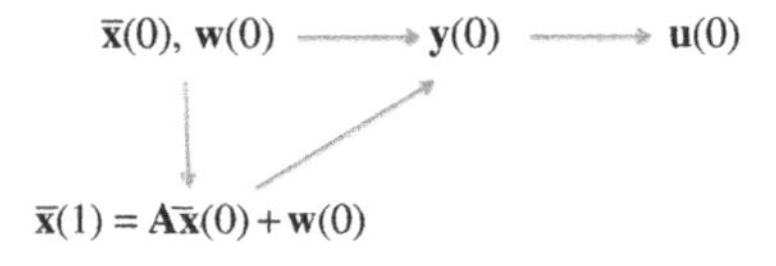

FIG. 5.12

Illustration of Markov chains in the noiseless digital channel.

Theorem 8. *The optimal control law for the linear system has the property of certainty equivalence if and only if it has no dual effect.*

The following lemma provides a sufficient condition for the no dual effect, which can be used to judge whether a control action with certainty equivalence is optimal:

Lemma 1. *Denote by $\bar{\mathbf{x}}(t)$ the uncontrolled system state. If $\sigma(\bar{y}(0:t)) \subset \sigma(y(0:t), \mathbf{u}(0:t-1))$ and $E[\bar{\mathbf{x}}|\bar{y}(0:t)] = E[\bar{\mathbf{x}}|y(0:t), \bar{y}(0:t), \mathbf{u}(0:t-1)]$, then the control has no dual effect.*

Based on this lemma, Ref. [14] discussed the no dual effect in the two communication channels introduced above:

- Noiseless digital channel: Since the input and the output of the communication channel are the same (while the number of bits is limited), the sensor can know exactly the control action that the controller will take. It is shown in Ref. [14] that the optimal encoder has the following form:

$$
\begin{aligned}
x(t) &= q_t(\mathbf{x}(t) - E[\mathbf{x}(t)|y(0:t-1), \mathbf{u}(0:t-1)]) \\
&= q_t(\bar{\mathbf{x}}(t) - E[\bar{\mathbf{x}}(t)|y(0:t-1), \mathbf{u}(0:t-1)]),
\end{aligned}
\tag{5.63}
$$

  where q_t is a quantizer. Note that the second equation is due to the sensor's perfect knowledge of the control action. By applying induction, we can prove the following lemma:

  Lemma 2. *For a noiseless digital channel, $\bar{y}(t) = y(t)$ and $\bar{\mathbf{x}}(t) \to y(0:t) \to \mathbf{u}(0:t-1)$ is a Markov chain.*

  A rigorous proof is given in Ref. [14]. Here we provide the proof for $t = 1$, as illustrated in Fig. 5.12. When $t = 0$, obviously we have $\bar{y}(0) = y(0)$. Since the control action is completely determined by the received message, $\bar{\mathbf{x}}(0), \mathbf{w}(0) \to y(0) \to \mathbf{u}(0)$ forms a Markov chain. Since $\bar{\mathbf{x}}(1) = \mathbf{A}\bar{\mathbf{x}}(0) + \mathbf{w}(0)$, $\bar{\mathbf{x}}(1) \to y(0) \to \mathbf{u}(0)$ also forms a Markov chain.

  By applying Lemma 1, we draw the conclusion that the control through a noiseless digital channel has no dual effect.

- Memoryless vector Gaussian channel: The optimal encoder for the vector Gaussian channel is still unknown. Ref. [14] restricted the encoder to be linear and deterministic:

$$
\begin{aligned}
x(t) = \mathbf{D}_{1t}\mathbf{x}(0:t) + \mathbf{D}_{2t}\mathbf{u}(0:t-1) \\
+ \mathbf{D}_{3t}x(0:t-1) + \mathbf{D}_{4t}y(0:t-1),
\end{aligned}
\tag{5.64}
$$

where $\{\mathbf{D}_{nt}\}_{n=1,2,3,4}$ are matrices of appropriate dimensions. Since the communication channel output is given by $y(t) = x(t) + v(t)$, it can also be written as a linear combination of the initial state, dynamics noise, communication noise, and control actions:

$$\begin{aligned} y(t) &= \mathbf{F}_{1t}\mathbf{x}(0) + \mathbf{F}_{2t}\mathbf{w}(0:t-1) \\ &\quad + \mathbf{F}_{3t}v(0:t) + \mathbf{F}_{4t}\mathbf{u}(0:t-1), \end{aligned} \tag{5.65}$$

where $\{\mathbf{F}_{nt}\}_{n=1,2,3,4}$ are the corresponding matrices. The uncontrolled channel output can also be written as

$$\bar{y}(t) = \mathbf{F}_{1t}\mathbf{x}(0) + \mathbf{F}_{2t}\mathbf{w}(0:t-1) + \mathbf{F}_{3t}v(0:t), \tag{5.66}$$

since the controls are all zero. Then we have $\bar{y}(0:t) = y(0:t) - \mathbf{G}_t\mathbf{u}(0:t-1)$, where $\mathbf{G}_t$ is an appropriate matrix. When estimating the uncontrolled system state $\bar{\mathbf{x}}(t)$, all the related information in $y(0;t)$ and $\mathbf{u}(0:t-1)$ is summarized in $\bar{y}(0;t)$. Hence the condition in Lemma 1 is valid, and thus the control through the memoryless vector Gaussian channel also has no dual effect.

Based on the above arguments, we reach the conclusion that, for both the noiseless digital communication channels and memoryless vector Gaussian channels, the structure of separated system state estimation and control is still optimal in linear dynamics.

5.3.3 SEQUENTIAL ESTIMATION

When the structure of separated system state estimation and control is optimal, the controller can use the same design as that in the case of perfect feedback (e.g., the LQG control). However, the system state estimation needs substantial revisions due to the limited communication capacity. We discuss the communication requirements for system state estimation at the controller in this subsection.

For the linear dynamics in Eq. (5.4), system state estimation can be considered as a lossy source coding problem, since the sensor encodes the observations and the decoder output always has errors except in trivial cases. The lossy source coding is traditionally studied using rate-distortion theory [33], as a branch of information theory, which discusses the minimum communication requirement to satisfy the requirement on the distortion at the decoder output. However, the information-theoretic argument is based on the assumption of infinite codeword length and thus infinite delays, which is reasonable for traditional data communications (e.g., one packet encoded as a codeword may have thousands of bits). Obviously such arguments are not suitable for CPSs, since the encoding procedure must have a limited delay. Hence in contrast to the block structure of traditional data communications, the encoding procedure in a CPS should have a sequential structure, as illustrated in Fig. 5.13.

Before we study sequential rate distortion, we provide a review of traditional rate-distortion theory. When the true information symbol is x while the recovered one

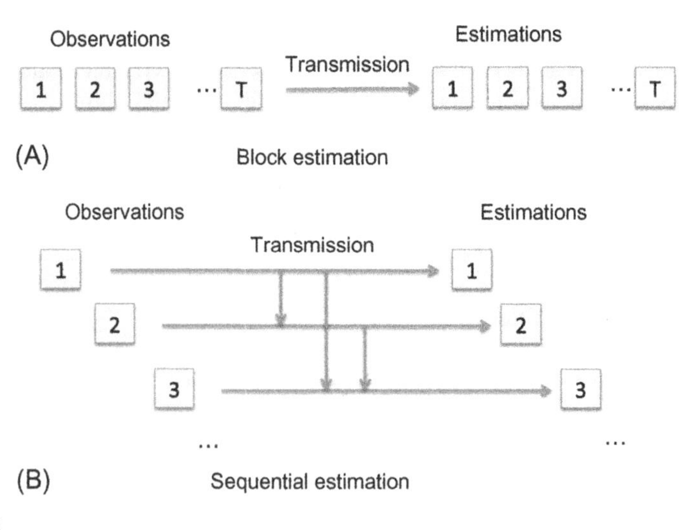

FIG. 5.13

Comparison between block and sequential estimation.

is y, we denote the distortion by $D(x, y)$. Then the rate-distortion function is defined as follows:

Definition 8. For a certain information source generating symbols $\{X(t)\}$, the rate-distortion function is defined as

$$R(D) = \inf_{Q_{Y|X}(y|x), D_Q \leq D} I(Y; X), \tag{5.67}$$

where $Q_{Y|X}$ is the conditional probability of generating the recovered symbol Y given the true information symbol X, D_Q is the distortion incurred by the conditional distribution $Q_{Y|X}$, and I is mutual information.

Intuitively, the rate-distortion function means the minimum average number of bits needed to achieve a distortion less than or equal to D. Meanwhile, the rate-distortion function can also determine whether a communication channel with a certain capacity can support the transmission of an information source with a given distortion bound. Roughly speaking, when transmitting the information of a Gauss-Markov process through a communication channel having capacity C, distortion D can be achieved if $R(D) < C$, while it cannot be achieved if $R(D) > C$. This dichotomy is illustrated in Fig. 5.14.

Similarly to the traditional definition of rate-distortion function with infinite delay, Ref. [14] defined the sequential rate-distortion function as follows:

Definition 9. For information source $\{X(t)\}$ (the recovered symbols are $\{Y(t)\}$), the sequential rate-distortion function is defined as

$$R_T^s(D) = \inf_{P \in \mathcal{P}_T(D)} \frac{1}{T} I(X(0:T-1); Y(0:T-1)), \tag{5.68}$$

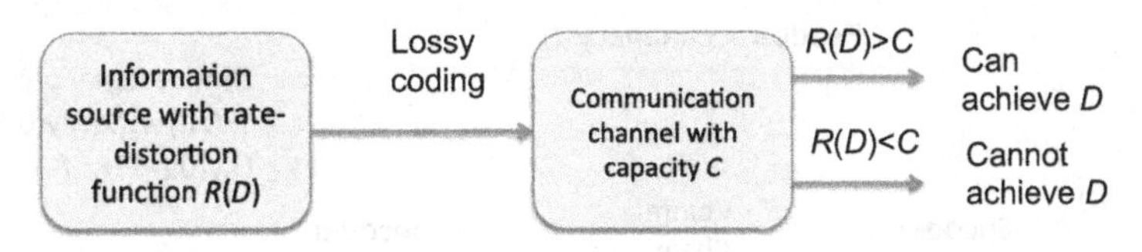

FIG. 5.14

Dichotomy of the rate-distortion function.

where the family of conditional probability $\mathcal{P}_T(D)$ is given by

$$\mathcal{P}_T(D) = \{\{P(dY(t)|x(0:t-1), y(0:t-1))\}_{t=0,\dots,T-1} \\ |E[d(X(t), Y(t))] \leq D, \forall t\}. \tag{5.69}$$

We notice that the sequential rate-distortion function has the following features:

- The necessary rate may be dependent on the time T.
- Only the causal information is considered in the mutual information, which is due to the requirement of sequential decoding.
- The expected distortion is no larger than D for each time slot, instead of being averaged over all time slots.

The following theorem shows the application of the sequential rate-distortion function in the lossy transmission of information sources [14]. Note that a sufficient condition is much more difficult to find.

Theorem 9. *Consider a memoryless communication channel with channel capacity C. A necessary condition for achieving the distortion D causally is* $R_T^s(D) \leq C$ *for all T.*

Gauss-Markov source

Since the sequential rate-distortion function provides a lower bound for the channel capacity of the communication channel, it is important to calculate the sequential rate-distortion function for various information sources for the purpose of online system state estimation. We follow the argument in Ref. [14] to discuss the sequential rate-distortion function for a d-dimensional Gauss-Markov source where

$$\mathbf{x}(t+1) = \mathbf{A}\mathbf{x}(t) + \mathbf{w}(t) \tag{5.70}$$

and

$$d(\mathbf{x}, \mathbf{y}) = \|\mathbf{x} - \mathbf{y}\|_2^2. \tag{5.71}$$

We assume a Gaussian communication channel with dimension d.

The following lemma shows the necessary property for the infimizing channel in Eq. (5.68) in the definition of a sequential rate-distortion function:

Lemma 3. *The infimizing channel in the Gauss-Markov information source in Eq. (5.70), namely* $P(dY(t)|x(0:t-1), y(0:t-1))$, *is a Gaussian channel of the form* $P(dY(t)|x(t), y(0:t-1))$.

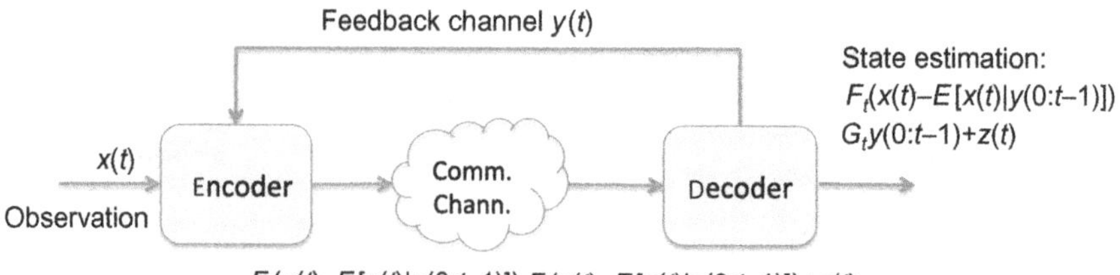

FIG. 5.15

Realization of channel $P(dY(t)|x(t), y(0 : t-1))$ in a memoryless Gaussian channel with perfect feedback.

Compared with the standard form $P(dY(t)|x(0 : t-1), y(0 : t-1))$, in the form $P(dY(t)|x(t), y(0 : t-1))$ the distribution of Y is dependent only on the current input of the communication channel, instead of all the history. Hence the recovered symbols have the following form:

$$\hat{\mathbf{x}}(t) = \mathbf{F}x(t) + \mathbf{G}y(0 : t-1) + \mathbf{z}(t), \tag{5.72}$$

where $\mathbf{F}$ is a $d \times d$ matrix, $\mathbf{G}$ is a $d \times (t-1)d$ matrix, and $\{\mathbf{z}(t)\}$ is a series of independent Gaussian random vectors of dimension d.

Such a channel can be realized for a memoryless Gaussian channel with perfect feedback, which is illustrated in Fig. 5.15. The encoder obtains the innovation, which is given by

$$x(t) = \mathbf{F}_t(\mathbf{x}(t) - E[\mathbf{x}(t)|y(0 : t-1)]). \tag{5.73}$$

Since the communication channel output $y(0 : t-1)$ can be sent to the encoder via the perfect feedback channel, the encoder can easily compute $x(t)$ as above. The decoder receives $y(t) = x(t) + \mathbf{z}(t)$ and then obtains the system state estimation:

$$\begin{aligned}\hat{\mathbf{x}}(t) = \mathbf{F}(\mathbf{x}(t) - E[\mathbf{x}(t)|y(0 : t-1)])\\ + \mathbf{F}_t E[\mathbf{x}(t)|y(0 : t-1)] + \mathbf{G}y(0 : t-1) + \mathbf{z}(t).\end{aligned} \tag{5.74}$$

The channel input should satisfy the power constraint:

$$P \geq E[x(t)x^T(t)] = \text{trace}[\mathbf{F}_t^T \Sigma_{\mathbf{x}(t)|y(0:t-1)} \mathbf{F}_t], \tag{5.75}$$

where $\Sigma_{\mathbf{x}(t)|y(0:t-1)}$ is the covariance matrix of the prediction error, namely

$$\Sigma_{\mathbf{x}(t)|y(0:t-1)} = \text{Cov}(\mathbf{x}(t) - E[\mathbf{x}(t)|y(0 : t-1)]). \tag{5.76}$$

For the realized channel, Ref. [14] has proved that

$$I(\mathbf{x}(0 : T-1); \hat{\mathbf{x}}(0 : T-1)) = \sum_{t=0}^{T-1} I(x(t); y(t)), \tag{5.77}$$

because

$$\begin{aligned} I(\mathbf{x}(0:T-1);\hat{\mathbf{x}}(0:T-1)) &= \sum_{t=0}^{T-1} I(\mathbf{x}(0:T-1);\hat{\mathbf{x}}(t)|\hat{\mathbf{x}}(0:t-1)) \\ &= \sum_{t=0}^{T-1} I(\mathbf{x}(t);\hat{\mathbf{x}}(t)|\hat{\mathbf{x}}(0:t-1)), \end{aligned} \tag{5.78}$$

where the second equation is due to the form of the estimation in Eq. (5.72), and

$$\begin{aligned} I(\mathbf{x}(t);\hat{\mathbf{x}}(t)|\hat{\mathbf{x}}(0:t-1)) &= h(\hat{\mathbf{x}}(t)|\hat{\mathbf{x}}(0:t-1)) - h(\hat{\mathbf{x}}(t)|\mathbf{x}(t),\hat{\mathbf{x}}(0:t-1)) \\ &= h(y(t)|\hat{\mathbf{x}}(0:t-1)) - h(y(t)|x(t),\hat{\mathbf{x}}(0:t-1)) \\ &= h(y(t)) - h(y(t)|x(t)) \\ &= I(x(t);y(t)), \end{aligned} \tag{5.79}$$

where the second equation is because

$$\hat{\mathbf{x}}(t) = y(t) + \mathbf{F}_t E[\mathbf{x}(t)|\hat{\mathbf{x}}(0:t-1)] + \mathbf{G}_t\hat{\mathbf{x}}(0:t-1). \tag{5.80}$$

Before we proceed to compute the sequential rate-distortion function, we provide a review of the traditional rate-distortion function. Consider a d-dimensional Gaussian source with independent samples and $d \times d$ covariance matrix Σ_x. The distortion function is given by $d(\mathbf{x},\hat{\mathbf{x}}) = \|\mathbf{x}-\hat{\mathbf{x}}\|^2$. Then we have [33]

$$R(D) = \frac{1}{2}\sum_{n=1}^{d} \log\frac{\lambda_n}{\delta_n}, \tag{5.81}$$

where $\{\lambda_n\}_{n=1,\ldots,d}$ are the eigenvalues of Σ_x and

$$\delta_n = \begin{cases} c, & \text{if } c \le \lambda_n, \\ \lambda_n, & \text{if } c > \lambda_n, \end{cases} \tag{5.82}$$

where the constant c is chosen such that $\sum_{n=1}^{d}\delta_n = D$. In particular, when $d = 1$, we have

$$R(D) = \frac{1}{2}\log\frac{\lambda_1}{D}. \tag{5.83}$$

Hence δ_n can be considered as the portion of distortion assigned to the nth eigenvector of Σ_x; then the assignment of $\{\delta_n\}$ can be considered as water filling, as illustrated in Fig. 5.16.

The optimal channel to minimize the rate R given the distortion D is given as follows [75]. Two channels are defined:

- Backward channel: $P(\mathbf{x}|\hat{\mathbf{x}})$. It is given by

$$\mathbf{x} = \hat{\mathbf{x}} + \delta, \tag{5.84}$$

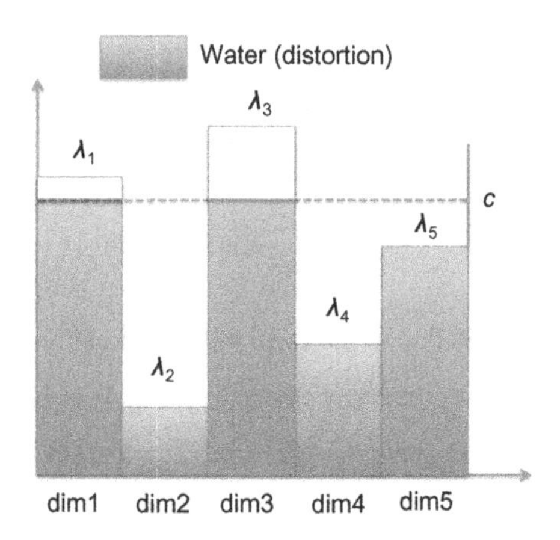

FIG. 5.16

Illustration of water filling in the rate-distortion function of a Gaussian information source.

where δ is the error, which is Gaussian distributed with zero mean and covariance matrix $\Sigma_{x|y}$. Hence the output $\hat{\mathbf{x}}$ is also Gaussian distributed with zero mean and expectation:

$$\Sigma_{\hat{x}} = \Sigma_x - \Sigma_{x|y}. \tag{5.85}$$

- Forward channel: $P(\hat{\mathbf{x}}|\mathbf{x})$. It is given by

$$\hat{\mathbf{x}} = \mathbf{H}\mathbf{x} + \mathbf{z}, \tag{5.86}$$

where $\mathbf{H} = E[\hat{\mathbf{x}}\mathbf{x}^T]E[(\mathbf{x}\mathbf{x}^T)^{-1}]$, and $\mathbf{z}$ is Gaussian with zero mean and covariance matrix

$$\Sigma_z = E[\hat{\mathbf{x}}\hat{\mathbf{x}}^T] - E[\hat{\mathbf{x}}\mathbf{x}^T]E[(\mathbf{x}\mathbf{x}^T)^{-1}]E[\mathbf{x}\hat{\mathbf{x}}^T]. \tag{5.87}$$

To realize the forward channel $\hat{\mathbf{x}} = \mathbf{H}\mathbf{x} + \mathbf{z}$, one can define an invertible matrix Γ such that $\Gamma\mathbf{H}\Gamma^{-1} = \mathbf{I}$. Define $\mathbf{b} = \Gamma\hat{\mathbf{x}}$, $\mathbf{a} = \Gamma\mathbf{x}$, and $\mathbf{n} = \Gamma\mathbf{z}$. Then the channel $\mathbf{b} = \mathbf{a} + \mathbf{n}$ with power constraint $E[\mathbf{a}^T\mathbf{a}] \leq trace(\Gamma\Sigma_x\Gamma^T)$ is matched to the information source.

We also need the concept of a matched channel. The stochastic kernel $P(d\hat{\mathbf{x}}(t)|\mathbf{x}(0 : t-1), \hat{\mathbf{x}}(t-1))$ can be considered as a communication channel with input $\mathbf{x}(t)$ and output $\hat{\mathbf{x}}(t)$ [76]. If the real communication channel is equal to the channel that minimizes the rate given the distortion, then the communication channel is said to be matched to the information source. When the communication channel is matched to the source, there is no need for the structure of separated source encoder and channel encoder. The information can be sent in an uncoded manner without any loss in performance.

Then for the sequential rate-distortion function, we can convert it to the traditional rate-distortion expression. We begin with the scalar Gaussian source case. The distortion is given by $\sigma_x^2 2^{-2R}$. The encoder computes the innovation information:

$$\begin{aligned} &\mathbf{x}(t) - E[\mathbf{x}(t)|\hat{\mathbf{x}}(0:t-1)] \\ &= \mathbf{A}\mathbf{x}(t-1) + \mathbf{w}(t) - \mathbf{A}\hat{\mathbf{x}}(t-1) \\ &= \mathbf{A}(\mathbf{x}(t-1) - \hat{\mathbf{x}}(t-1)) + \mathbf{w}(t) \\ &= \mathbf{A}\delta(t-1) + \mathbf{w}(t), \end{aligned} \tag{5.88}$$

where $\delta(t) = \mathbf{x} - \hat{\mathbf{x}}(t)$ is the estimation error. Since the error variance is equal to the distortion, namely $D(t) = E[\delta^2(t)]$, then the variance of the innovation information is given by

$$E[\|x(t)\|^2] = A^2 D(t-1) + \sigma_w^2, \tag{5.89}$$

where σ_w^2 is the scalar notation of Σ_w.

In the above argument for the traditional rate-distortion function, we have shown that the scalar Gaussian variable has variance σ_x^2 when a Gaussian channel having capacity R is matched to the source. Hence the distortions at different times can be obtained recursively as follows:

$$\begin{cases} D(t) = (A^2 D(t-1) + \sigma_w^2) 2^{-2R(t)}, & \forall t \geq 1, \\ D(0) = \sigma_{x_0}^2 2^{-2R(0)}. \end{cases} \tag{5.90}$$

For the sequential rate-distortion function, we have $D(t) = D$. Hence we have

$$\begin{cases} R(t) = \max\{0, 1/2 \log(A^2 + (\sigma_w^2/D))\}, & \forall t \geq 1, \\ R(0) = \max\{0, 1/2 \log(\sigma_w^2/D)\}. \end{cases} \tag{5.91}$$

When T is sufficiently large, the impact of $R(0)$ will vanish. Hence we have

$$\lim_{T\to\infty} R_T^s(D) = \max\left\{0, \frac{1}{2}\log\left(A^2 + \frac{\sigma_w^2}{D}\right)\right\}. \tag{5.92}$$

Now we handle the generic case of a d-dimensional Gauss-Markov source. The covariance matrix of the distortion error $\delta(t)$ is denoted by $\Sigma(t) = E[\delta(t)\delta^T(t)]$. The evolution of the covariance matrix can be easily shown to be

$$\Sigma(t) = \mathbf{A}\Sigma(t-1)\mathbf{A}^T + \Sigma_w. \tag{5.93}$$

Consider the unitary matrix $\mathbf{U}(t)$ that diagonalizes the matrix $\mathbf{A}\Sigma(t-1)\mathbf{A}^T + \Sigma_w$:

$$\mathbf{U}(\mathbf{A}\Sigma(t-1)\mathbf{A}^T + \Sigma_w)\mathbf{U}^T = \text{diag}[\lambda_1(t), \ldots, \lambda_d(t)]. \tag{5.94}$$

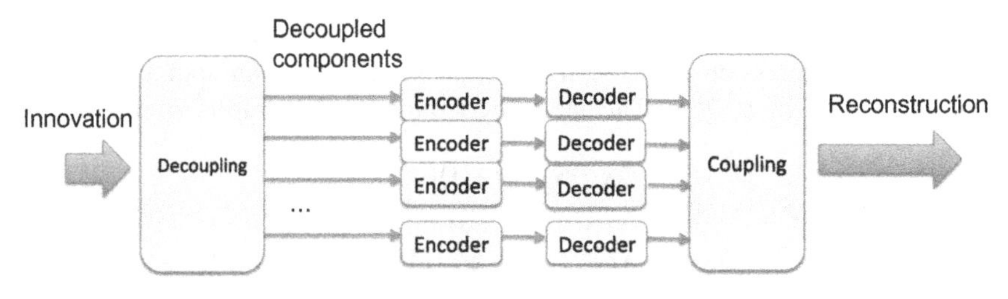

FIG. 5.17

The coding and decoding procedure for a d-dimensional Gaussian source.

From the previous water-filling argument, to achieve the distortion D, the coding rate needs to be

$$R(t) = \sum_{n=1}^{d} \frac{1}{2} \log \left(\frac{\lambda_n(t)}{\sigma_n(t)} \right), \tag{5.95}$$

where

$$\sigma_n(t) = \begin{cases} c(t), & \text{if } c(t) \leq \lambda_n(t), \\ \lambda_n(t), & \text{if } c(t) > \lambda_n(t), \end{cases} \tag{5.96}$$

where the constant $c(t)$ is chosen such that $\sum_{n=1}^{d} \sigma_n(t) = D$.

Then by using the rate-distortion function for the scalar Gaussian source, the error covariance matrix of the d-dimensional source at time slot t is given by

$$\begin{aligned} \Sigma(t) &= \mathbf{U}^T(t)\mathbf{U}(t)(\mathbf{A}\Sigma(t-1)\mathbf{A}^T + \Sigma_w)\mathbf{U}^T(t) \\ &\times \begin{pmatrix} 2^{-2R_1(t)} & \cdots & \cdots & \\ \ddots & \vdots & & \ddots \\ \cdots & \cdots & 2^{-2R_d(t)} & \end{pmatrix} \mathbf{U}(t). \end{aligned} \tag{5.97}$$

The evolution of the error covariance matrix is explained in Fig. 5.17. The encoding and decoding procedures are given as follows:

1. The sensor receives $\mathbf{x}(t)$ and then computes the innovation $\mathbf{e}(t) = \mathbf{x}(t) - E[\mathbf{x}(t)|y(0:t-1)]$.
2. The sensor computes the decoupling unitary matrix $\mathbf{U}$ and obtains the decoupled components $\mathbf{x}'(t) = \mathbf{U}\mathbf{e}(t)$.
3. The sensor uses the encoding approach in the scalar Gaussian source to encode each component of $\mathbf{e}'(t)$ and sends out the signal through the communication channel.
4. The estimator receives the signals of each component and reconstructs the transferred innovation. The estimation is denoted by $\hat{\mathbf{e}'}(t)$.
5. The estimator estimates the innovation by $\hat{\mathbf{e}}(t) = \mathbf{U}^T\hat{\mathbf{e}'}(t)$ and then obtains the estimation $\hat{\mathbf{x}}(t) = \hat{\mathbf{e}}(t) + E[\mathbf{x}(t)|y(0:t-1)]$.

When the distortion is sufficiently small, we have

$$\mathbf{U}\Sigma(t)\mathbf{U}^T = \mathrm{diag}(D/d, \ldots, D/d), \tag{5.98}$$

and thus the error covariance matrix evolution is given by

$$\begin{aligned}\Sigma(t) &= \mathbf{U}^T(t)\mathbf{U}(t)\left(\frac{D}{d}\mathbf{A}\mathbf{A}^T + \Sigma_w\right)\mathbf{U}^T(t)\\ &\quad\times\begin{pmatrix} 2^{-2R_1(t)} & \cdots & \cdots \\ \ddots & \vdots & \ddots \\ \cdots & \cdots & 2^{-2R_d(t)}\end{pmatrix}\mathbf{U}(t)\\ &= \frac{D}{d}\mathbf{I}.\end{aligned} \tag{5.99}$$

Based on the above equality, we have

$$\begin{aligned}R_n(t) &= \frac{1}{2}\log\frac{\lambda_n\left(\frac{D}{d}\mathbf{A}\mathbf{A}^T + \Sigma_w\right)}{\frac{D}{d}}\\ &= \frac{1}{2}\log\lambda_n\left(\mathbf{A}\mathbf{A}^T + \frac{d}{D}\Sigma_w\right).\end{aligned} \tag{5.100}$$

Hence the total rate is given by

$$\begin{aligned}\sum_{n=1}^{d} R_n(t) &= \sum_{n=1}^{d}\frac{1}{2}\log\lambda_n\left(\mathbf{A}\mathbf{A}^T + \frac{d}{D}\Sigma_w\right)\\ &= \frac{1}{2}\log\left|\mathbf{A}\mathbf{A}^T + \frac{d}{D}\Sigma_w\right|.\end{aligned} \tag{5.101}$$

Then when the distortion D is small and as T tends to infinity, we have

$$\lim_{T\to\infty} R_T^s(D) = \frac{1}{2}\log\left|\mathbf{A}\mathbf{A}^T + \frac{d}{D}\Sigma_w\right|. \tag{5.102}$$

Noiseless digital channel

When the communication channel is digital and noiseless, it is not matched to the information source. Since the communication channel is digital, it is necessary to quantize the information source and then transmit the discrete bits through the digital channel. For the sequential rate-distortion function, we need a sequential quantizer as follows [14]:

Definition 10. A sequential quantizer is a series of functions mapping from $R^{(t+1)\times d} \times R^{t\times d}$ to R^d, and the corresponding range is at most countable.

For practical applications, we consider the following "operational sequential rate-distortion function," which explicitly considers the quantizer structures:

Definition 11. The operational sequential rate-distortion function is defined as

$$R_T^{S,o}(D) = \inf_{q_0,\ldots,q_{T-1}\in\mathcal{F}_T^o} \frac{1}{T} H(\hat{\mathbf{x}}(0:T-1)), \tag{5.103}$$

where $\mathcal{F}_T^o$ is the set of quantizers satisfying the distortion constraint, namely

$$\mathcal{F}_T^o = \{(q_0,\ldots,q_{T-1}) | E[d(\mathbf{x}(t),\hat{\mathbf{x}}(t))] \le D\}. \tag{5.104}$$

Intuitively, the operational sequential rate-distortion function is to minimize the uncertainty of reconstruction while keeping the expected distortion at each time below D. However, it is more difficult to compute the exact value of the operational sequential rate-distortion function since it involves the structures of the quantizers. We can use the sequential rate-distortion function as an approximation. It is easy to verify that the sequential rate-distortion function provides a lower bound for the operational sequential rate-distortion function since

$$H(\hat{\mathbf{x}}(0:T-1)) \ge I(\mathbf{x}(0:T-1), \hat{\mathbf{x}}(0:T-1)). \tag{5.105}$$

Moreover, Ref. [14] has shown that, for very low distortion (thus very high coding rate), the operational sequential rate-distortion and sequential rate-distortion functions are infinitesimally close to each other:

Theorem 10. *For each T, we have*

$$\lim_{D\to 0} \frac{R_T^{S,o}(D)}{R_T^S(D)} = 1. \tag{5.106}$$

5.4 STOCHASTIC SYSTEMS: STABILITY

In the previous section, we studied the communication requirements for controllers with separated state estimation and control. However, there are two limitations to this setup:

- For generic cases, it is not clear whether separated state estimation and control are optimal.
- For the communication channel, we used the channel capacity to measure the capability of conveying information. However, the concept of channel capacity is reasonable for communications with long delays, not for the case of CPSs with very limited delays.
- The possible transmission errors, due to the noisy channel, are not considered.

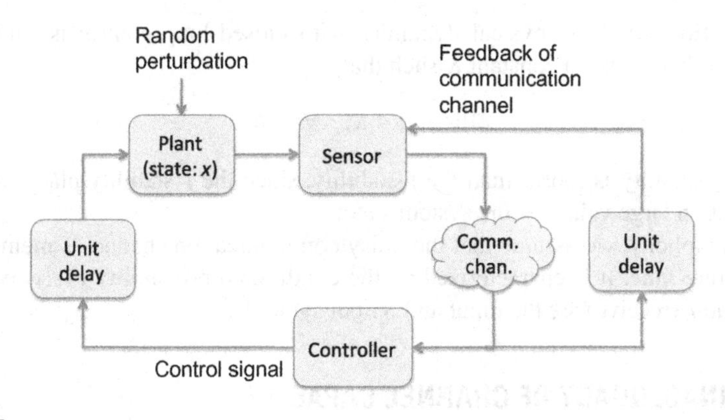

FIG. 5.18

System model for studying the communication requirements to stabilize the physical dynamics [13].

Hence in this section, we will study a novel metric for the communication channel in the context of stabilizing stochastic linear physical dynamics, namely the anytime capacity, which was proposed by Sahai and Mitter in 2006 [13].

5.4.1 SYSTEM MODEL

For simplicity, we consider a scalar system state in a discrete time framework, whose evolution law is given by

$$x(t+1) = Ax(t) + u(t) + w(t), \tag{5.107}$$

where x is the system state, u is the control action, and w is time-independent random perturbation with bounded amplitude (i.e., $|w(t)| < \Omega/2$), $E[w] = 0$, and $E[w^2] = \sigma_w^2$. We assume that $A > 1$; otherwise, there is no need for communications since the system is inherently stable.

The overall system under study is illustrated in Fig. 5.18. A sensor observes the system state of the physical dynamics and then sends messages through a noisy communication channel. There could be a feedback channel for the forward communication channel, through which the sensor knows the output of the communication channel. It is also possible for the sensor to know the control action taken by the controller, if the communication channel output is fed back to the sensor and the sensor shares the strategy of the controller.

The following two concepts of system stability will be used in the subsequent analysis:

Definition 12. The physical dynamics with closed-loop control is said to be f-stable if

$$P(|x(t)| > m) < f(m), \quad \forall t, m \geq 0, \tag{5.108}$$

where f is a positive function of m.

Definition 13. The physical dynamics with closed-loop control is said to be η-stable if there exists a constant K such that

$$E[|x(t)|^{\eta}] \leq K, \quad \forall t \geq 0. \tag{5.109}$$

The η-stability is looser than the f-stability, since the f-stability places a direct constraint on large values of the system state.

For simplicity, we assume that the noisy communication channel is memoryless and discrete-time. It is characterized by the conditional probability $P(r|s)$, where s (send) and r (receive) are the input and output symbols.

5.4.2 INADEQUACY OF CHANNEL CAPACITY

For a communication channel, the traditional channel capacity is given by

$$C = \sup_{P(S)} I(R; S), \tag{5.110}$$

which is optimized over the possible input symbol probability $P(S)$. As explained in Chapter 2, the channel capacity measures the capability of conveying information with asymptotically long codeword lengths and arbitrarily small error rates. However, in the context of controlling physical dynamics, the infinite delay of codewords in the traditional setup of information is intolerable, thus making the traditional concept of channel capacity inadequate.

Below is an example proposed in Ref. [13] showing the inadequacy of the traditional concept of channel capacity. Consider the binary erasure channel illustrated in Fig. 5.19. The probability that the transmitted symbol, 0 or 1, is correctly received is $1 - \delta$, while the probability of the symbol being erased (thus receiving a common symbol e) is given by δ. Similarly, we can have an L-bit erasure channel, in which the alphabets of inputs and outputs are both $\{0, 1\}^L$ (hence the input and output are L-dimensional binary vectors, or L-bit packets). The correct transmission probability is $p(s|s) = \delta$, while the erasure probability $p(e|s)$ is δ. It has been shown that the channel capacity of an L-bit erasure channel is given by $(1 - \delta)L$.

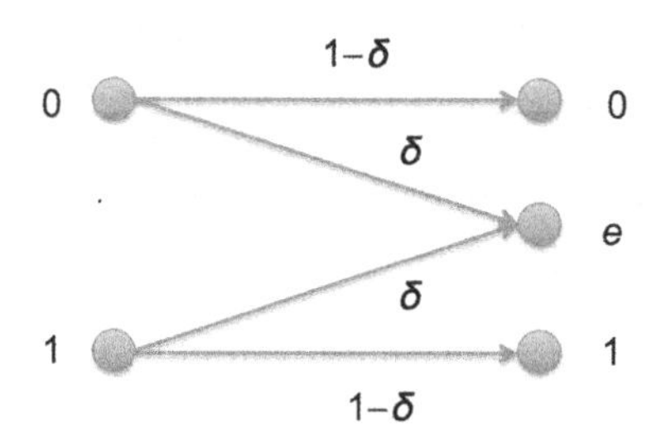

FIG. 5.19

Illustration of erasure channel.

Furthermore, we can have a real erasure channel, in which the input and output alphabets are real numbers R, and the conditional probabilities are $p(x|x) = 1-\delta$ and $p(0|x) = \delta$.

For a counterexample of traditional channel capacity, we consider the system in Eq. (5.107) with $A = 1.5$ and $\Omega = 1$ (hence the perturbation satisfies $|w(t)| \leq 0.5$). We assume that the communication channel is the real erasure case with $\delta = 0.5$. Hence with probability 0.5, the system state $x(t)$ can be perfectly transmitted to the controller, while the transmission is blocked with probability 0.5. The optimal control strategy is to transmit the system state measured by the sensor directly, thus $s(t) = x(t)$; the control action is $u(t) = -Ar(t)$. Hence when the transmitted symbol is successfully passed through the communication channel, the previous system state can be completely canceled by the control action and only the random perturbation remains. However, if the transmitted symbol is erased, no control action is carried out (since $u(t) = 0$).

In the event that all the transmissions after time $t-i-1$ are erased while the first $t-i$ transmissions are correct, the system state at time $t+1$ is given by

$$x(t+1) = \sum_{j=0}^{i} 1.5^j w(t-j). \tag{5.111}$$

Since the set of events (denoted by $\mathcal{E}$) that all the transmissions after time $t-i-1$ (i from 0 to t) are erased while the first $t-i$ transmissions are correct is only a subset of the whole set of events (e.g., there could be alternative erasures and transmission successes), we have

$$\begin{aligned}
E[x^2(t+1)] &> E[x^2(t+1)|\mathcal{E}]P(\mathcal{E}) \\
&= \sum_{i=0}^{t}\left(\frac{1}{2}\right)^{i+1} E\left[\left|\sum_{j=0}^{i} 1.5^j w(t-j)\right|^2\right] \\
&= \sum_{i=0}^{t}\left(\frac{1}{2}\right)^{i+1} \sum_{j=0}^{t}\sum_{k=0}^{t} 1.5^{j+k} E[w(t-j)w(t-k)] \\
&= \sum_{i=0}^{t}\left(\frac{1}{2}\right)^{i+1} \sum_{j=0}^{t} 1.5^{2j}\sigma_w^2 \\
&= \frac{4\sigma^2}{5}\sum_{i=0}^{t}\left(1.125^{i+1} - 0.5^{i+1}\right),
\end{aligned} \tag{5.112}$$

where the first inequality is due to the fact that $\mathcal{E}$ is only a subset of the events, and the third equation is due to the assumption that the noise is white. It is easy to verify that $E[x^2(t+1)]$ diverges as $t \to \infty$.

According to the above argument, the communication channel cannot stabilize the physical dynamics. Meanwhile, the real erasure channel has an infinite channel capacity, since one successfully transmitted real-valued symbol can convey infinite

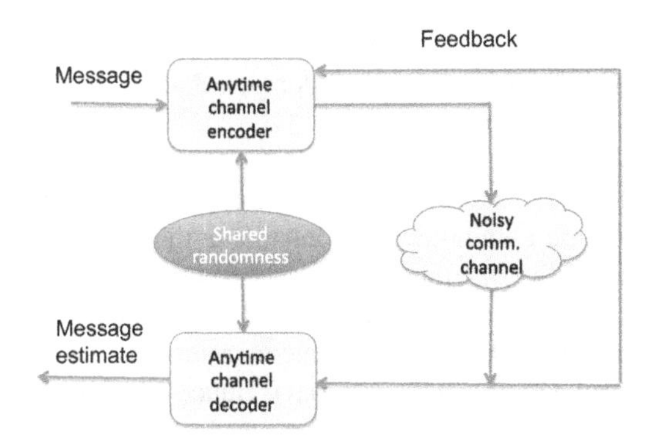

FIG. 5.20

Structure of the encoder and decoder in the anytime capacity setup.

information. Hence the channel capacity in the traditional sense does not necessarily characterize the capability of real-time transmission and system stabilization. Ref. [13] realized this deficiency and thus proposed the concept of anytime capacity, which will be explained subsequently.

5.4.3 ANYTIME CAPACITY

The focus of anytime capacity is "on the maximum rate achievable for a given sense of reliability rather than the maximum reliability possible at a given rate" [13]. The structure of the anytime capacity setup is illustrated in Fig. 5.20.

We denote by M_τ the R-bit observation received by the sensor's encoder at time τ. Based on the output of the communication channel, the decoder obtains a reconstructed message $\hat{M}_\tau(t)$ ($\tau \leq t$), namely the estimation of M_τ at time t. The error probability is $P(\hat{M}_{1:t-d}(t) \neq M_{1:t-d})$, namely the probability that not all the messages $M_1, \ldots, M_{t-d}$ are correctly decoded at time t. The procedure is illustrated in Fig. 5.21.

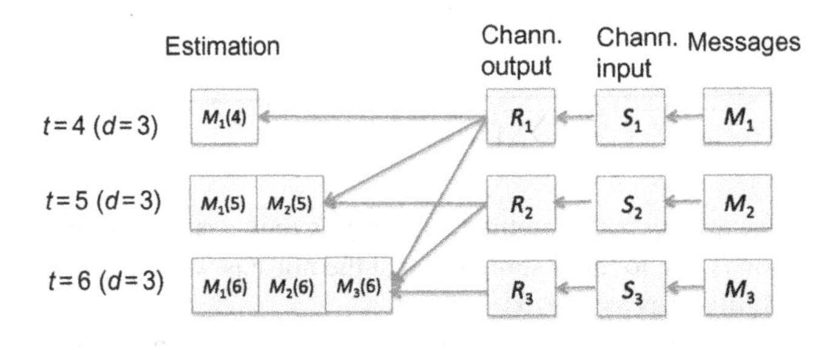

FIG. 5.21

Procedure of encoding and decoding in the anytime capacity setup.

Based on this context, the anytime capacity of a communication channel is defined as follows [13]:

Definition 14. A communication system over a noisy channel with rate R has an encoder and decoder pair, which work as follows:

- At time τ, the encoder receives an R-bit message M_τ.
- The encoder generates a channel input based on all the previously received messages and feedbacks from the decoder. There could be a delay $1+\theta$ in the feedback; hence the information of the feedback mechanism is $r(1:t-1-\theta)$ at time t.
- The decoder generates estimates for the messages, namely $\{\hat{M}_\tau(t)\}_{\tau\le t}$, at time t based on all the received channel outputs.

The system with rate R achieves anytime reliability α if there exists a constant K such that

$$P(\hat{M}_{1:\tau}(t) \neq M_{1:\tau}) \le K2^{-\alpha(t-\tau)}, \tag{5.113}$$

for every τ and t. If Eq. (5.113) is valid for every possible message M, then we say that the system achieves uniform anytime reliability α. The corresponding encoding mechanisms are called anytime codes (uniform anytime codes respectively).

The α-anytime capacity $C_{\text{any}}(\alpha)$ of a communication channel is the maximum of the rate R at which a communication system can be designed for the channel to achieve uniform anytime reliability α.

It is easy to verify the following inequality [13]:

$$C_{\text{any}}(E_r(R)) \ge R, \tag{5.114}$$

where $E_r(R)$ is the traditional error exponent for the transmission rate R of the same communication channel [75]. Intuitively, the anytime capacity has a higher requirement than the traditional channel capacity.

Necessity

The following theorem shows the necessity of the anytime capacity for the purpose of stabilizing the linear physical dynamics in a CPS.

Theorem 11. *Consider $\eta > 0$. If there exists a communication system and controller such that $E[|x(t)|^\eta < K]$ for noise sequence $|w(t)| \le \frac{\Omega}{2}$, then the communication channel satisfies*

$$C_{\text{any}}(\eta) \ge \log_2 A. \tag{5.115}$$

The rigorous proof of the theorem is given in Ref. [13]. Here, we provide a sketch of the proof. The key point is the simulated plant as shown in Fig. 5.22. Due to the perfect feedback channel, the sensor can simulate the plant observer and controller. The idea is to link communication reliability with control stability.

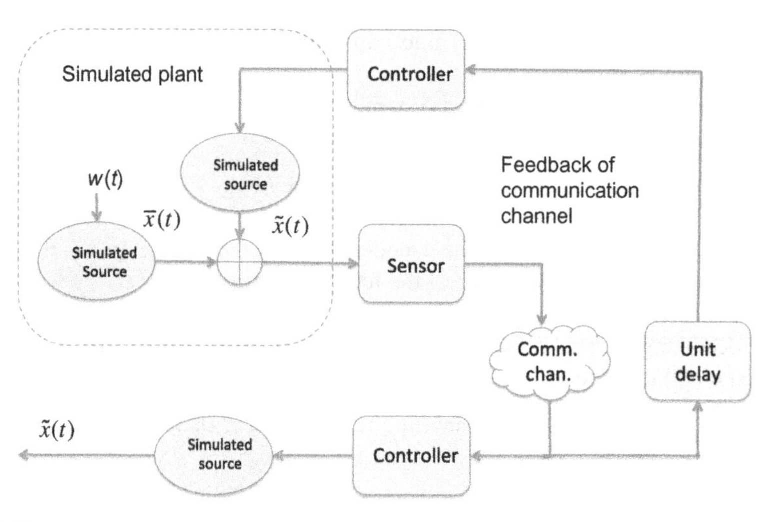

FIG. 5.22

Anytime code structure with simulated plant.

Below is the basic setup of the proof:

- We can consider the system illustrated in Fig. 5.22 as a special communication system. The information source is a sequence of i.i.d. bits $\{s_i(t)\}_{i,t}$ at the sensor. The bit sequence will be regenerated at the controller.
- The bit sequence $\{s_i(t)\}_{i,t}$ is encoded as follows. The sensor uses the bit sequence to generate time-independent random perturbations $\{w(t)\}$, which correspond to the random perturbations in real physical dynamics. Hence the sensor can generate a simulated plant with simulated random perturbation and simulate control actions. The system state in a simulated plant x consists of two parts $\bar{x}$, which is driven by the generated random perturbation w, and $\tilde{x}$, which is driven by the control action.
- We assume that the simulated physical dynamics can be stabilized by the control action. As we will see, the simulated dynamics $x(t)$, as the sum of the simulated sources $\bar{x}(t)$ and $\tilde{x}(t)$, can be reconstructed at the controller with an error of a particular magnitude. The simulated dynamics $\tilde{x}(t)$ in Fig. 5.22 can also be simulated at the controller due to the shared initial state and the control actions at both the sensor and the controller. Then we can use the bit sequence $s(t)$ to convey information about $\bar{x}$. Since x is small due to the stabilization, $\bar{x}$ can be approximated by $\tilde{x}$. Thus $\bar{x}$ can be reconstructed by the controller with small error (since $\tilde{x}$ can be generated by the controller), and the bit sequence $s(t)$ can be decoded with a particular error rate, which can be controlled by the system stability characterized by η.
- Essentially, this communication system conveys information via the simulated physical dynamics, thus linking the communication capacity and system stability.

The details are given below. The simulated system state $\tilde{x}(t)$ has the following dynamics:

$$\tilde{x}(t+1) = A\tilde{x}(t) + u(t), \tag{5.116}$$

where $u(t)$ can be perfectly simulated by the sensor since it knows the output of the communication channel. In the other simulated dynamics, the system state is driven by the simulated noise, namely

$$\bar{x}(t+1) = A\bar{x}(t) + w(t). \tag{5.117}$$

Then the sum of $\bar{x}$ and $\tilde{x}$ is a simulated version of the true dynamics, since

$$\begin{aligned} x(t+1) &= \bar{x}(t+1) + \tilde{x}(t+1) \\ &= A\bar{x}(t) + A\tilde{x}(t) + u(t) + w(t) \\ &= Ax(t) + u(t) + w(t). \end{aligned} \tag{5.118}$$

Since the original physical dynamics can be stabilized and $|x(t)|$ will be sufficiently small, then $\tilde{x}(t)$ will be sufficiently close to $-\bar{x}(t)$.

In Fig. 5.22, we assume that the observer transmits bit i at time $\frac{i}{R}$ where R is the data rate. It is easy to verify that

$$\begin{aligned} \bar{x}(t+1) &= \sum_{i=0}^{t} A^i w(t-1-i) \\ &= A^t \sum_{j=0}^{t} A^{-j} w(j). \end{aligned} \tag{5.119}$$

The sum $\sum_{j=0}^{t} A^{-j} w(j)$ is similar to the representation of a factional number with base A. Then, using induction, we can prove that $\bar{x}$ can be written as

$$\bar{x}(t) = \gamma A^t \sum_{k=0}^{\lfloor Rt \rfloor} (2+\epsilon_1)^{-k} S_k, \tag{5.120}$$

where S_k is the kth bit, and γ and ϵ_1 are constants whose expressions can be found in Ref. [13]. This can be achieved by representing the tth random perturbation $w(t)$ as

$$w(t) = \gamma A^{t+1} \sum_{k=\lfloor Rt \rfloor + 1}^{\lfloor R(t+1) \rfloor} (2+\epsilon_1)^{-k} S_k. \tag{5.121}$$

Assuming that the rate R is smaller than $\log_2 A$, it was shown in Ref. [13] that bits can be encoded into the simulated plant. At the output of the communication channel, estimates $\bar{s}_i(t)$ can be extracted from the ith bit such that

$$P(\hat{s}_{1:j}(t) \neq s_{1:j}(t)) \leq P\left(|x(t)| \geq A^{t-\frac{j}{R}} \left(\frac{\gamma \epsilon_1}{1+\epsilon_1} \right) \right). \tag{5.122}$$

Here the key idea is that, since we have assumed that the physical dynamics are stabilized and thus $x(t)$ is very small, $\bar{x}(t)$ is very close to $\tilde{x}(t)$ and thus can be estimated from $\tilde{x}(t)$ known at the controller (since the initial state and control actions are shared by both the controller and sensor).

Due to the assumption that the system is η-stable, we have

$$P(|x(t)| > m) \leq \frac{E[|x(t)|\eta]}{m^{\eta}} \leq Km^{-\eta}, \tag{5.123}$$

where the first inequality is due to the Markov inequality, which measures the probability of deviation from the expectation, and the second one is due to the assumption of η-stability.

Substituting Eq. (5.123) into Eq. (5.122), we have

$$P(\hat{s}_{1:j}(t) \neq s_{1:j}(t)) \leq \left(K\left(\frac{1}{\gamma} + \frac{1}{\gamma\epsilon_1}\right)\right) 2^{-\eta \log_2 A\left(t-\frac{i}{R}\right)}. \tag{5.124}$$

Notice that $\frac{i}{R}$ is the time that the ith bit is sent and thus $t - \frac{i}{R}$ is the (continuous-time) delay of decoding. This concludes the proof by checking the definition of anytime capacity.

Sufficiency

We first extend the anytime capacity to a broader meaning:

Definition 15. We say that a communication system with rate R achieves g-anytime reliability given a function $g(d)$ if

$$P(\hat{M}_{1:t-d}(t) \neq M_{1:t-d}(t)) < g(d), \tag{5.125}$$

where $g(d) = 1, \forall d < 0$.

Similarly to C_{any} in the previous discussion, we can define $C_{g\text{-any}}$. Note that the definition of anytime capacity in Definition 14 is simply a special case of the g-anytime capacity, if the function g is exponential; i.e.,

$$g(d) = K2^{-\alpha d}. \tag{5.126}$$

Now we discuss the following sufficient condition for the stability of physical dynamics:

Theorem 12. *Suppose that the noisy communication channel satisfies* $C_{g\text{-any}}(g) \geq \log_2 A$. *Then the physical dynamics in Eq. (5.107) can be stabilized such that*

$$P(|x(t)| > m) \leq g(K + \log_A m). \tag{5.127}$$

Encoding procedure

Due to the assumption of perfect feedback of communication channels, the sensor can perfectly know the control action taken by the controller, namely $u(t)$. On the other hand, the sensor also perfectly knows the system state $x(t)$. Hence it also has perfect estimation of the random perturbation by using

$$w(t) = x(t+1) - Ax(t) - u(t). \tag{5.128}$$

Hence one natural idea is to encode $w(t)$ and send it to the controller, since w is the only randomness in the system. However, Ref. [13] proposed a more effective approach in which "the observer will act as though it is working with a virtual controller through a noiseless channel of finite rate R..." The detailed encoding procedure is given as follows.

The sensor simulates a virtual process $\bar{x}(t)$ whose dynamics are given by

$$\bar{x}(t+1) = A\bar{x}(t) + w(t) + \bar{u}(t), \tag{5.129}$$

where $\bar{u}(t)$ is the computed action of a virtual controller simulated at the sensor. It also keeps updating the following two virtual dynamics:

$$x_u(t+1) = Ax_u(t) + \bar{u}(t) \tag{5.130}$$

and

$$\tilde{x}(t) = A\tilde{x}(t) + w(t), \tag{5.131}$$

which results in $\bar{x}(t) = x_u(t) + \tilde{x}(t)$. Since the sensor can stabilize the simulated dynamics $\bar{x}(t)$ within an interval, $-x_u(t)$ should be very close to $\tilde{x}(t)$. Hence the actual controller will try to take actions close to $x_u(t)$. This can be accomplished by the sensor sending an approximation of the computed control action $\bar{u}$ to the actual controller, which is illustrated in Fig. 5.23.

In summary, the whole system works as follows. Since the sensor knows all the information of the physical dynamics, it simulates the stabilization of the dynamics by using computed control actions; the sensor sends its computed control actions to the controller after quantization; the controller estimates all the previous control actions with an exponentially decreasing error rate (guaranteed by the anytime capacity); then the controller estimates the current system state of the simulated dynamics at the sensor and takes control actions to force the true system state to be close to the simulated one. Note that, although the sensors sends its computed control actions to the controller, it does not directly inform the controller about the control action that the controller should take; it simply informs the controller where the simulated system state is and thus the controller can try to catch up with the simulated system state, which is destined to be stable (Fig. 5.24).

Due to the communication constraint R, the sensor can take one of 2^R values (where we assume that 2^R is an integer). If $\bar{x}(t)$ is within the interval $\left[-\frac{\Delta}{2}, \frac{\Delta}{2}\right]$, $A\bar{x}(t)$ lies in $\left[-\frac{A\Delta}{2}, \frac{A\Delta}{2}\right]$. The sensor can choose one of the possible control actions

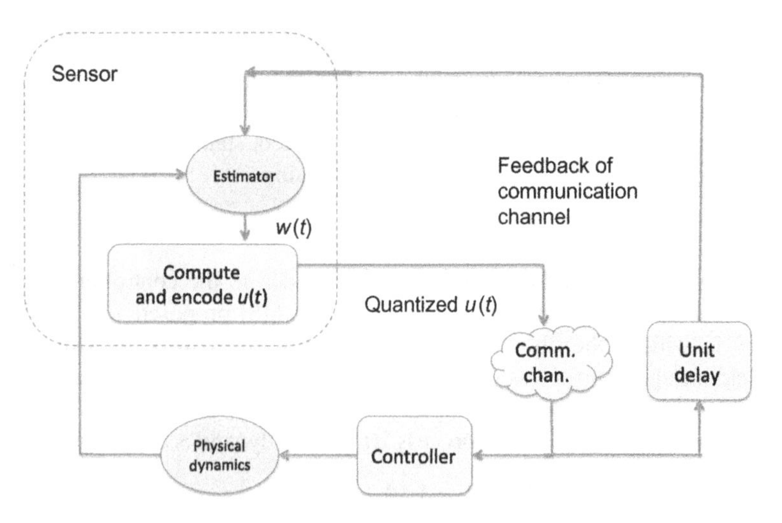

FIG. 5.23

Coding procedure for sufficiency.

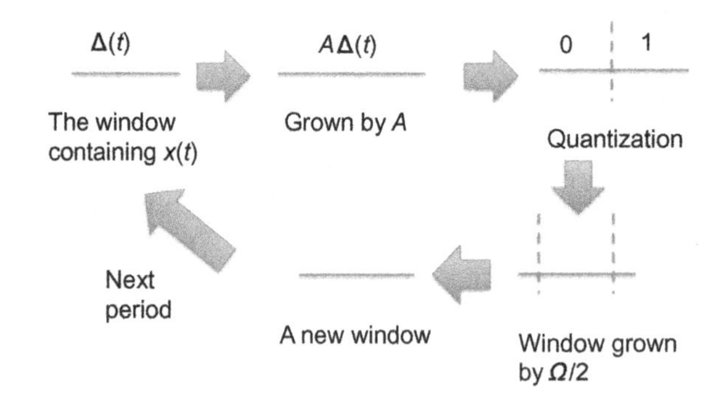

FIG. 5.24

Analysis of coding procedure.

uniformly aligned in the interval $\left[-\frac{A\Delta}{2}, \frac{A\Delta}{2}\right]$, which is within the distance $\frac{A\Delta}{2^{R+1}}$ of $A\bar{x}(t)$, such that

$$A\bar{x}(t) + \bar{u}(t) \in \left[-\frac{A\Delta}{2^{R+1}}, \frac{A\Delta}{2^{R+1}}\right]. \tag{5.132}$$

After the perturbation by the random noise, we have

$$\bar{x}(t+1) \in \left[-\frac{A\Delta}{2^{R+1}} - \frac{\Omega}{2}, \frac{A\Delta}{2^{R+1}} + \frac{\Omega}{2}\right]. \tag{5.133}$$

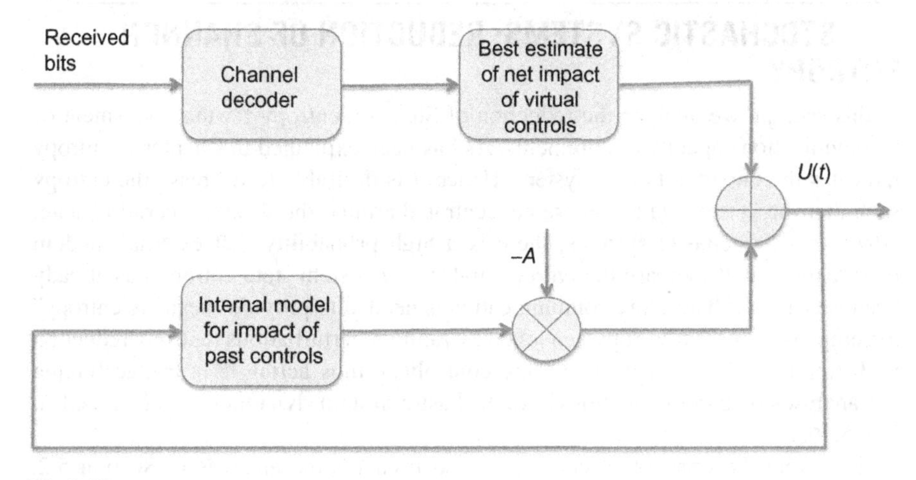

FIG. 5.25

Procedure for calculating the control action.

To make $\bar{x}(t)$ always within $\left[-\frac{\Delta}{2}, \frac{\Delta}{2}\right]$, we require

$$\frac{A\Delta}{2^{R+1}} \leq \frac{\Delta}{2}, \tag{5.134}$$

which can be achieved if

$$\Delta = \frac{\Omega}{1 - A2^{-R}}, \tag{5.135}$$

when $R > \log_2 A$.

Then the sensor can simply send out R bits to indicate the virtual control it takes to limit the virtual state $\bar{x}(t)$ within $\left[-\frac{\Delta}{2}, \frac{\Delta}{2}\right]$. Upon receiving the bits from the sensor, the controller chooses a control to make the true system state $x(t)$ as close to the virtually simulated state $\bar{x}$ as possible.

The strategy of the control action computed at the controller is illustrated in Fig. 5.25. The controller obtains the estimate of the system state using

$$\hat{\mathbf{x}}_{t+1}(t) = \sum_{i=0}^{t} A^i \hat{u}_{t-i}(t), \tag{5.136}$$

where $\hat{u} + t - i(t)$ is the estimation of the computed control action at time $t - i$ (i.e., $\bar{u}(t - i)$) given all the received channel outputs before time $t + 1$.

Then the controller chooses the control action $u(t)$ such that $\bar{x}(t + 1)$ equals $\hat{x}(t)$; i.e.,

$$u(t) = \hat{x}_{t+1}(t) - A\bar{x}(t). \tag{5.137}$$

Given the above strategies of coding, decoding, and control action, the stability of the dynamics is rigorously proved in Ref. [13].

5.5 STOCHASTIC SYSTEMS: REDUCTION OF SHANNON ENTROPY

In this section, we analyze the reduction of Shannon entropy for the assessment of communication capacity requirements. As has been explained in Chapter 2, entropy indicates the uncertainty of a system. Hence it is desirable to decrease the entropy such that the system state is more concentrated around the desired operation state. However, in stochastic systems, there is a high probability that external random perturbations will increase the entropy, unless the system state entropy has already been very large.[2] Therefore communication is needed to provide "negative entropy" to compensate for the entropy generated by random perturbations (entropy reduction will eventually be carried out by the controller), thus acting as a bridge linking the analyses of communications and stochastic system dynamics, as illustrated in Fig. 5.26.

Note that the concept of entropy is also used for the analysis in Section 5.2, where the system is deterministic and the only uncertainty stems from the unknown initial state. In this section, the system is stochastic; hence the Shannon entropy, instead of the topological entropy, is used to measure the system uncertainty. The relationship between the topological entropy and the Shannon-type entropy (more precisely, the Shannon-Sinai entropy of dynamical systems [12]) can be described by the Variational Principle (Theorem 6.8.1 in Ref. [12]), which will not be explained in detail in this book.

In this section, we will provide a comprehensive introduction to communication capacity analysis based on the reduction of Shannon entropy in physical dynamics. We will first provide a qualitative explanation of entropy reduction from the viewpoint of cybernetics. Then we begin from discrete state systems and extend to continuous state systems.

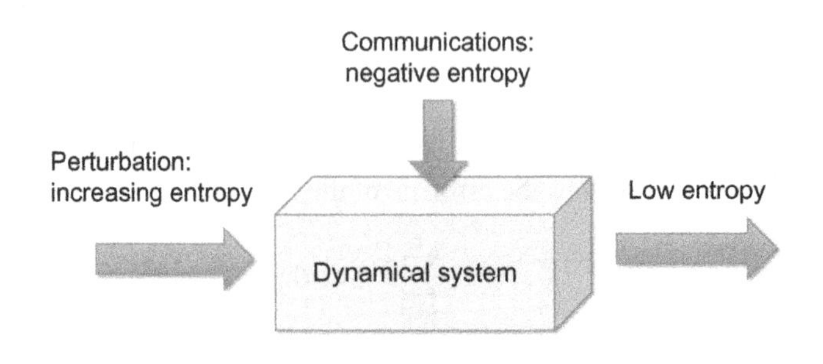

FIG. 5.26

An illustration of entropy reduction.

[2]When the system state entropy is very large, it is possible that random perturbations can decrease the entropy, since the entropy does not necessarily decrease in a generic stochastic system [33].

5.5.1 CYBERNETICS ARGUMENT

We first provide an analysis of Shannon entropy in dynamics, using the arguments of cybernetics [77,78]. The analysis does not explicitly concern communications; however, it provides insights for future discussions.

Law of requisite variety

The relationship between communications and control was originally considered by Wiener [78]. Then Ashby [77] proposed the Law of Requisite Variety in his celebrated book. Here the term "variety" means the number of states of the corresponding system, which is actually closely related to the concept of entropy (if the system state is uniformly distributed, then entropy is equal to log(variety)). Based on the concept of variety, Ashby proposed the follow law of requisite variety:

> To control a system, the variety of the control mechanism should be greater than or equal to that of the system being controlled.

Without a rigorous mathematical formulation and proof, Ashby claimed "only variety in the regulator can force down the variety due to the disturbances; only variety can destroy variety." This is illustrated by the example in Fig. 5.27. We consider two players, R (regulator) and D (disturbance). The varieties of their actions are denoted by V_R and V_D, respectively. We assume that player D takes action first and then R follows. This is similar to practical control systems: the system state is first randomly perturbed and then controlled by the controller. In Fig. 5.27A, each player has three possible actions (α, β and γ for R, and 1, 2, 3 for D). The outcomes of different action pairs are given in the table. The goal of the game is: if the outcome is a, R wins; otherwise, R loses. It is obvious that, regardless of the action taken by

R \ D	1	2	3
α	b	a	c
β	a	c	b
γ	c	b	a

(A) Variety_regulator = Variety_disturbance

R \ D	1	2	3	4	5	6	7	8	9
α	f	k	m	b	c	h	j	d	l
β	f	e	k	b	q	h	d	p	n
γ	k	f	a	b	c	m	d	j	h

(B) Variety_regulator < Variety_disturbance

FIG. 5.27

Examples to illustrate the law of requisite variety.

player D, player R can always choose its action adaptively and achieve the goal a. In this case, we claim that player R can control the game. However, in Fig. 5.27B, player D has more options than player R, thus making R unable to control the game.

We now consider a more generic case. If two elements in the same row are identical, then player R need not distinguish the corresponding actions of player D, which is too favorable to R. Hence we assume that no two elements in the same row are identical. Then it is easy to prove that the variety of the outcome (denoted by V_O), given the strategy of R, cannot be less than V_D/V_R.

Then if we use a logarithmic scale to measure the variety, the above conclusion can be translated into the following inequality:

$$\log(V_O) \geq \log(V_D) - \log(V_R). \tag{5.138}$$

If the distributions of O, R, and D are all uniform, then we have

$$H(O) \geq H(D) - H(R). \tag{5.139}$$

This inequality implies that a larger $H(R)$ (i.e., having more states of the controller) helps to better reduce the entropy of the output; otherwise, if $H(R)$ is small, the output may have a larger uncertainty than the disturbance. Although communication is not explicitly mentioned in the argument, we can consider the communication network as part of the controller. If the communication capacity is too small, then the controller cannot have much variety (since it does not have many options due to the limited number of reliable messages sent from the sensor), thus causing large entropy (or uncertainty) at the system output. Note that these arguments are merely qualitative. More rigorous analysis will be provided subsequently.

Shannon entropy in discrete-value dynamics

Based on the qualitative argument in the law of requisite variety, we provide a detailed analysis of the entropy change in discrete physical dynamics by following the argument of Conant [79]. Consider a controller R (called a regulator in Ref. [79]) regulating a variable Z (which can be considered as the system state) subject to random perturbation S. R and S jointly determine the output Z. It is assumed that R, S, and Z have finite alphabets and the system evolves in discrete time, given by

$$Z(t) = \phi(S(t), R(t)), \tag{5.140}$$

where ϕ is the evolution law.

We further assume that S is independent of R. Notice that the system dynamics is memoryless. Two types of control strategies are considered, as illustrated in Fig. 5.28:

- Point regulation: The goal is to minimize the changes of output Z (i.e., to make the regulated variable Z as constant as possible).
- Path regulation: The goal is to minimize the unpredictability of the outcomes. Hence the outcome Z could change; however, the change should be as predictable as possible.

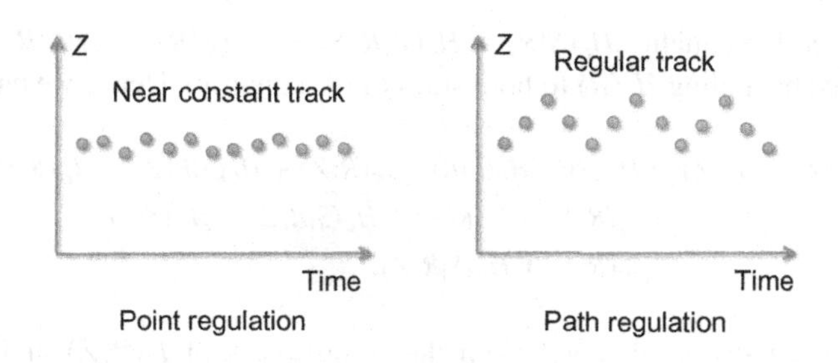

FIG. 5.28

Illustrations of point and path regulations.

We analyze the point regulation first. Since the regulator wants to minimize the change (or the uncertainty) of Z, it targets minimizing the entropy of Z. When R is active, the action selected by R is a function of Z (i.e., closed-loop control). The corresponding entropy of Z is denoted by $H_C(Z)$, where the subscript c means closed loop. When R is idle and uses a fixed action i (i.e., open-loop control), the corresponding entropy of Z is denoted by $H_o^i(Z)$, where the subscript o means open loop. The minimum entropy subject to the open-loop control is given by

$$H_o^*(Z) = \min_i H_o^i(Z). \tag{5.141}$$

Then we obtain the following theorem on the difference between $H_o^*(Z)$ and $H_C(Z)$, i.e., the entropy reduced by the controller R when compared with the optimal open-loop control.

Theorem 13. *The gap between the open-loop and closed-loop entropies satisfies*

$$H_o^*(Z) - H_C(Z) \leq I_C(R;S) + K, \tag{5.142}$$

where $K = \log_2$*(the largest number of repeated entries in* $\phi(\cdot, R)$*).*

Proof. For open-loop control, we have

$$\begin{aligned} H_o^*(Z) &= H_o^*(S,Z) - H_o^*(S|Z) \\ &= H_o^*(S) + H_o^*(Z|S) - H_o^*(S|Z) \\ &= H_C(S) - H_o^*(S|Z), \end{aligned} \tag{5.143}$$

where the superscript $*$ means using open-loop control, which minimizes the entropy. Note that we used the facts $H_C(S) = H_o^*(S)$ (since S is not dependent on the control action) and $H_o^*(Z|S) = 0$ (Z is deterministic given S and R).

For closed-loop control, we have

$$\begin{aligned} H_C(Z) &= H_C(R,Z) - H_C(R|Z) \\ &= H_C(Z|R) + H_C(R) - H_C(R|Z) \\ &= H_C(Z|R) + I_C(R;Z) \\ &= H_C(S|R) + H_C(Z|R,S) - H_C(S|R,Z) + I_C(R;Z), \end{aligned} \tag{5.144}$$

where we used the equality $H_c(S|R) + H_c(Z|R,S) = H_c(Z|R) + H_c(S|R,Z)$ (which can be proved by adding $H_c(R)$ to both sides of the equality). Hence we have

$$\begin{aligned} H_o(Z) - H_c(Z) &= H_c(S) - H_c(S|R) - I_c(R;Z) + H_c(S|R,Z) - H_o^*(S|Z) \\ &= I_c(R;S) - I_c(R;Z) + H_c(S|R,Z) - H_o^*(S|Z) \\ &\leq I_c(R;S) + H_c(S|R,Z), \end{aligned} \tag{5.145}$$

where the inequality is obtained from the positiveness of $I_c(R;Z)$ and $H_o^*(S|Z)$. Notice that, given R and Z, the uncertainty comes from the repeated outcomes in $\phi(\cdot, R)$ for a fixed R. Hence $H_c(S|R,Z) \leq K$ according to the definition of K. This concludes the proof. □

From the definition of K, we have the following corollary:

Corollary 1. *If there are no repeated outcomes in $\phi(\cdot, R)$ for a fixed R, we have*

$$H_o^*(Z) - H_c(Z) \leq I_c(R;S). \tag{5.146}$$

Remark 1. From the theorem, we observe that, when $K = 0$, the reduction of entropy (when compared with open-loop control) is bounded by the mutual information between the controller and perturbation. It can be explained as the information about the system dynamics provided by the sensor to the controller, or equivalently the effective information communicated by the sensor. Hence the communication capacity provides an upper bound for the reduction of entropy.

On the other hand, it is possible that the entropy reduction is larger than the communication capacity. This requires $K > 0$; i.e., the dynamics itself has the capability of reducing entropy when open-loop control is used. One extreme case is that, regardless of R and S, the output Z is always the same, thus making the entropy zero.

We then consider path regulation, in which we have a series of perturbations $S(1:T)$, control actions $R(1:T)$, and outputs $Z(1:T)$. The argument is similar. Again, we consider the open-loop and closed-loop cases. When the controller R is active (i.e., closed-loop control), we add a subscript c to the quantities; when open-loop control is used, we use the subscript o.

Using the same argument as in the point regulation, we can prove a similar conclusion:

$$H_o(Z(1:T)) - H_c(Z(1:T)) \leq I(R(1:T); S(1:T)) + KT, \tag{5.147}$$

which also implies

$$h_o(Z) - h_c(Z) \leq i(R;S) + K, \tag{5.148}$$

where h is the entropy rate, defined as $h(X) = \lim_{T\to\infty} \frac{1}{T} H(X(1:T))$, and the average mutual information i is similarly defined.

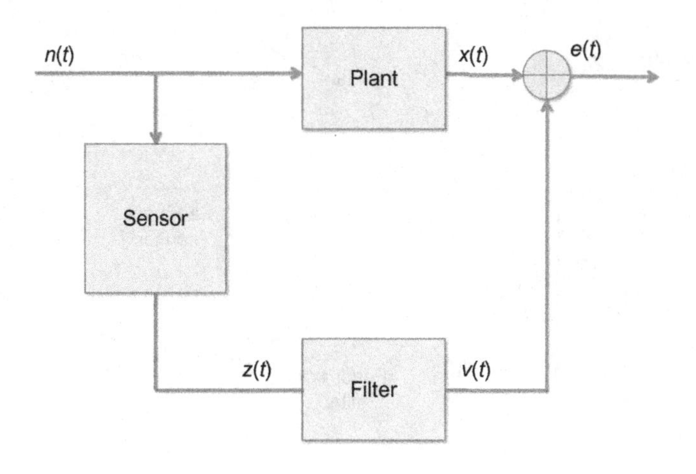

FIG. 5.29

System structure for estimation.

Shannon entropy in continuous-value dynamics

In the previous discussion, we considered discrete systems with very simple architectures (in particular, they are memoryless). We now provide a detailed analysis of the entropy change in continuous physical dynamics having more detailed and practical structures by following the argument in Ref. [80].

First we consider the entropy in the procedure of system state estimation, since in many situations (such as the optimal control of linear systems), the controller estimates the system state first and then computes the corresponding control action. The estimation procedure is illustrated in Fig. 5.29, where $\mathbf{n}$ is the random perturbation, $\mathbf{x}$ is the system state, and $\mathbf{e}$ is the estimation error. The sensor observes the random perturbation directly and the filter provides an estimation about $\mathbf{x}$. We assume

$$\begin{cases} \mathbf{x} = \mathbf{D}\mathbf{n}, \\ \mathbf{v} = \mathbf{F}(\mathbf{z}), \\ \mathbf{e} = \mathbf{x} - \mathbf{v}. \end{cases} \tag{5.149}$$

The entropy relationships in the estimation problem are disclosed in the following theorem:

Theorem 14. *For the above estimation system, we have the following conclusions:*

- *Regardless of* $\mathbf{F}$ *and* $\mathbf{D}$, *the entropy of the error vector* $\mathbf{e}$ *is bounded by*

$$h(\mathbf{e}) \geq h(\mathbf{x}) - I(\mathbf{x}; \mathbf{z}). \tag{5.150}$$

- *Minimizing* $h(\mathbf{e})$ *is equivalent to minimizing* $I(\mathbf{x}; \mathbf{z})$. *If* $\mathbf{x}$ *and* $\mathbf{z}$ *are mutually independent,* $h(\mathbf{e})$ *achieves the minimum.*

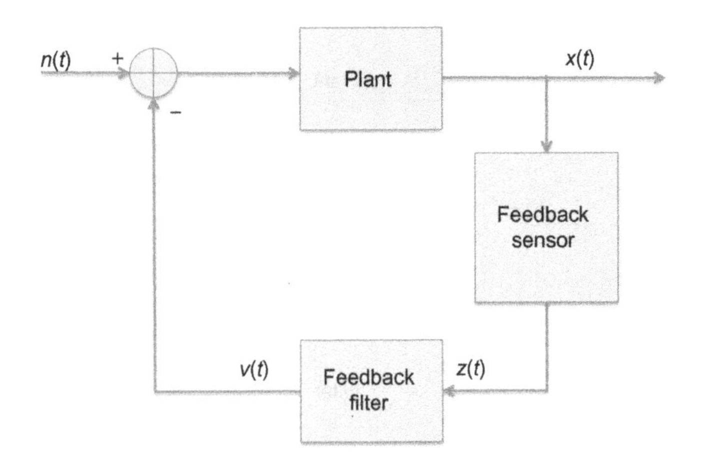

FIG. 5.30

Architecture of disturbance rejecting feedback control system.

Remark 2. A smaller $h(\mathbf{e})$ means less uncertainty in $\mathbf{e}$, i.e., a more precise estimation. If $\mathbf{e}$ is a scalar and is Gaussian with zero expectation (thus being unbiased), the minimization of error entropy is equivalent to minimizing the mean square error of the estimation. Since $\mathbf{x}$ is entirely determined by $\mathbf{n}$, then $h(\mathbf{x})$ is fixed and cannot be changed by the estimation mechanism. Hence according to Eq. (5.150), we can only increase the mutual information between $\mathbf{x}$ and $\mathbf{z}$ (i.e., the sensor report contains more information about the system state $\mathbf{x}$) in order to decrease the lower bound of $h(\mathbf{e})$. In the second conclusion, making $\mathbf{e}$ and $\mathbf{z}$ independent is equivalent to all the information in $\mathbf{x}$ being contained in $\mathbf{z}$. Actually, in the estimation of linear systems, this is the celebrated orthogonality principle in linear estimations [81].

We then consider the entropy change in the disturbance rejection feedback control, whose architecture is illustrated in Fig. 5.30. Here, $\mathbf{n}(t)$ is the random perturbation and $\mathbf{x}(t)$ is the system state. The feedback consists of a linear prefilter $\mathbf{B}$, a measurement error $\mathbf{w}$, and a post filter $\mathbf{C}$. The output of the sensor is given by

$$\mathbf{z} = \mathbf{C}(\mathbf{B}\mathbf{x} + \mathbf{w}). \tag{5.151}$$

Then the feedback signal is given by

$$\mathbf{v}(t) = F(\mathbf{z}(t), t), \tag{5.152}$$

where F is a generic function which can be nonlinear and time varying.

The input of the plant is the error between the perturbation $\mathbf{n}$ and the estimation $\mathbf{v}$; i.e.,

$$\mathbf{e}(t) = \mathbf{n}(t) - \mathbf{v}(t). \tag{5.153}$$

The linear output of the plant $\mathbf{x}$ is then given by

$$\mathbf{x}(t) = \mathbf{D}\mathbf{e}(t), \tag{5.154}$$

where $\mathbf{D}$ is a fixed matrix.

Our goal is to compare the entropy of the system state in closed-loop and open-loop control systems. In open-loop control, where we set $F = 0$, we denote the output by $\mathbf{x}_o(t)$, given by $\mathbf{x}_o(t) = \mathbf{D}\mathbf{n}(t)$. The following theorem shows the relationship between $h(\mathbf{x})$ and $h(\mathbf{x}_o)$:

Theorem 15. *Regardless of the feedback filter F, the following equality always holds:*

$$h(\mathbf{x}_o) - h(\mathbf{x}) = I(\mathbf{x}_o; \mathbf{z}) - I(\mathbf{x}; \mathbf{z}). \tag{5.155}$$

Immediate conclusions can be drawn in the following corollary:

Corollary 2. *The following inequalities hold:*

$$h(\mathbf{x}_o) - h(\mathbf{x}) \leq I(\mathbf{x}_o; \mathbf{z}), \tag{5.156}$$

$$h(\mathbf{x}) \geq h(\mathbf{x}_o|\mathbf{z}), \tag{5.157}$$

and

$$h(\mathbf{x}_o) - h(\mathbf{x}) \leq I(\mathbf{x}_o; \mathbf{C}(\mathbf{B}\mathbf{D}\mathbf{n} + \mathbf{w})). \tag{5.158}$$

Remark 3. Note that we do not specify the control mechanism since function F is arbitrary. Eqs. (5.156) and (5.158) provide bounds for the entropy reduction brought about by the feedback control. Eq. (5.157) shows that, to minimize the lower bound of $h(\mathbf{x}_o)$, we should minimize the uncertainty of $\mathbf{x}_o$ given $\mathbf{z}$; i.e., letting $\mathbf{z}$ bring as much information about $\mathbf{x}_o$ as possible.

Controller design based on entropy

Since entropy measures the uncertainty of system dynamics, it can be used as the criterion of controller synthesis. Here, we briefly introduce the formulation of the entropy-based control system [82]. We assume that $\mathbf{y}(t) \in [a, b]$ is the output of a stochastic system at time slot t. The control action is denoted by $\mathbf{u}(t)$. The distribution of $\mathbf{y}$ and $\mathbf{u}$ is denoted by γ. The B-spline expansion of the distribution is given in [83]

$$\gamma(\mathbf{y}, \mathbf{u}) = \sum_{i=1}^{n} w_i(\mathbf{u}) B_i(\mathbf{y}), \tag{5.159}$$

where B_i is the ith basis function and w_i is the expansion coefficient. The cost function is defined as

$$J(\mathbf{u}) = -\int_a^b \gamma(\mathbf{y}, \mathbf{u}) \log \gamma(\mathbf{y}, \mathbf{u}) d\mathbf{y} + R\|\mathbf{u}\|^2, \tag{5.160}$$

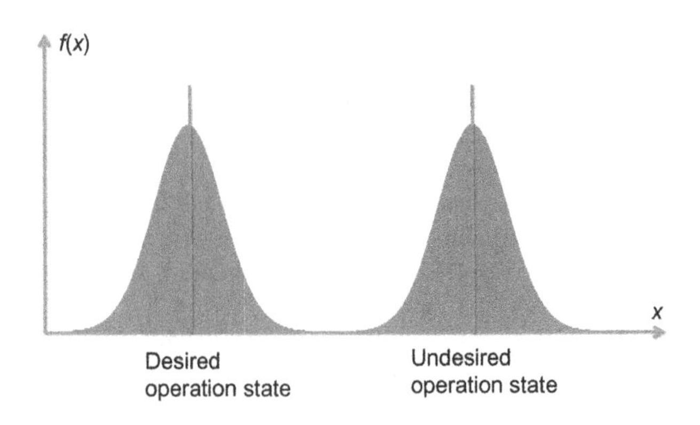

FIG. 5.31

The same distribution shape with different relative locations.

which is a combination of the output entropy and the cost of action. The optimal control action can then be obtained by taking the derivative of J with respect to $\mathbf{u}$. The details can be found in Ref. [82].

Criticisms on entropy-based control

Although entropy provides a good measure of the uncertainty of dynamics and low entropy is a necessary condition for a good operation status of a system, there have been criticisms of entropy-based control [84]:

- Entropy-based control approaches do not solve the problems that traditional control theory cannot solve. Moreover, entropy is only valid for stochastic systems and cannot handle deterministic systems.
- The criterion of entropy minimization is questionable. Entropy is dependent on the relative shape of the distribution and is independent of the relative locations. For example, the two distributions illustrated in Fig. 5.31 have the same distribution since they have the same shape. However, the one on the left is centered around the desired operation status while the other is not. Hence it should be desirable to achieve the distribution on the left; however, entropy-based control cannot distinguish between the two distributions and may provide the undesired one. A remedy is to add a constraint on the expectation of the distribution when minimizing the entropy.
- The entropy of the output in the sense of an alphabet without the definition of distances may not represent the variance of the output in its numerical meaning. This can be well illustrated in the following example [84]. Suppose that the controller and perturbation each have three possible actions (denoted by (q, r, p) and (a, b, c), respectively). The outcomes of the system are given in the following matrix, where the columns and rows represent the actions of controller and perturbation:

$$\begin{pmatrix} 1 & 9 & 4 \\ 5 & 2 & 8 \\ 7 & 6 & 3 \end{pmatrix}$$

Consider two strategies of the controller: (A) fix the action p; (B) adaptive control action: $a \to q$, $b \to r$, and $c \to p$. It is easy to check that the first strategy gives the outputs $\{1, 4, 9\}$ while the second strategy results in $\{4, 5, 6\}$. If we interpret the outputs as abstract alphabets, the outputs of the two strategies have the same entropy. However, if the outputs are explained as numbers, obviously the second strategy results in much less numerical variance.

5.5.2 DOES PRACTICAL CONTROL REALLY REDUCE ENTROPY?

Many existing control strategies are not based on the performance metric of entropy. However, they can also indirectly reduce the entropy, since uncertainty is usually undesirable. In this chapter, we take LQG control [72] as an example. The system dynamics are linear with Gaussian noise (both in the system state evolution and observation). The cost of the control is given by

$$J = \sum_{t=1}^{T} \mathbf{x}^T(t)\mathbf{Q}\mathbf{x}(t) + \mathbf{u}^T(t)\mathbf{R}\mathbf{u}(t), \tag{5.161}$$

where $\mathbf{Q}$ and $\mathbf{R}$ are both nonnegative definite matrices, and T is the final time under consideration. Intuitively, when $\mathbf{R} = 0$ and $\mathbf{Q} = \mathbf{I}$, all the efforts are used to reduce the expected square norm of $\mathbf{x}$. When $E[\|\mathbf{x}\|^2]$ is small, the uncertainty is small, thus making the entropy small. Note that reduction in $E[\|\mathbf{x}\|^2]$ does not necessarily imply reduction in the entropy; however, if $E[\|\mathbf{x}\|^2]$ is substantially reduced, there is a high probability that the entropy will also decrease.

Here, we use numerical results to demonstrate the entropy reduction. We consider a power network with N_g generators. For generator n, its dynamics are described by the following swing equation [4]:

$$M_n\ddot{\delta}(t) + D_n\dot{\delta}(t) = P^n_{\text{m}}(t) - P^n_{\text{e}}(t), \tag{5.162}$$

where δ is the phase, P^n_{m} is the mechanical power, and P^n_{e} is the electric power. M_n is the rotor inertia constant and D_n is the mechanical damping constant. We denote $f = \dot{\delta}$, which is the frequency of rotation.

Similarly to the seminal work by Thorp [56], we ignore the connection of loads. We assume that the system state is close to an equilibrium point. The standard frequency is denoted by f_0 (e.g., 60 Hz in the United States) and the frequency deviation of generator i is denoted by Δf_i. The angle deviation $\delta_i - f_0 t - \theta_i$ (where θ_i is the initial phase of generator i) is denoted by $\Delta\delta_i$. Then when Δf_i and $\Delta\delta_i$, $i = 1, \ldots, N_g$, are both sufficiently small, the dynamics can be linearized to

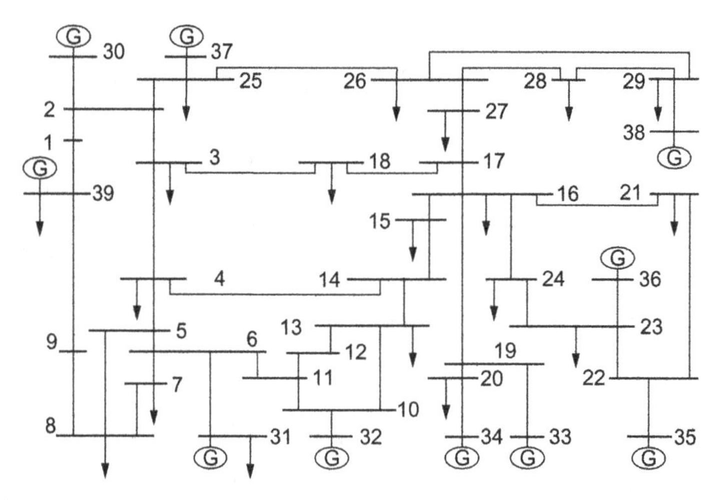

FIG. 5.32

Illustration of the IEEE 39-bus model.

$$
\begin{cases}
\Delta\dot{\delta}_i(t) = \Delta f_i(t), \\
M_i\Delta\dot{f}_i(t) + D_i\Delta f_i(t) = \Delta P^i_{\mathrm{m}}(t) - \sum_{k\sim i} c_{ik}(\Delta\delta_i - \Delta\delta_k),
\end{cases}
\tag{5.163}
$$

where ΔP^i_{m} is the difference between mechanical power and stable power, which is assumed to be the control action. The coefficients $\{c_{ik}\}_{ik}$ can be obtained from the analysis of real power. The details are omitted due to limitations of space, and can be found in Ref. [56]. Obviously, the state of each node is two-dimensional $(\Delta\delta_i, \Delta f_i)$.

We use the IEEE New England 39-bus model, which is illustrated in Fig. 5.32. The parameters of the transmission lines are obtained from the model. We assume that all generators have the same parameters: momentum $M = 6$ and damping $D = 0$ (i.e., we ignore damping). The feedback gain matrix $\mathbf{K}$ is obtained from the linear quadratic regulation (LQR) controller synthesis. The state of each bus is given by (f, δ), where f is the frequency and phase.

First we assume that there is no observation noise and the system state can be observed directly (i.e., $\mathbf{y}(t) = \mathbf{x}(t)$). We use LQG control by assuming $\mathbf{Q} = r\mathbf{I}$, $\mathbf{R} = \mathbf{I}$. We also add noise to the system state evolution with variance σ.

We consider the dynamics of Bus 1 by severing the connection to all other buses. We choose 10 random starting points. The 10 corresponding traces are illustrated in Fig. 5.33. We observe that the uncertainty is eliminated with time.

We then consider all the 39 buses. The traces of system entropy in four cases, where r is the ratio between $\mathbf{Q}$ and $\mathbf{R}$, are shown in Fig. 5.34. We observe that, in all these four cases, the entropy is a monotonically decreasing function of time. Moreover, a larger noise power or a smaller r will decrease the rate of entropy reduction.

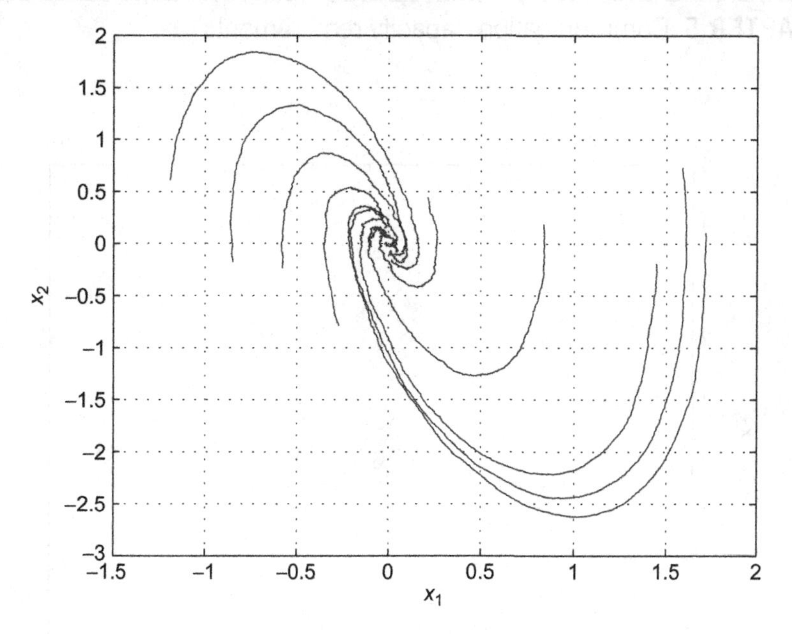

FIG. 5.33

Traces of the system state in two-dimensional space.

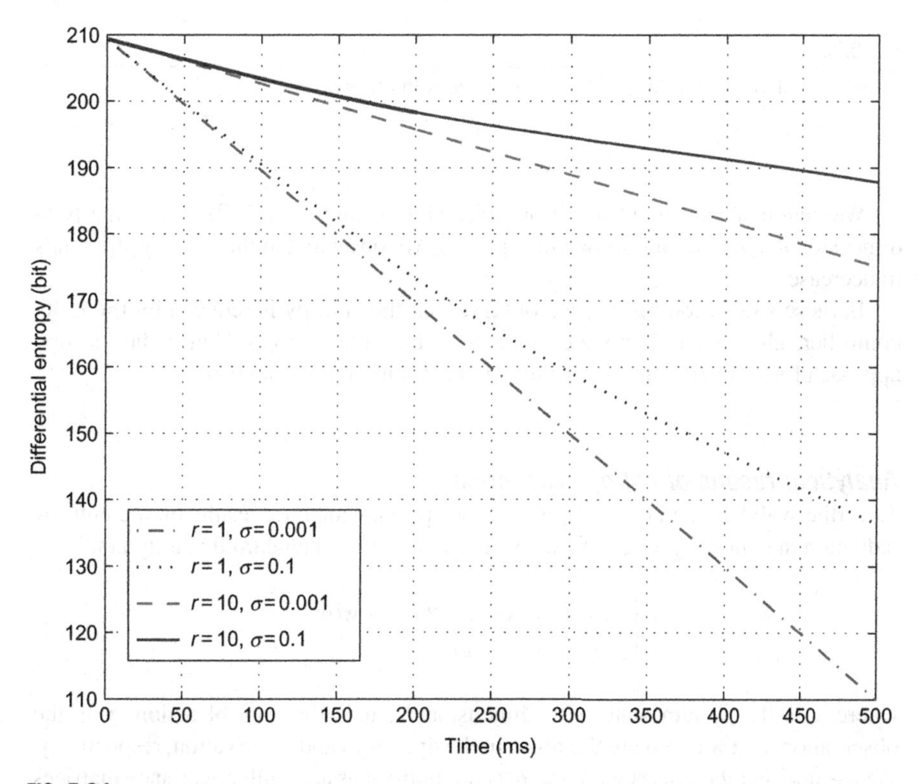

FIG. 5.34

Entropy evolution of the 39-bus power network.

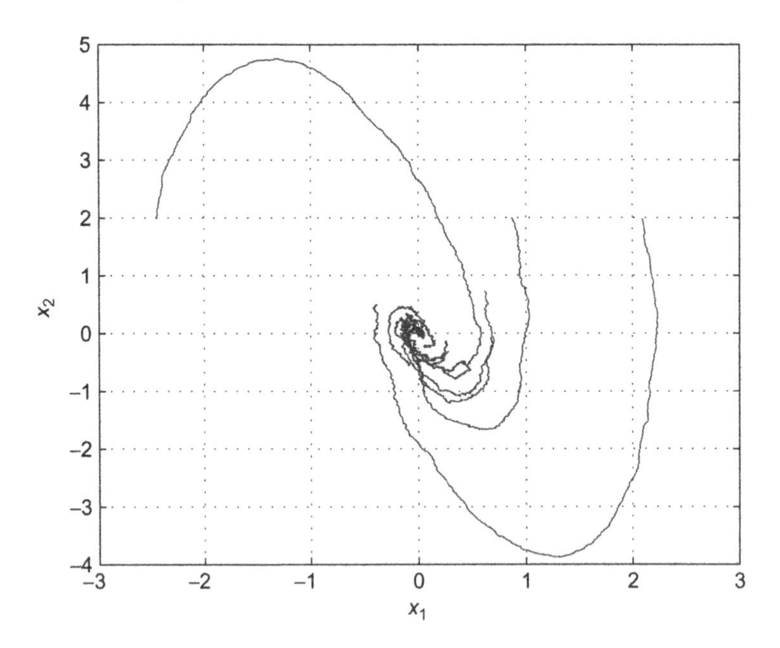

FIG. 5.35

Traces of system state in two-dimensional space with observation noise.

We then consider the observation noise and assume $\sigma = 1$. Traces of the two-dimensional dynamics are shown in Fig. 5.35. We observe that the entropy still tends to decrease.

In these simulation results, we observe that the entropy is reduced by the LQG controller, although it is not designed to reduce the entropy. Hence the entropy approach for analyzing the communication capacity requirements is valid.

Analytical results of entropy reduction

To further validate the entropy approach, we provide analytic results on the entropy reduction in control systems. We consider the following standard linear dynamics:

$$\begin{cases} \mathbf{x}(t+1) = \mathbf{A}\mathbf{x}(t) + \mathbf{B}\mathbf{u}(t) + \mathbf{w}(t), \\ \mathbf{y}(t) = \mathbf{C}\mathbf{x}(t) + \mathbf{n}(t), \end{cases} \tag{5.164}$$

where $\mathbf{x}$ is the system state with dimension N, $\mathbf{u}$ is the control action, $\mathbf{y}$ is the observation, and $\mathbf{w}$ and $\mathbf{n}$ are the noise in the dynamics and observation, respectively. We assume that the noises $\mathbf{w}(t)$ and $\mathbf{n}(t)$ are both Gaussian with covariance matrices Σ_w and Σ_n.

For the LQG control, the following theorem shows the exact evolution of the covariance matrix:

Theorem 16. *The evolution of the covariance matrix of the system state, Σ_x, satisfies*

$$\Sigma_x(t+1) = (\mathbf{A} - \mathbf{BK}(t))(\Sigma_x(t) + \Sigma(t|t))(\mathbf{A} - \mathbf{BK}(t))^T - \mathbf{A}\Sigma_x(t|t)\mathbf{A}^T + \Sigma_w, \tag{5.165}$$

where $\Sigma(t|t)$ is the covariance matrix estimation in Kalman filtering given all observations before time $t+1$.

Based on the evolution law of the covariance matrix in LQG control, we obtain the following sufficient condition of the temporal reduction of entropy:

Theorem 17. *For a time slot t, we assume that the following regulation assumptions hold:*

- *The eigenvalues of $\Sigma_x(t)$ and $\Sigma_x(t+1)$ satisfy*

$$\lambda(\Sigma_x(t)), \lambda(\Sigma_x(t+1)) \in [\lambda_{\text{low}}, \lambda_{\text{upp}}]. \tag{5.166}$$

- *There exists a $\theta > 0$ such that*

$$\text{trace}(\Sigma_x(t|t)) \leq \theta \, \text{trace}(\Sigma_x(t)). \tag{5.167}$$

Consider a positive number a such that

$$a \leq \sqrt[N]{\frac{N^N \lambda_{\text{low}}^N}{\lambda_{\text{low}}^N + (N^N - 1)\lambda_{\text{upp}}^N N}}. \tag{5.168}$$

Then if

$$\begin{aligned} &\text{trace}[\mathbf{P}(a'\Sigma(t|t-1) - \Sigma(t+1|t))] \\ &+ \text{trace}[\mathbf{P}^*(a'\Sigma(t|t) - \Sigma(t+1|t+1))] \\ &- \text{trace}[\mathbf{P}(a'\Sigma(t+1|t) - \Sigma(t+2|t+1))] \\ &- \text{trace}[(1-a')\mathbf{P}\Sigma_w] \geq 0, \end{aligned} \tag{5.169}$$

where $\mathbf{P}(t)$ satisfies

$$\mathbf{P}(t-1) = \mathbf{Q} + \mathbf{A}^T\mathbf{P}(t)\mathbf{A} - \mathbf{P}^*(t+1), \tag{5.170}$$

with

$$\mathbf{P}^*(t+1) = \mathbf{A}^T\mathbf{P}(t)\mathbf{B}(\mathbf{R} + \mathbf{B}^T\mathbf{P}(t)\mathbf{B})^{-1}\mathbf{B}^T\mathbf{P}(t)\mathbf{A} \tag{5.171}$$

and $\mathbf{P}(T) = \mathbf{Q}$, *and*

$$a' = \frac{a(\lambda_{\min}(\mathbf{KRK}^T) + \lambda_{\min}(\mathbf{Q}))}{\theta\lambda_{\max}(\mathbf{KRK}^T) + \lambda_{\max}(\mathbf{Q})}, \tag{5.172}$$

then we have

$$h(\mathbf{x}(t+1)) \leq h(\mathbf{x}(t)). \tag{5.173}$$

Remark 4. The proof of the theorem involves a series of inequalities and the details can be found in Ref. [85]. Essentially, it states that if the LQG cost function decreases sufficiently quickly, the entropy of the system state also decreases with time.

It is also possible that LQG control may increase the entropy. However, a small LQG cost function implies low entropy of the final system state. Take the simple case of $\mathbf{Q} = \mathbf{I}$ and $\mathbf{R} = 0$, for example. Since J is small, then $\sum_{n=1}^{N} \lambda_n(\Sigma_x(T))$, namely the sum of the eigenvalues of the covariance matrix $\Sigma_x(T)$ is small, where we apply the following fact:

$$\begin{aligned} E[\|\mathbf{x}\|^2] &= \text{trace}(\Sigma_x(T)) \\ &= \sum_{n=1}^{N} \lambda_n(\Sigma_x(T)). \end{aligned} \tag{5.174}$$

Then the entropy is dominated by a monotonic function of J, since

$$\begin{aligned} h_x(T) &\propto \det(\Sigma_x(T)) \\ &= \prod_{n=1}^{N} \lambda_n(\Sigma_x(T)) \\ &\leq \left(\frac{\sum_{n=1}^{N} \lambda_n(\Sigma_x(T))}{N}\right)^N \\ &\leq \left(\frac{J}{N}\right)^N. \end{aligned} \tag{5.175}$$

5.5.3 DISCRETE-STATE CPS: ENTROPY AND COMMUNICATIONS

In the previous subsections, we discussed the entropy change in the control system, which does not explicitly involve communications; hence it is still not applicable to a CPS. In this subsection, we will study a CPS with communications and discrete-state physical dynamics using the arguments in Ref. [15].

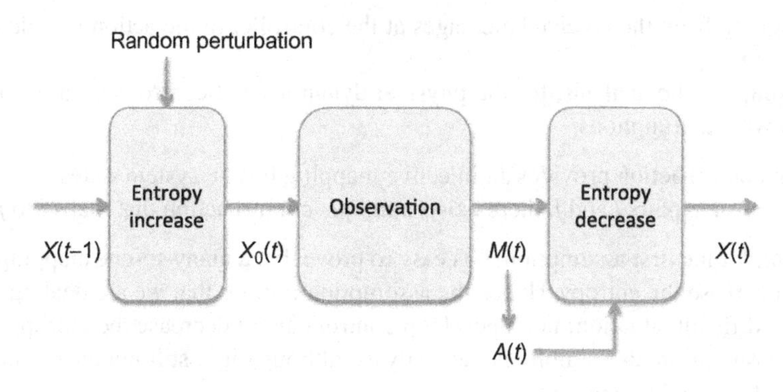

FIG. 5.36

Illustration of the three-stage model.

System model

We consider finite state physical dynamics with discrete timing structure. The N states of the physical dynamics are denoted by $x_1, \ldots, x_N$. In each time slot, the CPS proceeds in the following three stages, as illustrated in Fig. 5.36:

- Entropy increase stage: At the beginning of the time slot (e.g., the tth time slot), the state of the CPS is denoted by $X(t-1)$ with distribution p_{t-1}, and the corresponding entropy is denoted by $H(t-1)$. The state is perturbed by random perturbations and is changed to $X_0(t)$ with conditional probability $Q_{mn} = P(X_0 = n|X = m)$ and entropy $H_0(t)(> H(t-1))$. Here the impact of random perturbation is represented by the transition probability Q.
- Observation stage: An external sensor makes an observation on the system state and sends a message $M(t)$ to the controller. We assume that the observation is noiseless. The coding scheme and the communication channel are not specified. For notational simplicity, we define $P(x_i|m_j) = P(X(t) = x_i|M(t) = m_j)$ and $P_0(x_i|m_j) = P(X_0(t) = x_i|M(t) = m_j)$. We denote by R the number of possible messages.
- Entropy decrease stage: A control action $A(t)$ is computed by the controller and then actuated. The system state is then changed to $X(t)$ with distribution p_t and entropy $H(t)$. Then the physical dynamics proceeds to the next time slot.

For simplicity, we assume that the sensor is an external device and thus do not consider the entropy at the sensor itself. We do not consider the cost of computing.

We assume that there are D possible control actions that can be taken by the controller, denoted by $a_1, \ldots, a_D$. Each control action maps $X_0(t)$ to $X(t)$. W assume that the controlled dynamics is deterministic; i.e., given the action A and the state X_0 after the entropy increase stage, the system state X is uniquely determined. We denote the mapping, indexed by the control action, by f_i if $A = a_i$; i.e., $X(t) = f_i(X_0(t))$.

The mapping from the received messages at the controller to the action is called the control strategy.

To simplify the analysis, for the physical dynamics and control action, we have the following assumptions:

- Each control action provides an injective mapping for the system states.
- For any state pairs i and j, there exists a unique control action that maps i to j.

Note that, for the first assumption, it is easy to prove that a many-to-one mapping can help to decrease the entropy. Hence the assumption implies that we are dealing with the most difficult situation; i.e., open-loop control cannot decrease the entropy. The second assumption also simplifies the analysis, although it is still not clear whether it is easy to extend to a more generic case.

Note that all these assumptions are reasonable in practical cases. Take linear dynamics with continuous-valued state $\mathbf{x}(t)$ and dynamics $\mathbf{x}(t+1) = \mathbf{A}\mathbf{x}(t) + \mathbf{B}\mathbf{u}(t)$, for instance. If $\mathbf{B}$ is a square matrix and invertible (hence $\dim(\mathbf{u}) = \dim(\mathbf{x})$), then the mapping between $\mathbf{x}(t+1)$ and $\mathbf{u}(t)$ is one-to-one, given the current system state $\mathbf{x}(t)$. Meanwhile, if $\mathbf{A}$ is invertible, the mapping between $\mathbf{x}(t+1)$ and $\mathbf{x}(t)$ is also one-to-one, given $\mathbf{u}(t)$. Note that here the system state is continuously valued, which is different from the assumption in this subsection that the system state is discrete; however, if we partition the state space with sufficient precision, the two assumptions hold approximately.

For the communication and control strategies, we have the following assumptions for the control strategy:

- The message M is determined by the current observation; i.e., $M(t) = h(X_0(t))$, where h is the mapping mechanism.
- The action is dependent on the current received message, thus making the control Markovian. A non-Markovian strategy, which may improve the performance, will be studied in the future.
- The strategy is deterministic and one-to-one. This is reasonable since a randomized strategy may increase the output entropy. We denote by g the mapping from M to A; i.e., $A = g(M)$.

Entropy reduction in one time slot

We first analyze the entropy reduction within a single time slot. For notational simplicity, we omit the time indices in the notation, since the time slot is fixed. The following lemma is key to bridging the communication requirement and entropy reduction.

Lemma 4. *Given the above assumptions, we have*

$$I(X_0; M) = H(X_0) - H(X) + I(X; M). \tag{5.176}$$

Proof. The key of Lemma 4 is to prove $H(X_0|M) = H(X|M)$. According to the definition of conditional entropy (recall that $P(x_i|m_j) = P(X = x_i|M = m_j)$ and $P_0(x_i|m_j) = P(X_0 = x_i|M = m_j)$), we have

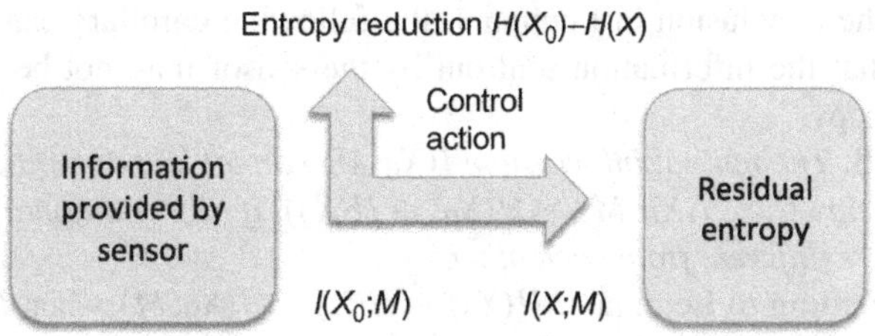

FIG. 5.37

Illustration of the entropy flow.

$$\begin{aligned}
H(X|M) &= -\sum_{i,j} P(x_i|m_j)p(m_j)\log P(x_i|m_j) \\
&= -\sum_j \sum_i P(x_i|m_j)p(m_j)\log P(x_i|m_j) \\
&= -\sum_j \sum_i P(f^{-1}_{g(m_j)}(x_i)|m_j)p(m_j) \\
&\times -\log P(f^{-1}_{g(m_j)}(x_i)|m_j) \\
&= -\sum_j \sum_k P_0(x_k|m_j)p(m_j)\log P_0(x_k|m_j) \\
&= H(X_0|M),
\end{aligned} \tag{5.177}$$

where the fourth equation is due to the assumptions of one-to-one mapping of the control output and the one-to-one mapping of the control strategy.

Then we have

$$\begin{aligned}
I(X_0;M) &= H(X_0) - H(X|M) \\
&= H(X_0) - (H(X) - I(X;M)).
\end{aligned} \tag{5.178}$$

This concludes the proof. □

Remark 5. The conclusion in Lemma 4 is equivalent to

$$H(X_0) - H(X) = I(X_0;M) - I(X;M), \tag{5.179}$$

where the left-hand side is the reduction of entropy of the system dynamics in this time slot, while the right-hand side is the difference between two mutual pieces of information. As illustrated in Fig. 5.37, the mutual information $I(X_0;M)$, namely the information on the system dynamics after the entropy increase stage, obtained by the controller, cannot be fully used to reduce the entropy; some residual, $I(X;M)$, will be unused, intuitively speaking. Interestingly, this is similar to the Carnot heat engine [86,87], in which the energy from a high-temperature source cannot be fully used to generate work, unless the absolute temperature of the colder source is zero.

Based on the conclusion in Lemma 4, the following corollary can be obtained, which states that the information sent out by the sensor may not be fully used to reduce the entropy.

Corollary 3. *The mutual information $I(X_0; M)$ cannot be fully utilized to reduce the system entropy (i.e., $I(X_0; M) \geq H(X_0) - H(X)$), if all conditional probabilities $\{P_0(x_i|m_j)\}_{i,j}$ are different from each other.*

Proof. According to Lemma 4, $H(X_0) - H(X) = I(X_0; M)$ when $I(X; M) = 0$; i.e., X and M are mutually independent. This implies

$$P(x_i|m_j) = P(x_i|m_k) = P(x_i), \tag{5.180}$$

$\forall i, j, k$. Furthermore, we have

$$P(x_i|m_j) = P_0(f^{-1}_{g(m_j)}(x_i)|m_j). \tag{5.181}$$

Hence the equation $H(X_0) - H(X) = I(X_0; M)$ can hold only when

$$P_0(f^{-1}_{g(m_j)}(x_i)|m_j) = P_0(f^{-1}_{g(m_k)}(x_i)|m_k), \tag{5.182}$$

for all i, j, and k. □

Based on Lemma 4, we obtain one of the main conclusions of this section in the following theorem, which provides upper and lower bounds for the entropy reduction in one cycle of the CPS. These bounds are independent of the controller design (or equivalently, valid for all possible controllers).

Theorem 18. *Given the finite-state CPS and the above assumptions, regardless of the controller design, the following bounds for entropy reduction are valid:*

- *Lower bound:*

$$\begin{aligned} H(X_0) - H(X) \geq\ & I(X_0; M) - H(X_0) \\ & + (1 - (\phi^*)^{-1}(H(X_0|M)))H^*(M), \end{aligned} \tag{5.183}$$

where the function ϕ^* is defined in Ref. [88], which is an increasing function and satisfies $\phi^*(0) = 0$, and $H^*(M)$ is defined as the entropy of the conditional probability of M given the event

$$\arg\max_x P(x_0|M) = X_0. \tag{5.184}$$

- *Upper bound:*

$$\begin{aligned} H(X_0) - H(X) \leq\ & I(X_0; M) - \frac{1}{2\ln 2}(\min_k P(M_j))^2 \\ & \times \min_{k_1 \neq k_2, j_1 \neq j_2} |P_0(x_{k1}|m_{j1}) - P_0(x_{k2}|m_{j2})|^2. \end{aligned} \tag{5.185}$$

Both bounds in Theorem 18 are tight under certain conditions, which is stated in the following corollary:

Corollary 4. *The lower bound in Eq. (5.183) is an equality when*

$$I(X_0; M) = H(X_0). \tag{5.186}$$

The upper bound becomes an equality when

$$P_0(x_{k1}|m_{j1}) = P_0(x_{k2}|m_{j2}) = \frac{1}{N}, \quad \forall k_1 \neq k_2, j_1 \neq j_2; \tag{5.187}$$

at this time, the information obtained by the controller is given by

$$I(X_0; M) = D(U||P_0), \tag{5.188}$$

where U is a uniform distribution among the N states; the gap between the information provided by M and the uncertainty in X_0 is given by

$$H(X_0) - I(X_0; M) = \log N. \tag{5.189}$$

Remark 6.

- Unfortunately, we are still unable to further simplify the lower bound; the major difficulty is the entropy $H^*(M)$, which is the entropy of M given that the estimation on X_0 is correct.
- When the upper bound becomes tight, although the residual entropy $I(X; M)$ becomes zero, the message M cannot provide all information about X_0; in particular, if X_0 is uniformly distributed, M provides no information about X_0 at all. Hence it may not maximize the entropy reduction if we minimize $I(X; M)$.

Entropy change in the long term

Since we assume that the number of states is finite and the system evolution (including the control policy, communication mechanism, and entropy increase mechanism) is time invariant, the system will converge to a stationary distribution of system states. This implies that the entropy will converge to a deterministic value H^*. It is an open problem to obtain an exact explicit expression for H^*, although we can compute the stationary distribution numerically. A lower bound of H^* is provided for a two-state system, when the communication channel is sufficiently good and the perturbation in the entropy increase stage is symmetric.

Proposition 1. *Consider a two-state physical dynamics, in which $Q_{11} = Q_{22} = \delta$. We assume that $I(X_0; M) = \alpha H(X_0)$, where $0 \leq \alpha \leq 1$; i.e., the proportion of the information contained in X_0 sent to the controller is α. When α is sufficiently close to 1 and δ is sufficiently close to 1, we have*

$$H^* \geq \frac{(1-\alpha)(H_{\alpha,\delta} - c_\delta(1-\alpha)H_\delta)}{1-(1-\alpha)c_\delta}, \tag{5.190}$$

where $H_\delta = -\delta \log \delta - (1-\delta)\log(1-\delta)$ *and*

$$c_\delta = (1-2\delta)\left.\frac{dH^{-1}(h)}{dh}\right|_{h=H_\delta} \times \log \frac{(1-\delta) - H^{-1}(H_\delta)(1-2\delta)}{H^{-1}(H_\delta)(1-2\delta)+\delta}, \tag{5.191}$$

where $H(x)$ $(x \leq 0.5)$ *is the entropy function for a binary distribution with probabilities* x *and* $1-x$, *and*

$$\begin{aligned} H_{\alpha,\delta} = &(H^{-1}((1-\alpha)H_\delta)(1-2\delta)+\delta) \\ &\times \log \frac{1}{H^{-1}((1-\alpha)H_\delta)(1-2\delta)+\delta} \\ &+ (1-\delta) - H^{-1}((1-\alpha)H_\delta)(1-2\delta)) \\ &\times \log \frac{1}{(1-\delta)-H^{-1}((1-\alpha)H_\delta)(1-2\delta)}. \end{aligned} \tag{5.192}$$

Remark 7. It is easy to observe that H_δ equals H_{X_0} when the system state X has entropy 0. Hence $H^* \geq (1-\alpha)H_\delta$ is a trivial lower bound. According to the proof of Proposition 1, we have $H_{\alpha,\delta} > H_\delta$. Hence the bound in Proposition 1 is tighter than the trivial one.

5.5.4 CONTINUOUS-STATE CPS: ENTROPY AND COMMUNICATIONS

In the previous discussion, we assumed that the dynamics is discretely valued. However, in practice, many systems are continuously valued. Here we follow Ref. [89] to study the relationship between entropy and communications in continuously valued dynamics.

System model

We consider the system illustrated in Fig. 5.38. Here **d** is the random disturbance. **x** is the state of the plant, while **y** is the output. **y** is fed back to a causal controller, which is also subject to noise **c**. The plant (physical dynamics) is described as follows:

$$\begin{cases} \mathbf{x}(t+1) = \mathbf{A}\mathbf{x}(t) + \mathbf{B}\mathbf{e}(t), \\ \mathbf{y}(t) = \mathbf{C}\mathbf{x}(t). \end{cases} \tag{5.193}$$

Here the feedback control can be considered as a noisy communication channel, which is continuous in both values and time. This is different from modern digital communication systems. However, the equivalent analog communication system can provide insight into digital ones.

The following assumptions are made:

- The noises in the observation **c**, the system disturbance **d**, and the initial state **x**(0) are mutually independent.

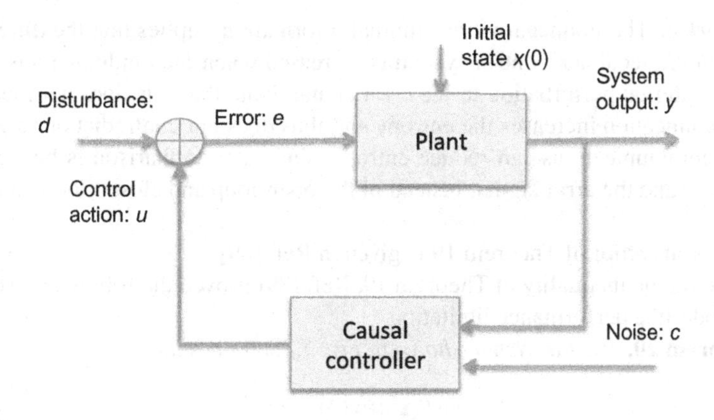

FIG. 5.38

A causal feedback control system.

- The control action is given by $\mathbf{u}(t) = K(t, \mathbf{y}(0:t), \mathbf{c}(0:t))$, where K can be a time-varying deterministic function and is dependent on the history of observations and observation noises, and is thus casual.

Traditional Bode's law

The fundamental limit of the control system in Fig. 5.38 can be characterized by Bode's Law, which will be explained subsequently. First we define the sensitivity function $S(z)$ as the transfer function between disturbance $\mathbf{d}$ and error $\mathbf{e}$. Since we expect small errors given the disturbance, we desire small $S(z)$. However, Bode's law states that for a strictly proper loop gain we have

$$\frac{1}{2\pi}\int_{-\pi}^{\pi} \log |S(e^{jw})| dw = \sum_{\lambda \in \Omega} \log |\lambda|, \tag{5.194}$$

where Ω is the set of unstable poles in open-loop control. Obviously, the right-hand term is independent of the feedback control scheme. Hence the sensitivity function S cannot be arbitrarily reduced over all the frequencies. Note that here we consider the system as deterministic (the perturbation is also deterministic). In the context of stochastic systems, we need to use the power spectrum density, instead of the signal spectrum. The extension to the stochastic system has been carried out in Ref. [89] and will be introduced subsequently.

Entropy reduction and Bode's law

In Ref. [89], the fundamental limit of the control system is studied from the viewpoint of entropy. The following theorem is of key importance in the analysis:

Theorem 19. *For the feedback control system illustrated in Fig. 5.38, we have*

$$h(\mathbf{e}(0:t)) \geq h(\mathbf{d}(0:t)) + I(\mathbf{x}(0); \mathbf{e}(0:t)), \quad \forall t = 0, 1, 2, \ldots \tag{5.195}$$

Remark 8. The nonnegativity of mutual information implies that the differential entropy in the feedback control system is increased when the randomness is passed from the external perturbation to the error signal. Note that this does not mean that the communication increases the entropy and thus does not contradict our argument that the communications can reduce entropy, since the comparison is between the disturbance and the error signal, instead of the open-loop and closed-loop controlled systems.

A rigorous proof of Theorem 19 is given in Ref. [89].

Based on the inequality in Theorem 19, Ref. [89] proved the following extension of the Bode-like performance limitation.

Theorem 20. *For the system shown in Fig. 5.38, we assume*

$$\sup_t E[\mathbf{x}^T(t)\mathbf{x}(t)] < \infty. \tag{5.196}$$

Then we have:

- *The following inequality holds:*

$$h_\infty(\mathbf{e}) \geq h_\infty(\mathbf{d}) + \sum_{i=1}^{N} \max\{0, \log(\lambda_i(\mathbf{A}))\}. \tag{5.197}$$

- *If* **e** *is asymptotically stationary, we have*

$$\frac{1}{4\pi}\int_{-\pi}^{\pi} \log(2\pi e F_{\mathbf{e}}(w))dw \geq h_\infty(\mathbf{d}) + \sum_{i=1}^{N} \max\{0, \log(\lambda_i(\mathbf{A}))\}, \tag{5.198}$$

  where $F_{\mathbf{e}}(w)$ is the power spectral density of **e**.
- *If* **e** *is asymptotically stationary and* **d** *is Gaussian autoregressive and asymptotically stationary, we have*

$$\frac{1}{2\pi}\int_{-\pi}^{\pi} \log(S_{\mathbf{d},\mathbf{e}}(w))dw \geq \sum_{i=1}^{N} \max\{0, \log(\lambda_i(\mathbf{A}))\}, \tag{5.199}$$

  where $S_{\mathbf{d},\mathbf{e}}(w)$ is the ratio between the power spectral densities of **e** *and* **d**, *which is the power spectrum density version of the sensitivity function.*

Remark 9. We notice that Eq. (5.199) is very similar to Eq. (5.194). The difference is that Eq. (5.199) is for stochastic systems while Eq. (5.194) is for deterministic systems. Moreover, Eq. (5.199) is an inequality while Eq. (5.194) is an equality.

Communication requirement

The following lemma was shown in Ref. [89] to illustrate the requirement of information flow in the feedback, or equivalently the required amount of communication in the feedback control:

Lemma 5. *In the feedback control system in Fig. 5.38, the following inequality holds:*

$$\limsup_{t\to\infty} \frac{I(\mathbf{x}(0), \mathbf{d}(0:t); \mathbf{u}(0:t))}{t} \geq \limsup_{t\to\infty} \frac{I(\mathbf{x}(0); \mathbf{e}(0:t))}{t} + I_{\infty}(\mathbf{d}; \mathbf{u}). \tag{5.200}$$

Remark 10. The left-hand side of Eq. (5.200) is the average amount of information (in terms of bits) hidden in the control action about the random initial state and random perturbation, or equivalently the effective information conveyed to the controller through the feedback. The right-hand side consists of two parts: the final information about the initial state remaining in the error signal and the information about the random perturbation carried by the control actions. This provides a lower bound for the communication requirement since the information flow through the feedback can be considered as communication.

5.6 NETWORKED STOCHASTIC SYSTEMS

In the previous discussions, we have assumed a single sensor and a single controller. However, in practice, there can be many sensors and controllers in a CPS. For example, in smart grids, there may be thousands of sensors (PMUs) and controllers (since each generator has a controller). Hence in this section, we discuss the relationship between entropy and communications in a networked CPS.

5.6.1 SYSTEM MODEL OF NETWORKED CPS

Physical dynamics

We consider N nodes with coupled physical dynamics in a cyber physical system (e.g., N generators in a power grid). These nodes form a network, in which $i \sim j$ means that nodes i and j are adjacent. We assume that the graph is connected; i.e., there are no isolated components. For simplicity, we consider linear dynamics, in which the state of the ith node is denoted by the vector $\mathbf{x}_i(t)$. The evolution of $\mathbf{x}_i(t)$ is given by

$$\dot{\mathbf{x}}_i(t) = \mathbf{A}_i\mathbf{x}_i(t) + \sum_{m\sim i} \mathbf{A}_{mi}\mathbf{x}_{\mathrm{m}}(t) + \mathbf{B}_i\mathbf{u}_i(t) + \mathbf{n}_i(t), \tag{5.201}$$

where $\mathbf{u}_i(t)$ is the control action taken by node i, and the matrices $\mathbf{A}_i$, $\{\mathbf{A}_{mi}\}$ and $\mathbf{B}_i$ are constants throughout time. The above dynamics of individual nodes can also be summarized into a higher dimensional form: $\dot{\mathbf{x}}(t) = \mathbf{A}\mathbf{x}(t) + \mathbf{B}\mathbf{u}(t) + \mathbf{n}(t)$, where the system states, control actions, and noises are stacked into single column vectors. The covariance matrix of $\mathbf{n}_i$ is denoted by $\Sigma_i^n(t)$ for vector states and $\sigma_i^2(t)$ for scalar states.

The differential entropy of the state of node i at time t is then defined as

$$h_i(t) = -\int \log(\rho_i(\mathbf{x}, t))\rho_i(\mathbf{x}, t)d\mathbf{x}, \tag{5.202}$$

where ρ_i is the probability density function of $\mathbf{x}_i(t)$, which is assumed to exist. It is well known that, when $\mathbf{x}_i$ is Gaussian distributed with covariance matrix $\Sigma_i(t)$, the closed form of the differential entropy is given by

$$h_i(t) = \frac{1}{2}\log[(2\pi e)^{n_i}|\Sigma_i(t)|], \tag{5.203}$$

where n_i is the dimension of $\mathbf{x}_i$.

Communication network

These physical dynamics nodes are also connected by a communication network. For simplicity, we assume that the communication network has the same topology as the physical dynamics network. The transmission rate is denoted by $R_{ij}(t)$ (nats/channel use) for the communications from node i to j. We ignore the time delay, which is reasonable if the physical dynamics are considerably slower than the communications (e.g., in smart grids), and also ignore transmission failures. For simplicity, we assume that the individual raw data on the system states is quantized and sent without any compression through distributed source coding. This is reasonable since most practical systems transmit individual raw data.

Feedback control

We assume that each physical dynamics node measures its system state and sends it to all its neighbors. Then each node can estimate its own state by using local measurement and the states of neighbors by using the reports from neighbors. We assume that linear control is used; i.e.,

$$\mathbf{u}_i = \mathbf{K}_i\hat{\mathbf{x}}_i + \sum_{j\sim i}\mathbf{K}_{ji}\hat{\mathbf{x}}_j, \tag{5.204}$$

where the superscript ˆ means estimation, and $\mathbf{K}_i$ and $\{\mathbf{K}_{ji}\}$ are the feedback gain matrices that are predetermined by the control policy.

5.6.2 ENTROPY PROPAGATION IN CPS

Motivating example

We need to realize that entropy can be propagated in the network of physical dynamics nodes. Take two linked generators in a power grid, for example, as illustrated in Fig. 5.39. Suppose that a random perturbation (e.g., the oscillation of mechanical power) occurs at generator 1 and makes its rotation slower or faster (thus causing uncertainty). If generator 1 rotates slower (or faster), it will pull (or push) power from (or to) generator 2, thus making generator 2 slower (or faster). Then state uncertainty of generator 2 is also increased. Hence the entropy is propagated from generator 1 to generator 2.

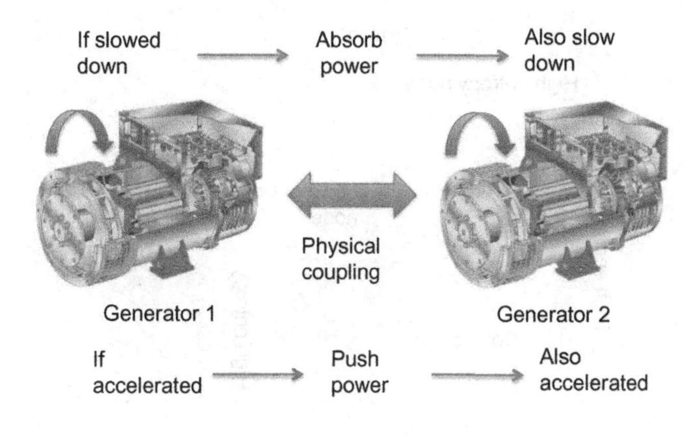

FIG. 5.39

Illustration of entropy propagation.

Entropy propagation with perfect communications

First we assume that the communication in the feedback is perfect. Then the overall system dynamics can be rewritten as $\dot{\mathbf{x}}(t) = \tilde{\mathbf{A}}\mathbf{x}(t) + \mathbf{n}(t)$, where $\tilde{\mathbf{A}} = \mathbf{A} + \mathbf{BK}$. We can prove the following theorem, which shows that the entropy increase at each node can be decomposed into two parts, namely local entropy generation and entropy propagated from neighbors.

Theorem 21. *Assume that each x_i is one-dimensional and $n_i(t) = 0$. Then for any node i, the change of its marginal entropy has the following decomposition form:*

$$\dot{h}_i(t) = E_{ii}(t) + \sum_{j\sim i} E_{ji}(t), \tag{5.205}$$

where $E_{ii}(t)$ is the local entropy generation rate given by

$$E_{ii}(t) = -\tilde{A}_{ii}h_i(t) + \tilde{A}_{ii}\int x_i \log \rho_i(x_i,t)\frac{\partial \rho_i(x_i,t)}{\partial x_i}dx_i, \tag{5.206}$$

and $E_{ji}(t)$ is the entropy generation at node i caused by neighbor node j, which is given by

$$E_{ji}(t) = \tilde{A}_{ji}\int \log \rho_i(x_i,t)\frac{\partial (E(x_j|x_i)\rho_i(x_i,t))}{\partial x_i}dx_i. \tag{5.207}$$

Notice that the expressions in Eqs. (5.206) and (5.207) are fairly complicated.

Corollary 5. *When* $\mathbf{x}$ *is Gaussian distributed, we have*

$$\begin{cases} E_{ii}(t) = \tilde{A}_{ii}, \\ E_{ji}(t) = \tilde{A}_{ji}r_{ij}(t)e^{h_j(t)-h_i(t)}, \end{cases} \tag{5.208}$$

where $r_{ij} = \frac{E[x_i x_j]}{\sqrt{E[x_i^2]E[x_j^2]}}$ is the correlation coefficient of x_i and x_j.

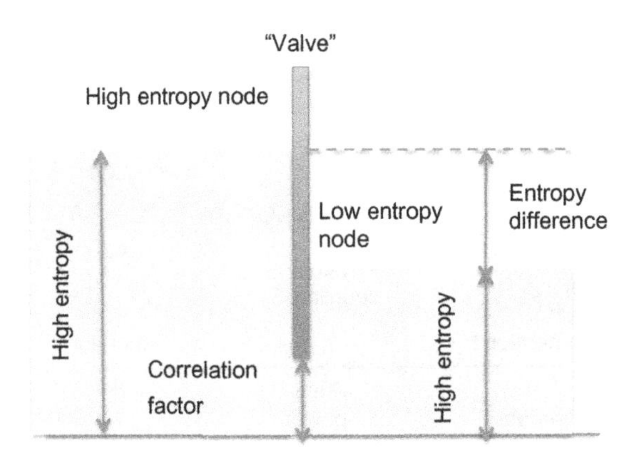

FIG. 5.40

Illustration of entropy propagation, similar to a valve.

Remark 11. The above conclusions provide not only a quantitative description on the entropy decimation, generation and transfer, but also the following interesting qualitative observations:

- The expression for E_{ii} in Eq. (5.208) denotes the rate of entropy decimation (if $\tilde{A}_{ii} < 0$) or generation (if $\tilde{A}_{ii} > 0$). The feature is that the local decimation/generation rate is fixed and is independent of the current value of entropy.
- E_{ji} is the entropy transfer rate between two adjacent nodes. We notice that once A_{ji}, A_{ij}, and r_{ij} have the same signs, the entropy transfers between nodes i and j are both positive. Hence the entropy may not be a conservative quantity in the network.
- For the case of vector system states at individual nodes, we can consider each dimension as a "virtual node" and apply the conclusion in Theorem 21.
- We notice that the entropy transfer rate is dependent on both the correlation factor r_{ij} and entropy difference $h_j(t) - h_i(t)$. Hence we can consider the entropy difference as the pressure and the correlation factor as a valve which determines the entropy transfer rate. In particular, consider the case in which the nodes have initial states close to zero, with low entropies and low correlations, and one node (say, node i) experiences a sudden entropy increase. Consider an adjacent node j. Then the entropy transfer from i to j is initially zero due to the low correlation factor. The state of node i will force that of node j to change to the same direction (if $\tilde{A}_{ij} > 0$) or the opposite direction (if $\tilde{A}_{ij} < 0$). Hence $|r_{ij}|$ will be substantially increased and $r_{ij}\tilde{A}_{ij}$ will become positive. The entropy transfer rate will be increased. This is illustrated in Fig. 5.40.
- When the entropy flows from a high entropy node to a low entropy one, the "valve" is first opened; i.e., the correlation coefficient r will increase to 1. Then the entropy difference will cause the entropy to flow from the high entropy node to the low entropy one, similar to the water flow caused by water level difference.

- Notice that the entropy evolutions in both the generic and Gaussian cases are dependent on the distributions $\{\rho_i\}_i$, conditional expectations $\{E[x_j|x_i]\}_{ij}$, and the correlation coefficients $\{r_{ij}\}_{ij}$. However, the corresponding differential equations (5.205)–(5.208) do not provide the evolution laws of $\{\rho_i\}_i$, $\{E[x_j|x_i]\}_{ij}$, and $\{r_{ij}\}_{ij}$. Hence these descriptions are incomplete.

Interdependency of entropy and communications

Now we take the limited communication bandwidth, which causes error or distortion in the feedback, into account. We define the system state estimation as $\mathbf{e}_{ji}(t) = \hat{\mathbf{x}}_{ji}(t) - \mathbf{x}_j(t)$, where $\hat{\mathbf{x}}_{ji}(t)$ is the version of $\mathbf{x}_j(t)$ conveyed by the communication link ji to node i.

We focus on the case of scalar system state, where the error variance e_j, denoted by σ^e_{ji}, is determined by both the variance of the information source $x_j(t)$ (denoted by $\sigma_j^2(t)$) and the communication rate R_{ji}; i.e.,

$$\sigma^e_{ji}(t) = g(\sigma_j^2(t), R_{ji}), \tag{5.209}$$

where g is the distortion function. It can be determined by either mathematical analysis or numerical simulation. For simplicity of analysis, we assume that $e_{ji}(t)$ is Gaussian, and is independent of $x_j(t)$ and $e_{jm}(t)$ $(m \neq i)$. Moreover, we assume

$$g(\sigma_j(t), R_{ji}(t)) = \sigma_j^2(t)\mathrm{e}^{-2R_{ji}(t)}, \tag{5.210}$$

which can be justified by the rate-distortion function of the Gaussian source.

5.6.3 JOINT EVOLUTION OF COMMUNICATION AND ENTROPY

Now we incorporate the communication rates into the evolution of the entropy. We assume that the time period for sampling and communication is ΔT. We also assume zero delay in the communications; i.e., the samples of observations are immediately sent to the destination. The transmission rate R_{ij} denotes the number of nats per sampling and communication.

Theorem 22. *For the case of scalar system states and Gaussian system state estimation errors, the entropy evolution law is given by*

$$\begin{aligned}\dot{h}_i(t) &= E_{ii}(t) + \sum_{j\sim i} E_{ji}(t) \\ &\quad + \frac{1}{2}\sum_{j\sim i}(B_iK_{ji})^2\mathrm{e}^{-2(h_i(t)-h_j(t)+R_{ji}(t))},\end{aligned} \tag{5.211}$$

as $\Delta T \to 0$.[3]

[3]Since R_{ji} is nats per channel use (sample), the transmission rate in bits/second tends to infinity when $\Delta T \to 0$. This infinite rate is due to the independent source coding for different times and different nodes. If the temporal and spatial redundancies in the system states are taken into account in the source coding, the rate can be bounded, which will be a topic of future study.

Remark 12. Both the entropy evolution and communication rates have been integrated in the same ordinary differential Eq. (5.211). This can be used to schedule the communication resource (e.g., the rates $\{R_{ij}\}$ must fall in a feasible region) and the control actions in continuous time.

5.6.4 CONTINUOUS SPACE LIMIT

Although both communications and physical dynamics can be described in the conclusion of Theorem 22, the description may be prohibitively complicated in a large-scale CPS. It is easy to verify that even in the Gaussian distribution case, we need to solve $\frac{10{,}000\times(10{,}000+1)}{2}$ differential equations for a 100×100 grid of CPS nodes with scalar states (since the covariance matrix of the overall system state is $10{,}000\times10{,}000$ and each element in the upper triangle of the matrix is described by one independent differential equation). For such a high-dimensional problem, usually we can consider the following two approaches:

- Dimension reduction: We can project the system dynamics to a lower dimensional subspace that we can handle. Usually the subspace selected for dimension reduction is the most informative one. Dimension reduction can be either linear (such as principal component analysis (PCA) [90]) or nonlinear (such as manifold learning [91]). We do not adopt this approach in this book, since it is difficult to evaluate the information in each subspace for such a dynamical system.
- Continuum limit: An alternative approach to handle the high-dimensional case is to increase the dimension to the uncountable case; i.e., we condense the network of nodes to a continuous space by shrinking the distances between adjacent nodes to zero. Then each point in the continuous space represents one node in the network, and thus the group of ordinary differential equations is converted to a single partial differential equation (PDE) [92]. Such an approach has been adopted to analyze the electromechanical waves in power grids [56]. Note that not every high-dimensional dynamical system can be converted to a PDE in continuous space, unless the network topology is geometrical.

In this book, we adopt the continuum limit approach.

Consensus dynamics

Not every type of dynamics has a continuum limit. We consider consensus dynamics in CPSs. We assume that each node has a scalar system state, whose evolution law is given by the following first-order differential equation:

$$\dot{x}_i(t) = A\sum_{j\sim i}(x_j(t) - x_i(t)), \qquad (5.212)$$

where A is common for all nodes. It actually describes the consensus dynamics of the states of different nodes (e.g., the consensus control of voltages in microgrids [93]).

To facilitate the continuum limit analysis, we make the following assumptions.

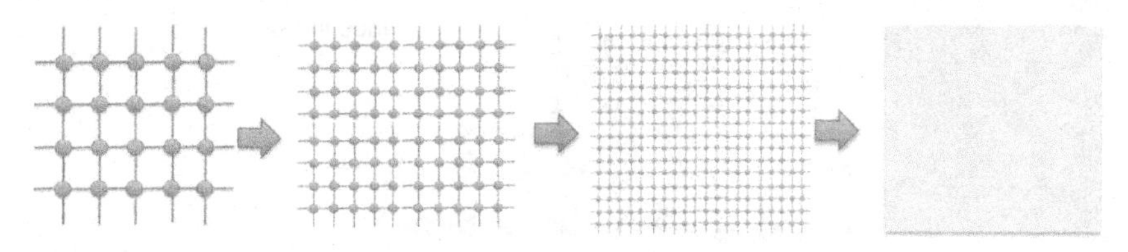

FIG. 5.41

Illustration of space condensation.

Assumption 1. The following assumptions are used for the physical dynamics:

- We assume that the network of physical dynamics is located in a plane; hence each node has four neighbors.
- We assume that the network has infinitely many nodes and all the nodes are symmetric (i.e., they have the same pattern of system state evolution).

Space condensation

We consider a planar networked CPS to explain the condensation procedure, which can be easily extended to the one-dimensional case (network on a line). In a planar network, each node is indexed by a two-tuple (i, j); its location is given by $(i\delta, j\delta)$, where δ is the distance between two adjacent nodes. Any function f (such as system state, or entropy) of the node (i, j) at time t is then denoted by $f_{ij}(t)$. Hence for a location in the plane $\mathbf{z} = (x, y)$, we define

$$f(\mathbf{z}, t) = f_{\lfloor \frac{x}{\delta} \rfloor, \lfloor \frac{y}{\delta} \rfloor}(t). \tag{5.213}$$

When $\delta \rightarrow 0$, the function defined in Eq. (5.213) will converge to a function in the continuous plane, since rational numbers are dense in real numbers. We assume that the limit function is continuously differentiable in both time and space, which has been demonstrated by our numerical simulations. This procedure is illustrated in Fig. 5.41.

Diffusion of uncertainty

We begin from the one-dimensional case; i.e., a network on a line. We have the following lemma describing how the covariance matrix of the dynamics evolves:

Lemma 6. *Denote by $\Sigma(t)$ the covariance matrix of the overall system state $\mathbf{x}(t) = (\mathbf{x}_1^T(t), \ldots, \mathbf{x}_N^T(t))^T$. Assume that the system state estimation errors are Gaussian distributed with covariance matrix $\Sigma_i^e(t)$ for node i. Then the evolution of $\Sigma(t)$ satisfies*

$$\dot{\Sigma}(t) = \mathrm{Sym}(\Sigma(t), \mathbf{A} + \mathbf{BK}) + \mathrm{Sym}(\Sigma_e(t), \mathbf{BK}) + \Sigma_n(t), \tag{5.214}$$

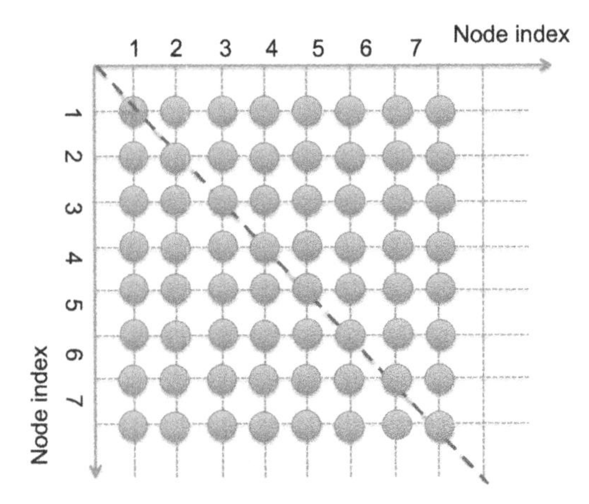

FIG. 5.42

Illustration of the covariance space.

where $\mathbf{A} = \{\mathbf{A}_{ij}\}_{ij}$, $\mathbf{B} = \text{diag}(\mathbf{B}_1, \ldots, \mathbf{B}_N)$, $\Sigma_e(t) = \text{diag}(\Sigma_1^e(t), \ldots, \Sigma_N^e(t))$ *is the covariance matrix of the error of system state estimation and we define*[4] $Sym(\mathbf{G}, \mathbf{H}) = \mathbf{G}\mathbf{H}^T + \mathbf{H}\mathbf{G}^T$ *for matrices* $\mathbf{G}$ *and* $\mathbf{H}$ *and* $\Sigma_n(t) = \text{diag}(\Sigma_1^n(t), \ldots, \Sigma_N^n(t))$.

We consider a two-dimensional space in which a virtual node (i,j) (which is not the physical dynamics node) is located at $(i\delta, j\delta)$ and represents $\Sigma_{ij} = E[x_i x_j]$ (i.e., the covariance of x_i and x_j), as illustrated in Fig. 5.42. Then we study the dynamics of covariance in this "virtual plane."

We then have

$$\begin{aligned}
\dot{\Sigma}_{ij}(t) &= \Sigma_{i,j-1}(t)A_{j-1,j} + \Sigma_{ij}(t)A_{ij} + \Sigma_{i,j+1}(t)A_{j+1,j} \\
&\quad + A_{i,i-1}\Sigma_{i-1,j}(t) + A_{ij}\Sigma_{ij}(t) + A_{i,i+1}\Sigma_{i+1,j}(t) \\
&= A\Sigma_{i,j-1}(t) + A\Sigma_{i,j+1}(t) + A\Sigma_{i-1,j}(t)(t) \\
&\quad + A\Sigma_{i+1,j}(t) - 4A\Sigma_{ij}(t) \\
&= A\delta^2 \frac{\frac{\Sigma_{i,j+1}(t) - \Sigma_{i,j}(t)}{\delta} - \frac{\Sigma_{i,j}(t) - \Sigma_{i,j-1}(t)}{\delta}}{\delta} \\
&\quad + A\delta^2 \frac{\frac{\Sigma_{i+1,j}(t) - \Sigma_{i,j}}{\delta} - \frac{\Sigma_{i,j}(t) - \Sigma_{i-1,j}(t)}{\delta}}{\delta}.
\end{aligned} \tag{5.215}$$

Then when $\lim_{\delta \to 0} A\delta^2 = D$, letting $\delta \to 0$ we have

$$\frac{\partial \Sigma(\mathbf{z}, t)}{\partial t} = D\Delta\Sigma(\mathbf{z}, t), \tag{5.216}$$

which is the heat equation in the plane.

[4]Note that this is similar to the symmetric product of tensors, which generates a symmetric tensor from two tensors.

Since the variances on the line are the diagonal elements in the virtual plane, namely $\sigma^2(x) = \Sigma(\mathbf{z})$ when $\mathbf{z} = (x, x)$, we obtain the following conclusion:

Proposition 2. *For one-dimensional and first-order dynamics with deterministic evolution law, as $\delta \to 0$, the dynamics of variances on the line is the same as the two-dimensional dynamics of heat diffusion along the diagonal line.*

Remark 13. Using a similar argument, the dynamics of variance in a planar CPS are the same as that of four-dimensional heat diffusion dynamics along the diagonal plane. It is easy to verify that the heat diffusion along the diagonal plane of a four-dimensional space is the same as the heat diffusion in the plane.

The main conclusion of the above continuum space limit analysis in continuous space is that the propagation of variance is the same as heat diffusion, which provides significant insight and conclusions on many useful properties (such as the speed of entropy propagation). It is well known that the solution to the heat equation in the plane is given by

$$\sigma^2(\mathbf{z}, t) = \int G(\mathbf{z} - \mathbf{s}, t) g(\mathbf{s}) ds, \tag{5.217}$$

where the Green function G is given by

$$G(\mathbf{z}, t) = \frac{1}{\sqrt{4\pi D t}} \exp\left(-\frac{\|\mathbf{z}\|^2}{4Dt}\right), \tag{5.218}$$

and $g(\mathbf{s})$ is the initial value of the variance. We assume the following initial values of variance:

$$g(\mathbf{z}) = \sigma_0^2 \delta(\mathbf{z}), \tag{5.219}$$

i.e., only the point of origin has uncertainty. Then the location $\mathbf{z}$ has variance at time t given by

$$\sigma^2(\mathbf{z}, t) = \frac{\sigma_0^2}{\sqrt{4\pi D t}} \exp\left(-\frac{\|\mathbf{z}\|^2}{4Dt}\right), \tag{5.220}$$

which is proportional to the Green function. Hence we have

$$h(\mathbf{z}, t) = h_0 - \frac{1}{4} \log 4\pi D t - \frac{\|\mathbf{z}\|^2}{8Dt}, \tag{5.221}$$

where h_0 is the differential entropy corresponding to σ_0^2.

We then obtain the following conclusion about the maximum entropy of a location:

Corollary 6. *For a planar first-order CPS with deterministic dynamics, given a location $\mathbf{z}$ and the initial value in Eq. (5.219), $h(\mathbf{z}, t)$ is maximized at time $t^* = \frac{\|\mathbf{z}\|^2}{2D}$ and*

$$h(\mathbf{z}, t^*) = h_0 - \frac{1}{4}\log\frac{\|\mathbf{z}\|^2}{2\pi} - \frac{1}{4}. \tag{5.222}$$

We define the arrival time of uncertainty as follows:

Definition 16. Given fixed h_0 and h_1, we define the (h_0, h_1)-arrival time at location $\mathbf{z}$ as

$$\tau_{h_0,h_1}(\mathbf{z}) = \inf\{\tau | h(\mathbf{z}, \tau) = h_1\}, \tag{5.223}$$

i.e., the first time that the entropy at location $\mathbf{z}$ reaches h_1. If $h(\mathbf{z})$ can never achieve h_1, then we define $\tau(h_0, h_1) = \infty$.

We then obtain the following conclusion about the arrival time of uncertainty:

Corollary 7. *For a planar first-order CPS with deterministic dynamics, given a location* $\mathbf{z}$ *and the initial value in Eq. (5.219), the arrival time of uncertainty is given by*

$$\tau_{h_0,h_1}(\mathbf{z}) = \inf\left\{\tau \left| -\frac{1}{4}\log 4\pi D\tau - \frac{\|\mathbf{z}\|^2}{8D\tau} = h_1 - h_0 \right.\right\}. \tag{5.224}$$

5.7 CONTROL COMMUNICATION COMPLEXITY

In all the previous discussions, we have considered the communications between sensors and controllers. In the context of distributed control systems, it is also possible for controllers to exchange information through communication channels with finite bandwidth. In this section, we follow the discussion in Ref. [70] to introduce the communication requirement between two controllers with coupled dynamics. Since the communication is for the computation of control actions at the two controllers, it can be considered as communications for computing functions at different agents. Hence the corresponding communication requirement is called Communication Complexity by following the terminology in the seminal paper by A. Yao in 1979 [94]. However, in contrast to the research in traditional communication complexity of distributed computing, in the situation under study there is no direct communication between the two agents; hence they communicate via the messages to and from the physical dynamics. Therefore the physical dynamics can be considered as a special communication channel, which brings special challenges to the analysis.

5.7.1 SYSTEM MODEL

The model for distributed control with a single agent is given by

$$\begin{cases} \mathbf{x}(t) = \mathbf{D}\mathbf{x}(t-1) + \mathbf{E}\mathbf{u}(t-1, \xi(t-1)), \\ \psi(t) = Q(t, \mathbf{C}\mathbf{x}(t)) \\ \xi(t) = K(t, \psi(t), \alpha). \end{cases} \tag{5.225}$$

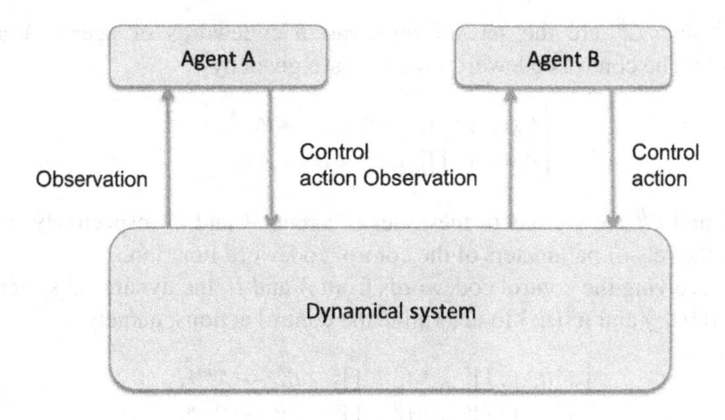

FIG. 5.43

The system model of distributed control with two agents.

The notation is explained below:

- $\mathbf{x}$ is the N-dimensional system state vector;
- $\mathbf{D}$ is the matrix characterizing the open-loop dynamics;
- the $N \times M$ matrix $\mathbf{E}$ is the control coefficient matrix;
- the function $Q(t, \cdot)$ is a time-variant coding function mapping from the L-dimensional observation $\mathbf{Cx}$ to a finite set L_t.
- $K(t, \cdot)$ is the feedback control law mapping from the encoded messages $\psi(t)$ to the control codeword $\xi(t)$.
- $\mathbf{u}$ is the control action mapping from the control codeword to the control action vector.
- α is of the parameter of control law, which belongs to set $\mathcal{A}$.

Here the sensor sends the quantized observation $\xi(t)$ to the decision maker; then the decision maker computes the control codeword and sends it to the physical plant to actuate the control action.

For the two-agent (denoted by A and B) case, the system is described by

$$\begin{cases} \mathbf{x}(t) = & \mathbf{Dx}(t-1) + \mathbf{E}_A\mathbf{u}_A(t, \xi^A(0:t-1), \psi^A(0:t)) \\ & +\mathbf{E}_B\mathbf{u}_B(t, \xi^B(0:t-1), \psi^B(0:t)), \\ \psi^A(t) = & Q_A(t, \mathbf{C}_A\mathbf{x}(0:t)), \quad \psi^B(t) = Q_B(t, \mathbf{C}_B\mathbf{x}(0:t)), \\ \xi^A(t) = & K_A(t, \psi^A(0:t), \alpha), \quad \xi^A(t) = K_B(t, \psi^B(0:t), \beta). \end{cases} \tag{5.226}$$

The system is illustrated in Fig. 5.43.

Similarly to the single-agent case, the quantization functions are mappings given by

$$\begin{cases} Q_A(t, \cdot) : \prod_{i=0}^{t} \mathcal{R}^{L_A} \to \mathcal{L}_t^A, \\ Q_B(t, \cdot) : \prod_{i=0}^{t} \mathcal{R}^{L_B} \to \mathcal{L}_t^B, \end{cases} \tag{5.227}$$

where $\mathcal{L}_t^A$ and $\mathcal{L}_t^B$ are the sets of observation codewords of agents A and B, respectively. The control codeword functions are given by

$$\begin{cases} K_A(t,\cdot) : \prod_{i=0}^{t} \mathcal{L}_t^A \times \mathcal{A} \to \mathcal{M}_t^A, \\ K_B(t,\cdot) : \prod_{i=0}^{t} \mathcal{L}_t^B \times \mathcal{B} \to \mathcal{M}_t^B, \end{cases} \tag{5.228}$$

where $\mathcal{L}_t^A$ and $\mathcal{L}_t^B$ are the sets of messages of agents A and B, respectively, while $\mathcal{A}$ and $\mathcal{B}$ are the sets of parameters of the control codeword function.

Upon receiving the control codewords from A and B, the dynamical system uses functions $\mathbf{u}^A(t,\cdot)$ and $\mathbf{u}^B(t,\cdot)$ to determine the control actions, namely

$$\begin{cases} \mathbf{u}^A(t,\cdot) : \prod_{i=0}^{t} \mathcal{M}_t^A \times \prod_{i=0}^{t} \mathcal{L}_t^A \to \mathcal{R}^{m_A}, \\ \mathbf{u}^B(t,\cdot) : \prod_{i=0}^{t} \mathcal{M}_t^B \times \prod_{i=0}^{t} \mathcal{L}_t^B \to \mathcal{R}^{m_B}, \end{cases} \tag{5.229}$$

where m_A and m_B are the dimensions of the control actions for A and B, respectively. Hence the protocol of distributed control is specified by the functions $\{Q^A(t,\cdot)\}_{t=0}^{\infty}$, $\{Q^B(t,\cdot)\}_{t=0}^{\infty}$, $\{K^A(t,\cdot)\}_{t=0}^{\infty}$, $\{K^B(t,\cdot)\}_{t=0}^{\infty}$, $\{\sqcap^A(t,\cdot)\}_{t=0}^{\infty}$, and $\{\sqcap^B(t,\cdot)\}_{t=0}^{\infty}$. The set of these parameters is denoted by Σ.

The information structure, namely the knowledge known to the agent, of each agent is given by

$$\begin{cases} \mathcal{I}^A(t) = (\psi_0^A, \ldots, \psi_t^A, \alpha), \\ \mathcal{I}^B(t) = (\psi_0^B, \ldots, \psi_t^B, \beta), \end{cases} \tag{5.230}$$

i.e., each agent only knows the observations sent to it and its own parameter.

5.7.2 CONTROL COMMUNICATION COMPLEXITY

Based on the above system model of distributed control, we discuss the communication requirement measured by control communication complexity.

Communication complexity

In generic distributed computing, the communication complexity is the amount of exchanged information among agents necessary to compute certain function(s). For example, agents A and B have private (thus unknown to each other) variables α and β, respectively. Now they want to compute function $f(\alpha,\beta)$ and the result should be known to both A and B. Note that the function f is known to both A and B. Then the communication complexity is the minimum number of bits that A and B need to exchange. A comprehensive introduction to communication complexity can be found in Ref. [95].

Consider the following example. The function is specified in Fig. 5.44A. The protocol of the computing procedure is illustrated in Fig. 5.44B. For each step, the left arrow denotes bit 0 while the right arrow means bit 1. Suppose that agent A transmits first. If its own bit is 1, $f(\alpha,\beta) = 1$ regardless of β; then agent A can simply inform B that the result is 1. Otherwise, agent A informs B its own bit α. In the second step,

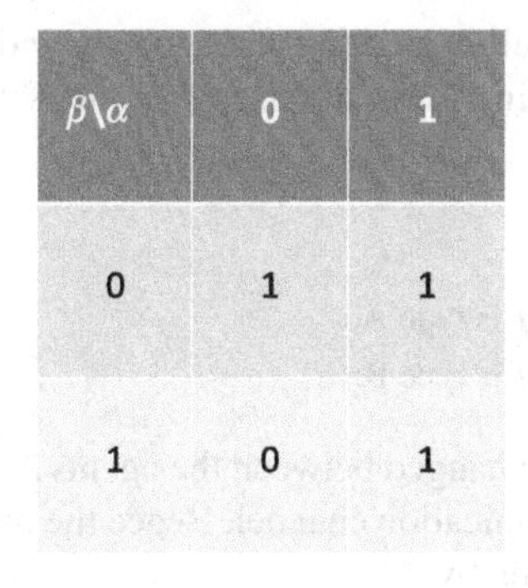

$\beta\backslash\alpha$	0	1
0	1	1
1	0	1

(A) Value of function f

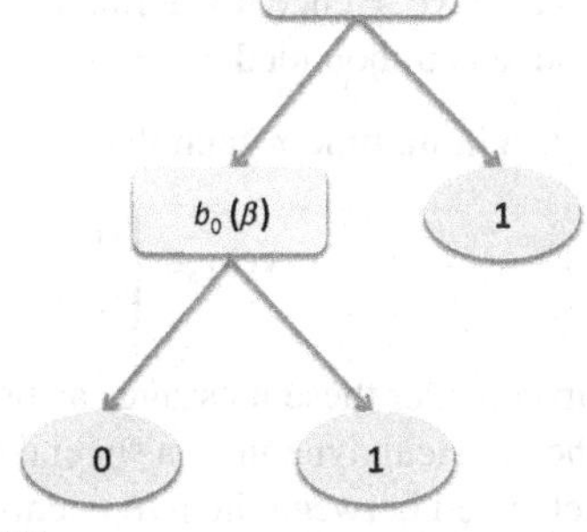

(B) Protocol for computing f

FIG. 5.44

Illustration of communication complexity.

B determines $f(\alpha, \beta)$ according to its own bit and the bit received from A. Hence after two steps the function is determined. The communication complexity, denoted by $D(f)$, is then defined as the depth of the tree. Hence in this example, we have $D(f) = 2$.

Control communication complexity

We assume that the two agents have a goal function f. Hence the whole system is represented by (f, Σ). The communication in the above distributed control model includes:

- The messages sent from the physical system to the two agents A and B; hence we define

$$\begin{cases} \lambda_t^A = \lceil \log_2 |\mathcal{L}_t^A| \rceil, \\ \lambda_t^B = \lceil \log_2 |\mathcal{L}_t^B| \rceil, \end{cases} \tag{5.231}$$

  namely the number of bits needed to describe the messages at time t.
- The messages sent from the agents to the physical systems (namely the control codewords); hence we define

$$\begin{cases} \mu_t^A = \lceil \log_2 |\mathcal{M}_t^A| \rceil, \\ \mu_t^B = \lceil \log_2 |\mathcal{M}_t^B| \rceil, \end{cases} \tag{5.232}$$

  namely the number of bits needed to describe the control codeword at time t.

  Two types of goal functions are considered:

- Type A: The goal function f represents the state that agents A and B desire to reach in finite steps. It is a function mapping from $\mathcal{A} \times \mathcal{B}$ to $\mathcal{R}^N$. Here it is assumed that the initial state is bounded.

- Type B: The goal function f represents the cell in a lattice that agents A and B want to achieve. Hence it is a function mapping from $\mathcal{A} \times \mathcal{B}$ to a cell in $\mathcal{R}^N$. The initial state is unbounded.

There exists a finite time τ such that

$$\mathbf{x}(\tau) = \begin{cases} f(\alpha, \beta), & \text{if } f \text{ is type A,} \\ \in f(\alpha, \beta), & \text{if } f \text{ is type B.} \end{cases} \tag{5.233}$$

We can consider these messages as being exchanged between the agents A and B through the physical dynamics, a special communication channel. Hence the number of bits exchanged between the two agents is given by

$$c(\Omega, \alpha, \beta, \mathbf{x}(0)) = \sum_{t=0}^{\infty} (\lambda_t^A + \lambda_t^B + \mu_t^A + \mu_t^B). \tag{5.234}$$

We further define

$$C(\Omega) = \sup_{\alpha, \beta, \mathbf{x}(0)} c(\Omega, \alpha, \beta, \mathbf{x}(0)), \tag{5.235}$$

which denotes the worst case communication cost.

Then the control communication complexity is defined as

$$Y(\Sigma, f) = \min_{\Omega} C(\Omega). \tag{5.236}$$

The following theorem provides a bound for the control communication complexity:

Theorem 23. *For any distributed control parameter Σ and goal function f, we have*

$$2D(f) \leq Y(\Sigma, f). \tag{5.237}$$

The proof is illustrated in Fig. 5.45. According to the definition of control communication complexity, for any local parameters and initial system state, we have

$$c(\Omega, \alpha, \beta, \mathbf{x}(0)) \leq Y(\Sigma, f), \tag{5.238}$$

for a fixed Ω.

We first consider agent A and the physical dynamics as a composite agent A'. Due to the definition of communication complexity in traditional distributed computing, we have

$$D(f) \leq \sum_{t=0}^{\infty} (\lambda_t^B + \mu_t^B). \tag{5.239}$$

Symmetrically, we can consider agent B and the physical dynamics as a composite agent B'. Similarly, we have

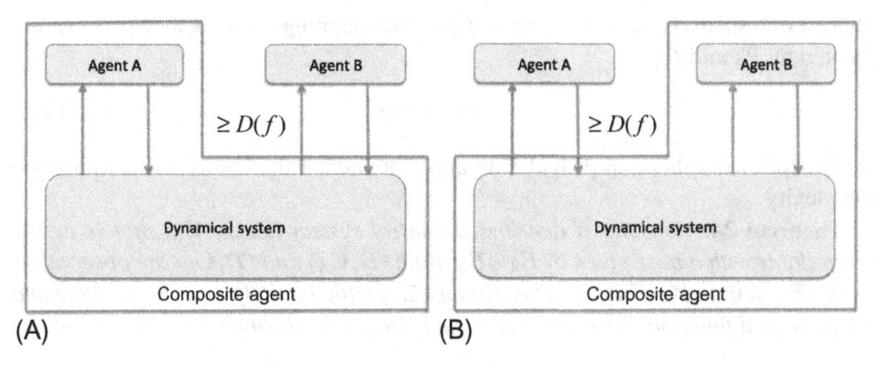

FIG. 5.45

Illustration of the proof of Theorem 23. (A) Agent *A* and physical dynamics as a composite agent. (B) Agent *B* and physical dynamics as a composite agent.

$$D(f) \leq \sum_{t=0}^{\infty}(\lambda_t^A + \mu_t^A). \tag{5.240}$$

Combining Eqs. (5.238)–(5.240), we obtain the conclusion in Theorem 23.

Definition 17. Assume that a subset $\mathcal{X}$ is bounded. Consider a lattice in $\mathcal{R}^N$, which is given by

$$\Lambda = \{\mathbf{Gi}, \mathbf{i} \in \mathbf{Z}^n\}, \tag{5.241}$$

where $\mathbf{Z} = (\ldots, -2, -1, 0, 1, 2, \ldots)$, n is the dimension of the space, and $\mathbf{G}$ is an $n \times n$ generator matrix. The order of $\mathcal{X}$ over the lattice Λ is the number of distinct cells of Λ that intersect $\mathcal{X}$. We denote it by $O(\mathcal{X})$.

The distributed control system $(\mathbf{D}, \mathbf{E}_A, \mathbf{E}_B)$ is said to be jointly controllable, if the following matrix satisfies

$$\text{rank}(\mathbf{E}_A, \mathbf{E}_B, \mathbf{D}\mathbf{E}_A, \mathbf{D}\mathbf{E}_B, \mathbf{D}^2\mathbf{E}_A, \mathbf{D}^2\mathbf{E}_B, \ldots, \mathbf{D}^{N-1}\mathbf{E}_A, \mathbf{D}^{N-1}\mathbf{E}_B) = N. \tag{5.242}$$

When the distributed control system is jointly controllable, let

$$\mathcal{B}_A = \{\mathbf{e}_1, \ldots, \mathbf{e}_{n_A}\} \tag{5.243}$$

be a set of linearly independent column vectors chosen from the matrix $(\mathbf{E}_A, \mathbf{D}\mathbf{E}_A, \ldots, \mathbf{D}^{N-1}\mathbf{E}_A)$, whose rank is n_A. Similarly we can define $\mathcal{B}_B$ as a set of independent column vectors from the matrix $(\mathbf{E}_B, \mathbf{D}\mathbf{E}_B, \ldots, \mathbf{D}^{N-1}\mathbf{E}_B)$, whose rank is n_B. Then any vector $\mathbf{x}$ in $\mathcal{R}^N$ can be uniquely represented by

$$\mathbf{x} = \mathbf{x}_A + \mathbf{x}_B, \tag{5.244}$$

where $\mathbf{x}_A \in \text{span}(\mathcal{B}_A)$ and $\mathbf{x}_B \in \text{span}(\mathcal{B}_B)$. The mappings from $\mathbf{x}$ to $\mathbf{x}_A$ and $\mathbf{x}_B$ are denoted by P_A and P_B:

$$\mathbf{x} = P_A\mathbf{x} + P_B\mathbf{x}. \tag{5.245}$$

The following theorem provides an upper bound for the control communication complexity:

Theorem 24. *Consider a distributed control system* $(\mathbf{E}, \mathbf{E}_A, \mathbf{E}_B)$ *that is jointly controllable with a base given by* $\mathcal{B}_A \cup \mathcal{B}_B$. *Both* $(\mathcal{D}, \mathbf{C}_A)$ *and* $(\mathcal{D}, \mathbf{C}_B)$ *are observable, and* $\mathbf{C}_B\mathbf{E}_A \neq 0$ *and* $\mathbf{C}_A\mathbf{E}_B \neq 0$. *Then for any* Σ, f *with a finitely large initial state and a Type A goal function, the control communication complexity is upper bounded by*

$$Y(\Sigma, f) \leq 4D(f) + 4\lceil \log_2 |\mathcal{X}_0| \rceil + 2\lceil \log_2 |\mathcal{F}_R| \rceil. \tag{5.246}$$

Since this theorem provides an upper bound for the minimum communication requirement, it suffices to construct an encoding and decoding scheme which achieves the goal function and has the same amount of communication as the right-hand side of Eq. (5.246). In the feasible communication and control scheme proposed in Ref. [70], the following three steps are carried out by both agents A and B, before which we fix a communication protocol Π that achieves the traditional communication complexity in the distributed computing problem.

- Step 1: The agents estimate the initial state and then communicate this to each other. During this period of time (suppose that it ends at time slot $n-1$), both agents set their control actions to zero. Then the observations at agent A are given by

$$(\mathcal{C}_A\mathbf{x}(0))^T, \ldots, (\mathcal{C}_A\mathbf{x}(n-1))^T)^T = \mathbf{H}_A\mathbf{x}(0), \tag{5.247}$$

where

$$\mathbf{H}_A = ((\mathcal{C}_A)^T, (\mathcal{C}_A\mathbf{D})^T, \ldots, (\mathcal{C}_A\mathbf{D}^{n-1})^T)^T. \tag{5.248}$$

Similarly, we have

$$((\mathcal{C}_B\mathbf{x}(0))^T, \ldots, (\mathcal{C}_B\mathbf{x}(n-1))^T)^T = \mathbf{H}_B\mathbf{x}(0), \tag{5.249}$$

where

$$\mathbf{H}_B = ((\mathcal{C}_B)^T, (\mathcal{C}_B\mathbf{D})^T, \ldots, (\mathcal{C}_B\mathbf{D}^{n-1})^T)^T. \tag{5.250}$$

Due to the assumption of observability of both agents A and B, $\mathbf{x}(0)$ can be uniquely determined by the observations at agent A (i.e., the left-hand side of Eq. (5.247)) and agent B (the left-hand side of Eq. (5.249)) separately. Since there are a total of $|\mathcal{X}_0|$ possible initial states (due to the definition of a Type A goal function), we can set

$$|\mathcal{L}_{n-1}^A| = |\mathcal{L}_{n-1}^B| = |\mathcal{X}_0|, \tag{5.251}$$

such that the observation codewords can identify the initial state for agents A and B. Hence in the first stage, the total number of bits exchanged between the physical dynamics and the agents is given by $2\lceil\log_2|\mathcal{X}_0|\rceil$.

- Step 2: A and B exchange information with each other in order to transcribe the transcripts of protocol Π. The total period of Step 2 is $D(f)$ time slots. Suppose that agent A sends the first bit ξ_{n-1}^A and agent B does not transmit. Then the system dynamics sets

$$\begin{cases}\mathbf{u}_A(n) = \xi_{n-1}^A\mathbf{v}_A,\\ \mathbf{u}_B(n) = 0,\end{cases} \tag{5.252}$$

where $\mathbf{v}_A$ is a preset vector, and the system state at time n is given by

$$\mathbf{x}(n) = \mathbf{D}^n\mathbf{x}(0) + \xi_{n-1}^A\mathbf{E}_A\mathbf{v}_A. \tag{5.253}$$

Then the quantization function Q_n^B can be defined such that

$$\psi_n^B(\mathbf{C}_B(\mathbf{x}(0)),\ldots,\mathbf{C}_B(\mathbf{x}(n))) = \xi_{n-1}^A. \tag{5.254}$$

This is feasible since ξ_{n-1}^A is a function of $\mathbf{x}(0)$ which can be uniquely determined from $\{\mathbf{C}_B(\mathbf{x}(0)),\ldots,\mathbf{C}_B(\mathbf{x}(n))\}$. In this manner, a single bit in the first step of Π is transmitted from agent A to agent B. Using the same approach, in the next round, agent B sends its first bit to agent A. Repeating this procedure, the two agents can complete the procedure of Π by exchanging $D(f)$ bits. Hence the procedure stops at time $n + D(f) - 1$. At most $2D(f)$ bits are transmitted, since each bit in the protocol Π requires two bits in the context of distributed control. Since each bit causes two possible system states, at time $n + D(f) - 1$ the number of possible system states (considering all possible initial states $\mathbf{x}(0)$) is given by

$$|\{\mathbf{x}(n + D(f) - 1)\}| = 2^{D(f)}|\mathcal{X}(0)|. \tag{5.255}$$

- Step 3: A and B steer the system state to the desired one, based on the initial state estimation. For notational simplicity, we define

$$\mathbf{x}_s = \mathbf{D}^n\mathbf{x}(n + D(f) - 1), \tag{5.256}$$

which is known to both agents at the end of the second step.
We can always find vector sequences $\{\mathbf{z}_A(1),\ldots,\mathbf{z}_A(N)\}$ and $\{\mathbf{z}_B(1),\ldots,\mathbf{z}_B(N)\}$ such that

$$P_Af(\alpha,\beta) = P_A\mathbf{x}_s + \mathbf{D}^{N-1}\mathbf{E}_A\mathbf{z}_A(1) + \cdots + \mathbf{E}_A\mathbf{z}_A(N) \tag{5.257}$$

and

$$P_Bf(\alpha,\beta) = P_B\mathbf{x}_s + \mathbf{D}^{N-1}\mathbf{E}_B\mathbf{z}_B(1) + \cdots + \mathbf{E}_B\mathbf{z}_B(N), \tag{5.258}$$

whose feasibility is guaranteed by the definitions of $\mathcal{B}_A$ and $\mathcal{B}_B$. Then we have

$$f(\alpha,\beta) = P_A\mathbf{x}_s + P_B\mathbf{x}_s + \mathbf{D}^{N-1}(\mathbf{E}_A\mathbf{z}_A(1) + \mathbf{E}_B\mathbf{z}_B(1)) + \cdots + \mathbf{E}_A(\mathbf{z}_A(N) + \mathbf{z}_B(N)). \quad (5.259)$$

We then define

$$\begin{cases} \mathbf{u}^A_{n+D(f)-1+k} = \mathbf{z}_A(k+1) \\ \mathbf{u}^B_{n+D(f)-1+k} = \mathbf{z}_B(k+1), \end{cases} \quad (5.260)$$

which can force the system state to reach $f(\alpha,\beta)$.

Finally, we can estimate the total number of bits that are transmitted during the above three steps:

- Step 1: To identify the initial state $\mathbf{x}(0)$, the following number of bits is needed:

$$B_1 = 2\lceil\log_2|\mathcal{X}_0|\rceil. \quad (5.261)$$

- Step 2: In the second step, at most $D(f)$ time slots are needed and two bits are transmitted in each time slot. Hence at most $2D(f)$ bits are exchanged, namely

$$B_2 = 2D(f). \quad (5.262)$$

- Step 3: At most $D(f) + \lceil\log_2|\mathcal{X}_0|\rceil + \lceil\log_2|\mathcal{F}_R|\rceil$ bits are needed for agent A to encode the control sequence. Symmetrically, the number of bits used for B is also $D(f) + \lceil\log_2|\mathcal{X}_0|\rceil + \lceil\log_2|\mathcal{F}_R|\rceil$. Hence in this step, the total number of bits is upper bounded by

$$B_3 = 2D(f) + 2\lceil\log_2|\mathcal{X}_0|\rceil + 2\lceil\log_2|\mathcal{F}_R|\rceil. \quad (5.263)$$

By adding B_1, B_2, and B_3, we can obtain the upper bound on the right-hand side of Eq. (5.246).

Using a similar argument, we can obtain an upper bound for the control communication complexity in the context of a Type B goal function, which is summarized in the following theorem:

Theorem 25. *Consider a distributed control system with configuration (Σ, f), where f is of Type B with lattice Λ. Suppose that $(\mathbf{D}, \mathbf{E}_A, \mathbf{E}_B)$ is jointly controllable with a base $\{\mathcal{B}_A, \mathcal{B}_B\}$, and $\mathbf{D}$ is invertible, while $(\mathbf{D}, \mathbf{C}_A)$ and $(\mathbf{D}, \mathbf{C}_B)$ are observable. Then the control communication complexity is upper bounded by*

$$Y(\Sigma, f) \leq 4D(f) + 4\lceil\log_2(O(\mathbf{D}^{2n+D(f)-1}\mathcal{X}_0, \Lambda))\rceil + 2\lceil\log_2|\mathcal{F}_R|\rceil. \quad (5.264)$$

The detailed proof is provided in Ref. [70]. Essentially, the proof is similar. The right-hand side of Eq. (5.264) can be achieved in three steps similar to those in the proof of the bound Eq. (5.246). The main difference is in the first step, since the number of possible initial states is uncountable. The system dynamics needs $N+D(f)$ time slots to estimate $\mathbf{x}(0)$ precisely and then use the lattice to encode the estimation.

Since both agents do not carry out controls, the system state at the end of step 1 is given by $\mathbf{D}^{N+D(f)-1}\mathbf{x}(0)$, which requires $\lceil \log_2(O(\mathbf{D}^{2n+D(f)-1}\mathcal{X}_0, \Lambda))\rceil$ bits to encode with respect to the lattice Λ. Hence a total of $2\lceil \log_2(O(\mathbf{D}^{2n+D(f)-1}\mathcal{X}_0, \Lambda))\rceil$ bits are needed in the first step. The procedures and analysis in the second and third steps are similar.

5.8 CONTROL AND INFORMATION IN PHYSICS

In the previous sections, our analysis was based on abstract mathematical models of physical dynamics. In practice, these physical dynamics must comply with the corresponding physics laws. Hence the relationship between information exchange and the control of systems has been an important topic in physics. As we have seen, the concept of entropy can be applied to provide bounds for the communication requirement in a CPS. Actually, entropy originates from research in thermodynamics and had been studied by physicists long before the inception of control and communication theories. In this section, we provide a brief introduction to the related research on the relationship between information and control in the physics community. Although their approaches and models are usually different from those of information scientists, the research may provide an alternative method for the analysis and design of CPSs.

5.8.1 ENTROPY AND CONTROL IN PHYSICS

Second law of thermodynamics

One of the links between information science and physics is the concept of entropy. As we have known, entropy is the fundamental concept in information theory. Originally it comes from thermodynamics. According to Clausius, entropy is a state function of a system, which is defined by

$$S_B - S_A = \int_A^B \frac{dQ}{T}, \tag{5.265}$$

where A and B are two equilibrium states of the system, dQ is the exchange of heat between the system and the heat source during the state change, and T is the temperature of the system.

The second law of thermodynamics states that, for an isolated system, the entropy does not decrease; i.e.,

$$S_B - S_A \geq 0, \tag{5.266}$$

if A is the initial state and B is the final state. Boltzmann found that the entropy is actually determined by the number of micro system states, namely

$$S = k \ln \Omega, \tag{5.267}$$

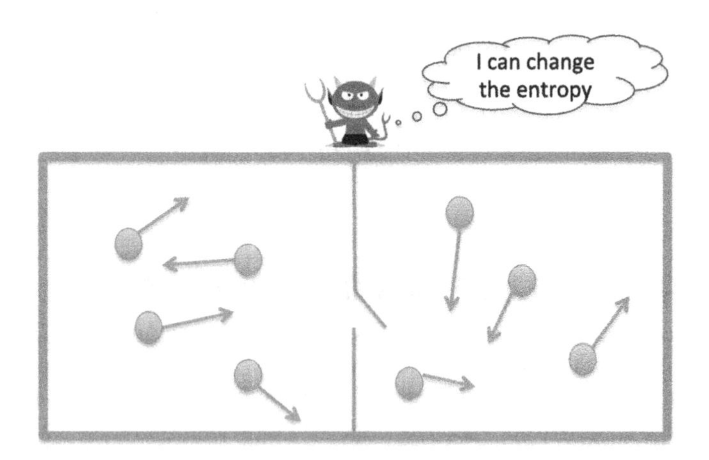

FIG. 5.46

Illustration of Maxwell's demon.

where k is the Boltzmann constant and Ω is the number of micro system states. Based on the insight of thermodynamics entropy, C. Shannon defined the information entropy as

$$S = -\sum_n p_n \log p_n, \tag{5.268}$$

where n is the index of the system state and p_n is the corresponding probability. It is easy to verify that when $p_n = \frac{1}{\Omega}$, namely the states are uniformly distributed, the information entropy is equivalent to the thermodynamic entropy, up to the Boltzmann constant. Essentially, entropy measures the uncertainty of the system.

Maxwell's demon

As has been stated in the second law of thermodynamics, the entropy of an isolated system cannot decrease spontaneously. However, in approximately 1897, J. Maxwell proposed a thought experiment in which the entropy of a system can be decreased. There has been much debate about this thought experiment, which is called "Maxwell's Demon." It has been found that the model is closely related to the feedback control of physical dynamics, as well as the fundamental limit of computing. A detailed discussion can be found in Ref. [96].

The proposed model is illustrated in Fig. 5.46. Consider a container filled with gas molecules. The gas is in the equilibrium state, which achieves the maximum entropy. Then it is divided by a wall into two parts, A and B. There is a small hole in the wall, which can be closed or opened. Now we suppose that a demon can observe the status of each gas molecule. It opens the door only in the following two situations:

- If a molecule in A coming to the door has a velocity less than the mean velocity of all gas molecules, the demon allows it to pass the hole.
- If a molecule in B coming to the door has a velocity greater than the mean velocity of all gas molecules, the demon allows it to pass the hole.

Then it is easy to see that, after a period of time, the average velocity of the gas molecules in B will be lower than that in A. Hence a disparity of velocity is created and the entropy of the whole system is decreased, which seems to contradict the second law of thermodynamics.

Soon, physicists pointed out that the action of the demon is not free, since it needs to "see" the molecules. For example, the French physicist Brillouin [97] assumed that the demon uses light signals to detect the velocity of the molecules. It has been argued that the detection of a molecule's status may not require extra energy; however, this information, which can also be considered as negative entropy, is necessary for the actions of the demon. Hence Maxwell's demon does not contradict the second law of thermodynamics.

We can also find an analogy between Maxwell's demon and a CPS:

- The demon can be considered as a controller which can change the status of the physical dynamics (e.g., in terms of entropy).
- Information is needed by the demon/controller to accomplish its task.

According to the second law of thermodynamics, the entropy of an isolated system cannot decrease spontaneously. Hence we can apply the second law of thermodynamics to informally analyze the entropy evolution in a CPS, which consists of the physical dynamics and the controller [98]. We denote by H_p and H_c the entropies of the physical dynamics and controller, respectively. The physical dynamics and the controller form an isolated system, whose entropy is given by $H_p + H_c$ because of the additivity of entropy. One possible evolution of the system entropy is given as follows.[5] Note that the process is based on some realistic assumptions in physics and is not generic.

- Initial stage: Without loss of generality, we assume that the state of the controller is completely known (e.g., its memory is all cleared to zero). The entropy of the physical dynamics is H_{p0}. Then the total system entropy is H_{p0}.
- Observation stage: We assume that the controller obtains an observation from the physical dynamics using a certain coupling mechanism.[6] The mutual information between the physical system state and the observation is denoted by I. The controller needs to memorize this observation. Hence its entropy is changed to I, due to the new uncertainty in its memory (from the view of an outsider).
- Control stage: The controller uses a certain physical process to decrease the entropy of the physical dynamics to a residual value R. Again, it is possible that no extra energy or entropy increase is needed by the controller [98]. Then the entropy of the controller is still I since it requires extra energy to erase the memory. The total system entropy is then given by $R + I$.

[5] Note that the entropy arises from the uncertainty from the view of an outsider.

[6] It has been realized by Bennett that such a coupling mechanism does not necessarily require energy [96]; hence the system is still isolated.

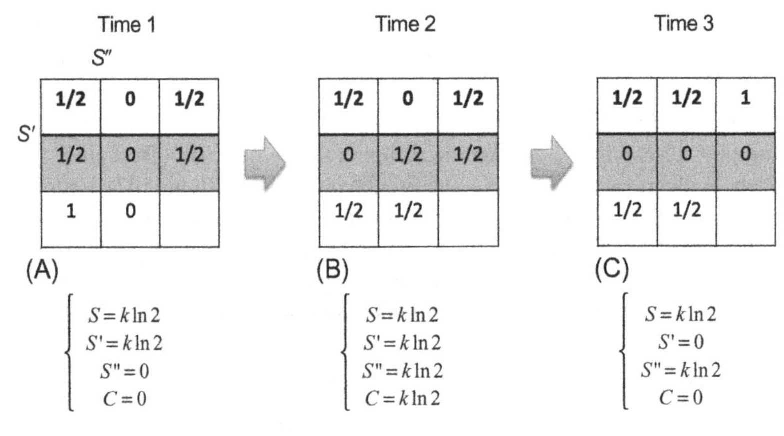

FIG. 5.47

An example of entropy evolution.

Then using the second law of thermodynamics, we have $R + I \geq H_{p0}$, since the entropy cannot be decreased after this cycle, thus resulting in

$$I \leq H_{p0} - R, \tag{5.269}$$

which implies that the information obtained by the controller should be greater than or equal to the decrease of entropy in the physical dynamics.

We illustrate the procedure using the example in Fig. 5.47. We consider two subsystems $\mathcal{S}'$ and $\mathcal{S}''$. $\mathcal{S}'$ is being controlled by $\mathcal{S}''$. The amount of information exchange is denoted by C. Each subsystem has two possible states 1 and 2. The above three steps in the generic process are illustrated as follows:

- Initial stage: At time 1, the state of $\mathcal{S}'$ is uniformly distributed. Hence the entropy of $\mathcal{S}'$ is $k \ln 2$. The state of $\mathcal{S}''$ is deterministic, thus having zero entropy. Then the entropy of the overall system S is $k \ln 2$.
- Observation stage: At time 2, the states of $\mathcal{S}'$ and $\mathcal{S}''$ affect each other. The state transitions are given in Fig. 5.48A. Essentially, the state of $\mathcal{S}''$ is synchronized to that of $\mathcal{S}'$, without affecting the state of $\mathcal{S}'$. This synchronization brings uncertainty to $\mathcal{S}''$, whose entropy is increased to $k \ln 2$. The information exchange is also $k \ln 2$.
- Control stage: At time 3, the state of $\mathcal{S}'$ is changed, based on the current state of $\mathcal{S}''$. The state transitions are given in Fig. 5.48B. Essentially, the state of $\mathcal{S}'$ is changed to 1 if the states of $\mathcal{S}'$ and $\mathcal{S}''$ are the same, and to 2 otherwise. According to the state of $\mathcal{S}''$ after the observation, the state of $\mathcal{S}'$ is fixed to 1, thus eliminating the entropy of $\mathcal{S}'$.

In the above analysis, we observe that, to reduce the entropy of $\mathcal{S}'$ by $k \ln 2$, the amount of communication is $k \ln 2$, and the entropy of the controller itself is also

1,1	1,2	1
2,2	2,1	2
1	2	

S' (middle row); S'' (bottom row)

State transition at the observation stage

(A)

1,1	2,2	1
2,1	1,2	2
1	2	

S' (middle row); S'' (bottom row)

State transition at the control stage

(B)

FIG. 5.48

State transitions at the observation and control stages.

increased to $k \ln 2$. Consider the overall system containing S' and S''; the overall system entropy does not decrease although the entropy of subsystem S' is decreased to zero. The state transition is illustrated in Fig. 5.48.

Szilard engine

In 1929, Leo Szilard refined the Maxwell's demon model and devised the so-called Szilard engine, which demonstrates that information is physical. The engine is illustrated in Fig. 5.49. Again, consider a container with pistons on both sides. There is a single gas molecule in the container. A removable partition is put in the middle of the container. The information on which half of the container the gas molecule lies in can be described by only one bit. If this bit is known, we can push the piston on the empty half of the container to the partition. Since there is no opposition to the piston, no work is needed to compress the empty half of the container. Then we release the partition and the piston will be pushed back to the original location due to

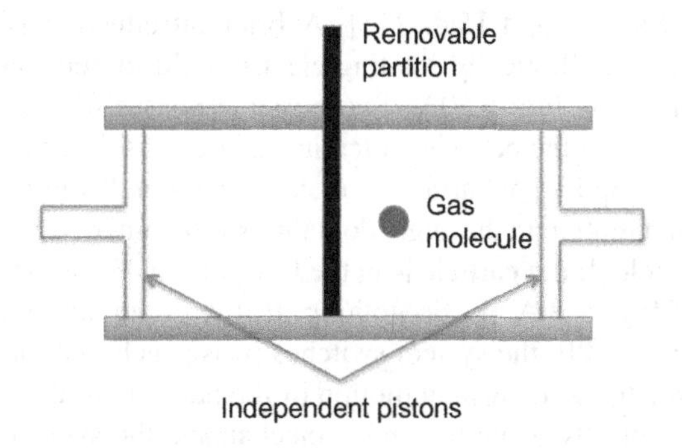

FIG. 5.49

Illustration of the Szilard engine.

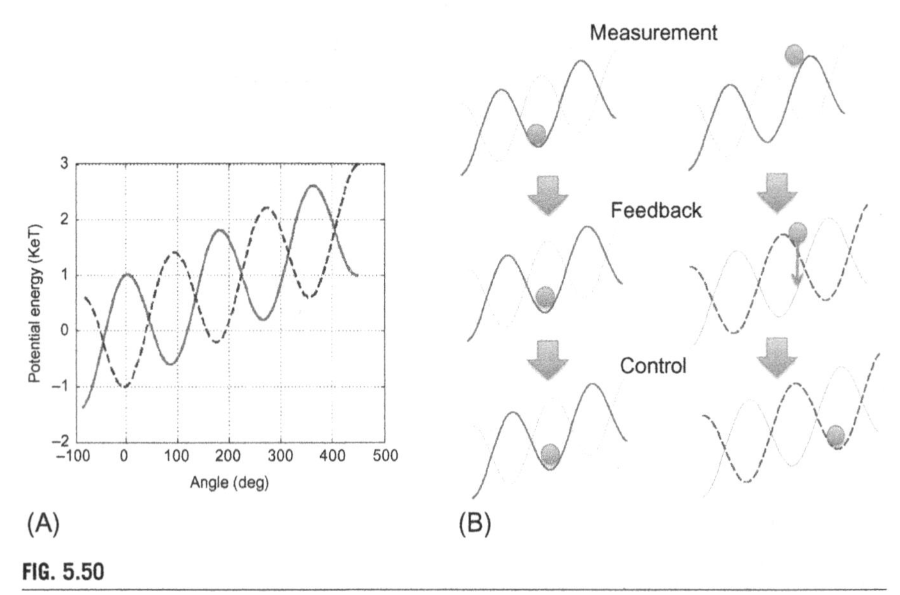

FIG. 5.50

Explanation of the Szilard experiment in Ref. [99]. (A) Potential energy as a function of angle. (B) Measurement and control.

the pressure caused by the gas molecule. It can be verified that we can extract energy of $kT \ln 2$ joules from this procedure. This procedure can be repeated by feeding the information of the gas to the controller again. In this manner, the information on the gas molecule location is converted to work, thus realizing the information-to-energy conversion. Therefore information is no longer an abstract concept. It should be considered as a physical concept similar to energy or power.

In 2010, the schematic experiment of the Szilard engine was reported to have been realized by Sagawa and Ueda [71]. A brief introduction is provided here. The physicists used an elliptically rotating electric field to generate tilted periodic potential as illustrated in Fig. 5.50A. Two possible phases of the potential can be generated. Each valley of the periodic potential can be considered as a step in a stair. A dimeric particle composed of particles with a diameter of 287 nm is put in the field. It exhibits Brownian motion in the angle domain. A fast camera is used to capture the location of the particle. If the particle is in the low potential location, as illustrated in the left column of Fig. 5.50A, we do nothing. If it is in the high potential location, as illustrated in Fig. 5.50B, the system switches phase such that the particle falls to a higher valley. Due to the Brownian motion of the particle, it may move to a lower valley or a higher one. By using the above mechanism, the system can prevent the particle from falling to a lower "step" and can only allow it to jump to a higher one. This procedure is illustrated in Fig. 5.51.

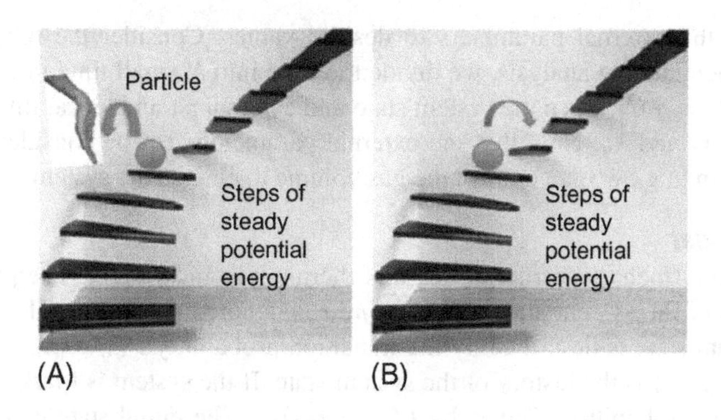

FIG. 5.51

Illustration of the Szilard experiment in Ref. [99]. (A) When the particle intends to jump to a lower step, the system prevents it. (B) When the particle intends to jump to a higher step, the system allows it.

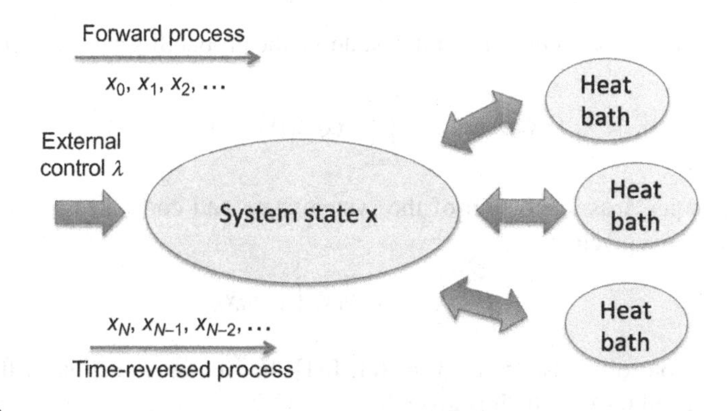

FIG. 5.52

Illustration of the thermodynamic system.

5.8.2 NONEQUILIBRIUM THERMODYNAMICS OF FEEDBACK CONTROL

In Ref. [99], a general theory is proposed to analyze the feedback control for thermodynamic systems. Note that the difference between this study and that of traditional control systems is that the physical dynamics in consideration is a thermodynamic system described by the laws of physics. Although the methodology is considerably different from those in a CPS or networked control, the approaches in the physics community can provide alternative ideas for researchers of controls or communications.

Consider a stochastic system in contact with M heat baths, denoted by $B_1, \ldots, B_M$, with temperatures $T_1, \ldots, T_M$, as illustrated in Fig. 5.52. Denote by x the point of the system in the phase space, while λ is the set of external parameters (e.g., the volume of gas is an external parameter). The feedback control aims to manipulate the external parameters to desired values. Consider the time interval $[0, \tau]$. To facilitate the analysis, we divide the time into N small time intervals with time span $\Delta t = \tau/N$. Then the system state and external parameters at time $n\Delta t$ are denoted by x_n and λ_n. Note that the external parameters can be considered as the control action (e.g., we can control the gas volume to change the system state).

System model

We define the stochastic structure for this thermodynamic system. The probability distribution of the system state x at time $n\Delta t$ is denoted by $P_n(x_n)$. The evolution of the system state is described by the transition probability $P(x_{n+1}|X_n, \lambda_n)$, where $X_n = \{x_0, \ldots, x_n\}$ is the history of the system state. If the system is Markovian, then the transition probability is given by $P(x_{n+1}|x_n, \lambda_n)$. The initial state is denoted by $P_0(x_0)$. Then the probability of trajectory X_n is given by

$$P(X_n|\Lambda_n) = P_0(x_0)\prod_{k=0}^{n-1} P(x_{k+1}|X_k, \lambda_k), \tag{5.270}$$

where $\Lambda_k = \{\lambda_0, \ldots, \lambda_k\}$. Given the initial state x_0, the probability can be rewritten as

$$P(X_n|\Lambda_n, x_0) = \prod_{k=0}^{n-1} P(x_{k+1}|X_k, \lambda_k). \tag{5.271}$$

For a quantity A as a function of the system state and control trajectories, the expectation of A is given by

$$\langle A\rangle = \int A(X_N, \Lambda_N)P(X_N|\Lambda_N)dX_N. \tag{5.272}$$

Backward control is also defined in Ref. [71]. For a control action λ, the time reversal is denoted by λ^*, which is given by

$$\lambda^* = -\lambda. \tag{5.273}$$

For example, if λ is to compress the volume of the gas, then λ^* has to enlarge the volume. The time reversed protocol of $\lambda(t)$ is then given by

$$\lambda^\dagger = \lambda^*(\tau - t), \tag{5.274}$$

where τ is the end of the time interval. The protocol can also be discretized to

$$\begin{aligned}\Lambda^\dagger &= (\lambda_0^\dagger, \ldots, \lambda_n^\dagger)\\ &= (\lambda_{N-1}^*, \ldots, \lambda_{N-n-1}^*).\end{aligned} \tag{5.275}$$

The probability of the time-reversed protocol is given by

$$P(X_n^\dagger|\Lambda_n^\dagger, x_0^\dagger) = \prod_{k=0}^{n-1} P(x_{k+1}^\dagger|X_k^\dagger, \lambda_k^\dagger). \tag{5.276}$$

Fluctuation theorem

Fluctuation theorems are conclusions on how the entropy of a system not in thermodynamic equilibrium will increase or decrease in a given time interval. The second law of thermodynamics claims that the entropy of an isolated system will always increase. However, in statistical mechanics, the probability that the entropy decreases is nonzero, although it is very small. Fluctuation theorems provide quantified conclusions on the probability. The first fluctuation theorem was found by computer simulations in 1993 and then mathematically proved in 1994. It is now a cornerstone of nonequilibrium statistical mechanics.

A general form of fluctuation theorems is given by

$$\frac{P(\sigma_t = A)}{P(\sigma_t = -A)} = \mathrm{e}^{At}, \tag{5.277}$$

where σ_t is the time-averaged generation of entropy during a time duration of t.

Theorem 26 (Detailed Fluctuation Theorem). *The following equality holds for the forward and time-reversed processes:*

$$\frac{P(X_n^\dagger | \Lambda_n^\dagger, x_0^\dagger)}{P(X_n | \Lambda_n, x_0)} = \mathrm{e}^{\sum_i \beta_i Q_i(X_N, \Lambda_N)}, \tag{5.278}$$

where $\beta_i = 1/T_i$ and $Q_i(X_N, \Lambda_N)$ is the heat absorbed from the ith heat bath.

The proof can be found in Ref. [71].

Eq. (5.278) can be rewritten as

$$\begin{aligned} \frac{P(X_n^\dagger | \Lambda_n^\dagger)}{P(X_n | \Lambda_n)} &= \frac{P(x_0)}{P^\dagger(x_0^\dagger)} \mathrm{e}^{\sum_i \beta_i Q_i(X_N, \Lambda_N)} \\ &= \mathrm{e}^{-\sigma(X_N)}, \end{aligned} \tag{5.279}$$

where

$$\sigma(X_N) = -\ln P_0^\dagger(x_0^\dagger) - (-\ln P_0(x_0)) - \sum_i \beta_i Q_i(X_N), \tag{5.280}$$

which consists of the following three terms:

- $-\ln P_0^\dagger(x_0^\dagger)$: this can be considered as the final thermodynamic entropy of the system.
- $-\ln P_0(x_0)$: this can be considered as the initial thermodynamic entropy of the system.
- $-\sum_i \beta_i Q_i(X_N)$: this can be considered as the entropy dissipated to the heat baths.

Therefore we can call $\sigma(X_N)$ the entropy production of the system.

The following theorem provides the foliation law for entropy production:

Theorem 27 (Crooke's Fluctuation Theorem). *The following equality for the forward and reverse-time processes holds:*

$$\frac{P^{\dagger}(-\sigma)}{P(\sigma)} = \mathrm{e}^{-\sigma}. \tag{5.281}$$

Note that Eq. (5.281) can be rewritten as

$$P^{\dagger}(-\sigma) = \mathrm{e}^{-\sigma} P(\sigma). \tag{5.282}$$

By integrating with respect to σ on both sides, we obtain the following form of fluctuation theorem:

Theorem 28 (Integral Fluctuation Theorem). *The following equality on entropy production is valid:*

$$\langle \mathrm{e}^{-\sigma} \rangle = 1. \tag{5.283}$$

Due to the concavity of function e^{-x}, we have

$$\mathrm{e}^{-\langle \sigma \rangle} \le \langle \mathrm{e}^{-\sigma} \rangle = 1, \tag{5.284}$$

which results in the following corollary as an expression of the second law of thermodynamics:

Corollary 8. *The following inequality on entropy production is valid:*

$$\langle \sigma \rangle \ge 0. \tag{5.285}$$

Consider an isothermal process with a single heat bath; the entropy production is given by

$$\sigma(X_N) = \beta(W(X_N) - \Delta F), \tag{5.286}$$

where $W(X_N)$ is the work done on the system and ΔF is the difference of free energy. Due to Eq. (5.283), we have

$$\langle \mathrm{e}^{-\beta W} \rangle = \mathrm{e}^{-\beta \Delta F}, \tag{5.287}$$

which is called the Jarzynski equality.

Feedback control

For the feedback control of a thermodynamic system, the controller needs measurements on the system state. We denote by $Y_n = (y_1, \ldots, y_n)$ the measurements on the thermodynamic system. It is assumed that the measurements are mutually independent given the true system states. Hence we have

$$P(Y_n|X_n) = \prod_k P(y_k|X_k). \tag{5.288}$$

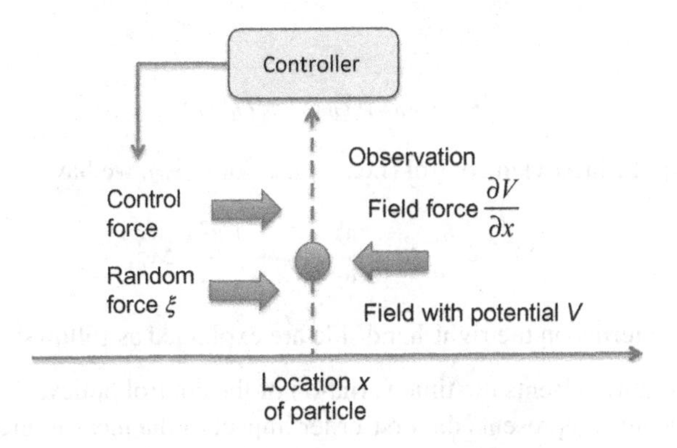

FIG. 5.53

Illustration of the thermodynamic model, in which a particle is moving in a one-dimensional space, subject to a field, a control external force, and random perturbation.

The control action is based on all the previous measurements. Hence we have

$$\Lambda_n(Y_{n-1}) = (\lambda_0(Y_0), \ldots, \lambda_{n-1}(Y_{n-1})). \tag{5.289}$$

We assume that the thermodynamic system is described by an overdamped Langevin system, whose evolution is given by

$$\eta \frac{dx(t)}{dt} = -\frac{\partial V(x, \lambda)}{\partial x} + f(\lambda) + \sqrt{2\eta k_B T}\xi(t), \tag{5.290}$$

whose terms and parameters are explained as follows:

- x is the system state (e.g., the location of a damped oscillator);
- η is the friction constant;
- $V(x, \lambda)$ is an external potential;
- $f(\lambda)$ is the external force as part of the control action;
- ξ is white Gaussian noise.

The model is illustrated in Fig. 5.53.

We also assume that the measurement is given by

$$y_n = x_n + \frac{\Delta R_n}{\Delta t}, \tag{5.291}$$

where ΔR_n is white Gaussian noise. Note that Eq. (5.290) is generally a nonlinear dynamic while Eq. (5.291) is a linear observation.

Since the control action is given by $\lambda_n(Y_n)$, the work performed by the controller on the thermodynamic system at time slot n is given by

$$\begin{aligned} W_n &= V(x_n, \lambda_{n+1}) - V(x_n, \lambda_n) \\ &= \frac{\partial V}{\partial \lambda}\delta\lambda_n + o(\Delta t), \end{aligned} \tag{5.292}$$

where

$$\delta\lambda_n = \lambda_{n+1}(Y_{n+1}) - \lambda_n(Y_n). \tag{5.293}$$

In the case of Markovian control (i.e., λ_n uses only y_n), we have

$$\Delta\lambda_n = \frac{\partial\lambda}{\partial t}\Delta t + \frac{\partial\lambda}{\partial y}\Delta y_n + \frac{1}{2}\frac{\partial^2\lambda}{\partial y^2}\Delta y_n^2, \tag{5.294}$$

where the three terms on the right-hand side are explained as follows:

- The first term represents the time variation of the control policy.
- The second term represents the first-order impact of the measurement change.
- The second term represents the second-order impact of the measurement change.

By discretizing and linearizing the Langevin equation, we have

$$\eta(x_{n+1} - x_n) = -Kx_n\Delta t + \lambda_n\Delta t + \sqrt{2\eta k_B T}\Delta W, \tag{5.295}$$

where K is a positive constant.

Fluctuation theorem for feedback control

As we have seen, various fluctuation theorems can be obtained for thermodynamic systems without control. Given the feedback control model in the above discussion, we can obtain a flotation theorem for feedback control [71].

For thermodynamic system with feedback, the joint probability of system state and observations is given by

$$\begin{aligned} P(X_n, Y_n) &= \prod_{k=0}^{n-1} P(y_{k+1}|X_{k+1})P(x_{k+1}|X_k, \lambda_k(Y_{k-1})) \\ &= P_c(Y_n|X_n)P(X_n|\Lambda(Y_{n-1})), \end{aligned} \tag{5.296}$$

where

$$P_c(Y_n|X_n) = \prod_{k=0}^{n-1} P(y_k|X_k) \tag{5.297}$$

and

$$P(X_n|\Lambda(Y_{n-1})) = \prod_{k=0}^{n-1} P(x_{k+1}|X_k, \lambda_k(Y_{k-1})). \tag{5.298}$$

Consider a variable A as a function of (X_n, Y_n). Its expectation is given by

$$\langle A\rangle = \int A(X_n, Y_n)P(X_n, Y_n)dX_n dY_n, \tag{5.299}$$

and the conditional expectation is given by

$$\langle A\rangle_{Y_n} = \int A(X_n, Y_n)P(X_n|Y_n)dX_n. \tag{5.300}$$

We define the "instantaneous" mutual information between the system state and observation with control as

$$\begin{aligned} I_c(Y_n; X_n) &= \sum_{k=1}^{N} I(y_n; X_n|Y_{n-1}) \\ &= \ln \frac{P_c(Y_n|X_n)}{P(Y_n)}, \end{aligned} \tag{5.301}$$

where $I(y_n; X_n|Y_{n-1})$ is the instantaneous mutual information between the observation at time n and the system state history, given the previous observations; i.e.,

$$\begin{aligned} I(y_n; X_n|Y_{n-1}) &= \ln \frac{P(y_n|X_n, Y_{n-1})}{P(y_n|Y_{n-1})} \\ &= \ln \frac{P(y_n|X_n)}{P(y_n|Y_{n-1})}. \end{aligned} \tag{5.302}$$

Note that I_c does not necessarily equal the "instantaneous" mutual information between the system state and observation trajectories equaling

$$I(X_n; Y_n) = \ln \frac{P(Y_n|X_n)}{P(Y_n)}, \tag{5.303}$$

because $P_c(Y_n|X_n)$ is not necessarily equal to $P(Y_n|X_n)$, since

$$P(Y_n|X_n) = \prod_{k=0}^{n-1} P(y_k|X_k, Y_k) \tag{5.304}$$

and $P(y_k|X_n, Y_n)$ is not necessarily equal to $P(y_k|X_k)$.

It is easy to verify that the following equality, which is similar to the integral fluctuation theorem in Theorem 28, holds:

$$\langle e^{-I_c}\rangle = 1, \tag{5.305}$$

since

$$\begin{aligned} \langle e^{-I_c}\rangle &= \int \frac{P(Y_N)}{P_c(Y_N|X_N)} P(X_N, Y_N)dX_N dY_N \\ &= \int P(Y_N)P(X_N|\Lambda(Y_{N_1}))dX_N dY_N \\ &= 1. \end{aligned} \tag{5.306}$$

Nonequilibrium equalities for feedback control

The following theorem provides an equality for feedback control:

Theorem 29 (Generalized Detailed Fluctuation Theorem With Feedback Control). *The following equality holds when feedback control exists:*

$$\frac{P^{\dagger}(X_N^{\dagger}, Y_N)}{P(X_N, Y_N)} = e^{-\sigma(X_N, \Lambda_N(Y_N) - I_c(X_N; Y_N))}, \tag{5.307}$$

where the "backward probability distribution" is defined as

$$P^{\dagger}(X_N^{\dagger}, Y_N) = P(X_N^{\dagger} \Lambda_N(Y_{N-1}))P(Y_N). \tag{5.308}$$

The proof is similar to that of Theorem 26. Note that the physical meaning of $P^{\dagger}(X_N^{\dagger}, Y_N)$ is the probability obtained from the following two-step experiments:

- Step 1: Carry out the forward experiment and obtain a series of observations Y_N.
- Step 2: Carry out the backward experiment with protocol $\Lambda_N^{\dagger}(Y_{N-1})$.

Based on Theorem 29, we also obtain the following theorems:

Theorem 30 (Generalized Integral Fluctuation Theorem With Feedback Control). *The following equality holds for the case of feedback control:*

$$\langle e^{-\sigma - I_c} \rangle = 1. \tag{5.309}$$

Theorem 31 (Generalized Second Law of Thermodynamics With Feedback). *The following inequality holds for the case of feedback control:*

$$\langle \sigma \rangle \geq -\langle I_c \rangle. \tag{5.310}$$

We can also obtain

$$\begin{aligned} \langle \sigma \rangle + \langle I_c \rangle &= \int P(X_N, Y_N) \ln \frac{P(X_N, Y_N)}{P^{\dagger}(X_N^{\dagger}, Y_N)} dX_N dY_N \\ &= D(P(X_N, Y_N) || P^{\dagger}(X_N^{\dagger}, Y_N)) \end{aligned} \tag{5.311}$$

by taking the expectation of the logarithm of both sides of Eq. (5.307). Note that $D(P(X_N, Y_N)||P^{\dagger}(X_N^{\dagger}, Y_N))$ is the Kullback-Leiber distance between the distributions $P(X_N, Y_N)$ and $P^{\dagger}(X_N^{\dagger}, Y_N)$. When the reversibility with feedback control holds, i.e.,

$$P(X_N, Y_N) = P^{\dagger}(X_N^{\dagger}, Y_N), \tag{5.312}$$

we have

$$\langle \sigma \rangle + \langle I_c \rangle = 0. \tag{5.313}$$

When $\sigma = \beta(W - \Delta F)$, the following generalized Jarzynski equality holds:

$$\langle e^{-\beta(W - \Delta F) - I_c} \rangle = 1, \tag{5.314}$$

which results in

$$\langle W\rangle \geq \Delta F - k_B T\langle I_c\rangle. \tag{5.315}$$

This inequality relates the work exerted by external sources, the reduction of free energy of the thermodynamic system, and information obtained from the feedback. Moreover, if $\Delta F = 0$ (namely the free energy of the thermodynamic system does not change), we define $W_{\text{ext}} = -W$ as the work extracted from the thermodynamics system, we have

$$\langle W_{\text{ext}}\rangle \leq k_B T\langle I_c\rangle. \tag{5.316}$$

Hence the work that we can extract from the information provided to the thermodynamic system through the feedback is upper bounded by the mutual information scaled by $k_B T$.

Take the Szilard engine discussed previously, for example. The information about the location of the particle is binary (0: the left half; 1: the right half). We assume that the measurement is passed through a binary symmetric channel with error probability ϵ; i.e., the controller receives the correct location of the particle with probability $1-\epsilon$. Using the argument in Ref. [71], we can obtain

$$W_{\text{ext}} = k_B T(\ln 2 + \epsilon\ln\epsilon + (1-\epsilon)\ln(1-\epsilon)). \tag{5.317}$$

Nonequilibrium equalities with efficacy

An alternative type of fluctuation theorem can also be derived. Consider a Markovian measurement; i.e.,

$$P(y_n|X_n) = P(y_n|x_n). \tag{5.318}$$

A forward experiment is carried out at times $t_1, \ldots, t_M$ with feedback control; a backward experiment is also carried out at times $t_{N-M}, \ldots, t_{N-1}$. The measurements in the backward experiment are denoted by $Y'_N = (y_{N-M}, \ldots, y'_{N-1})$. Given $X^\dagger_N$, we define

$$P_c(Y'_N|X_N\dagger) = \prod_{k=1}^{M} P(y_{N-k}|x^\dagger_{N-k}), \tag{5.319}$$

which denotes the probability that Y'_N is obtained given $X^\dagger_N$. Given the protocol $\Lambda(Y_N)^\dagger$, the probability of observing Y'_N is given by

$$P(Y'_N|\Lambda_N(Y_{N-1})^\dagger) = \int P_c(Y'_N|X_N\dagger)P(X^\dagger_N|\Lambda_N(Y_{N-1})^\dagger)dX^\dagger_N. \tag{5.320}$$

The time-reversed sequence of Y_N, denoted by $Y_N^\dagger$, is given by

$$Y_N^\dagger = (-y_{N-M}, \ldots, -y(N-1)). \quad (5.321)$$

Then the probability that Y_N' equals $Y_N^\dagger$ is given by

$$P(Y_N^\dagger | \Lambda_N(Y_{N-1})^\dagger) = \int P_c(Y_N^\dagger | X_N \dagger) P(X_N^\dagger | \Lambda_N(Y_{N-1})^\dagger) dX_N^\dagger. \quad (5.322)$$

We further assume time-reversed symmetry; i.e.,

$$P(y_n^* | x_n^*) = P(y_n | x_n). \quad (5.323)$$

We then obtain the following theorem, whose detailed proof is similar to that of Theorem 26 and can be found in Ref. [71].

Theorem 32 (Renormalized Detailed Fluctuation Theorem). *The following equality holds:*

$$\frac{P(Y_N^\dagger | \Lambda(Y_N)^\dagger)}{P(Y_N)} = \mathrm{e}^{-\sigma'(Y_N)}, \quad (5.324)$$

where the renormalized entropy production σ' is defined as

$$\sigma'(Y_N) = -\ln \int \mathrm{e}^{-\sigma(X_N, \Lambda_N(Y_{N-1}))} P(X_N | Y_N) dX_N. \quad (5.325)$$

Taking the expectation of both sides of Eq. (5.324), we obtain the following corollary:

Corollary 9. *The following equality holds:*

$$\langle \mathrm{e}^{-\sigma} \rangle = \gamma, \quad (5.326)$$

where γ is called the efficacy parameter of feedback control given by

$$\gamma = \int P(Y_N^\dagger | \Lambda_N(Y_{N-1})^\dagger) dY_N^\dagger. \quad (5.327)$$

For the case $\sigma = \beta(W - \Delta F)$, we have

$$\langle \mathrm{e}^{-\beta(W - \Delta F)} \rangle = \gamma, \quad (5.328)$$

which is a generalization of the Jarazynski equality.

The values of the efficacy parameter γ are as follows:

- When the feedback control is perfect (namely, the system is expected to return to the initial state with probability 1 in the time-reserved process), γ equals the number of possible observations of Y_N.
- In the Szilard engine, $\gamma = 2$.
- When there is no feedback control, $\gamma = 1$.

5.9 CONCLUSIONS

We have discussed various approaches of evaluating the communication requirements for controlling CPSs in different situations. As we have seen, several approaches are closely related to the entropy, either deterministic or stochastic, of the physical dynamics. Hence it is a common and generic approach to analyze the increase of the entropy of the physical dynamics and then study how the control can alleviate the entropy increase. We are still awaiting an elegant framework to unify these approaches.

CHAPTER

Network topology design

6

6.1 INTRODUCTION

In this chapter, we introduce the network topology design for CPSs, which is the first step of detailed communication network design. Essentially, the topology design of a communication network determines the connections between the communication nodes.

The following assumptions are used throughout this chapter:

- We consider wired networks, in particular wavelength-division multiplexing (WDM) networks, which use optical fibers. The topology design is also an important issue for wireless networks, e.g., in mobile ad hoc networks. However, large area wireless networks are usually not used in typical CPSs such as smart grids.
- We assume that the network is designed in one batch, rather than in incremental manner.
- We assume that the physical infrastructure of the communication network already exists. Our goal is to design the logical topology of the network.

Currently, there is no systematic framework for communication network topology design in CPSs. Most CPS communication networks use existing design paradigms of data communication networks, or heuristic design methodologies. Our main goal is to design the network topology taking into consideration the physical dynamics.

In the subsequent discussion, we will first provide a brief introduction to WDM networks. Then we introduce an algorithm for optimizing the topology taking into consideration the physical dynamics of the CPS. Finally, we briefly introduce team decision theory, which is related to network topology design in a CPS.

6.2 WDM NETWORKS AND DESIGN CONSTRAINTS

In this section, we provide a brief introduction to WDM networks and the corresponding constraints in topology design. More details can be found in Ref. [43].

Communications for Control in Cyber Physical Systems. http://dx.doi.org/10.1016/B978-0-12-801950-4.00006-8

6.2.1 WDM NETWORKS

WDM networks provide circuit switched end-to-end optical channels, which are also called light paths. Usually a light path uses one wavelength used in the optical fiber. WDM networks consist of the following two layers, which are illustrated in Fig. 6.1:

- Physical layer: In this layer, each node consists of an optical cross-connect (OXC) router. The OXCs are connected by optical fibers. When a packet reaches an OXC, it either passes through it without stopping or is taken out of the optical domain to local clients.
- Logical layer: In this layer, each node is an IP router, which corresponds to an OXC. Two adjacent routers are connected by a direct light path (thus the delay will be very small), or equivalently a logical link. Note that one logical link may actually pass through another OXC in the physical layer.

Take Fig. 6.1 for instance. In the logical layer, nodes R1 and R5 are adjacent. However, in the physical layer, the traffic between R1 and R5 will pass through node OXC3 without stopping. The detailed traffic situation is illustrated in Fig. 6.2.

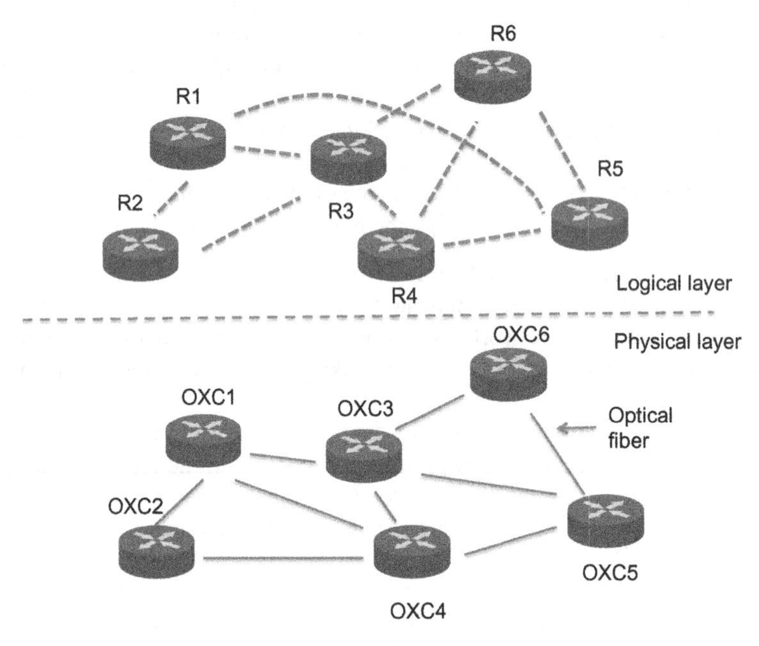

FIG. 6.1

Illustration of the logical and physical layers in WDM networks.

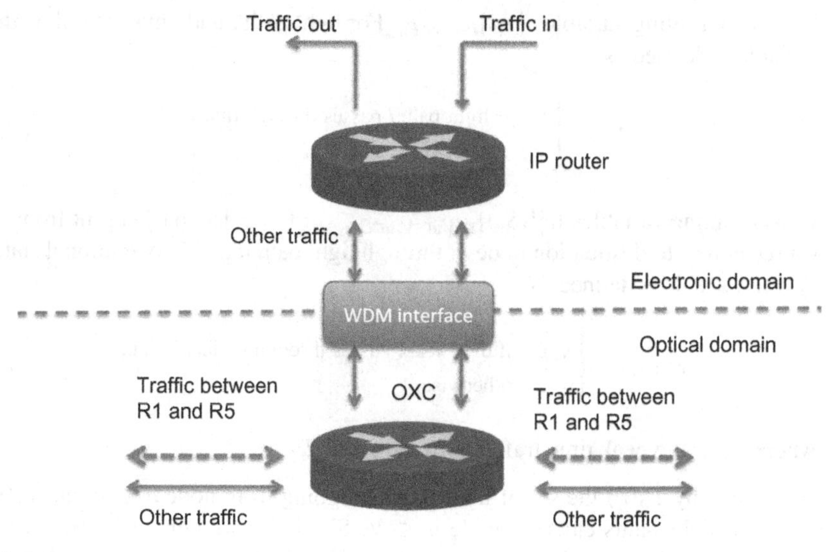

FIG. 6.2
Illustration of the traffic at OXC3.

6.2.2 DESIGN CONSTRAINTS

The design of a WDM network includes the following sequential steps [43]: (1) topology design, (2) using multiple parallel links to increase the resilience, (3) routing optimization, and (4) capacity planning. We follow Ref. [43] to study the design of topology in data WDM networks. Usually, the topology design is formulated as a constrained optimization problem. Here we provide the constraints in a WDM network as the foundation of the subsequent discussion.

Note that our goal is to design the logical topology of the WDM network. We represent the logical topology using a graph, in which each node is a router while each edge is a light path (or a logical link). We denote by $G = (V, E)$ the graph of logical topology, where V is the set of nodes and E is the set of edges. We also denote the set of possible light paths (or logical links) by $\mathcal{E}$. The physical topology is denoted by $G' = (V, E_0)$, where V is the same as defined in the logical topology and E_0 denotes the physical links. $\mathcal{S}$ is the set of operational states. For topology design, we have the following three parameters to manipulate:

- Light path variables $\{X_i\}_{i \in \mathcal{E}}$: For each possible light path i, the corresponding variable X_i is defined as

$$X_i = \begin{cases} 1, & \text{if } i \text{ is an edge,} \\ 0, & \text{otherwise.} \end{cases} \tag{6.1}$$

- Light path routing variables $\{Y_{iv}\}_{i\in\mathcal{E},v\in E_0}$: For light path i and physical link v, the variable is defined as

$$Y_{iv} = \begin{cases} 1, & \text{if light path } i \text{ passes through link } v, \\ 0, & \text{otherwise.} \end{cases} \tag{6.2}$$

- Packet routing variables $\{t_u^{sd}(S_v)\}_{s\in V,d\in V,u\in\mathcal{E},S_v\in\mathcal{S}}$: For a data packet sent from source node s to destination node d, through light path u and in operational state S_v, the variable is defined as

$$t_u^{sd}(S_v) = \begin{cases} \lambda_{sd}, & \text{if the packet passes through logical link } u, \\ 0, & \text{otherwise,} \end{cases} \tag{6.3}$$

where λ_{sd} is the peak time traffic between s and d.

We denote by $\Gamma_o(n)$ the set of light paths outgoing from node $n \in V$ and $\Gamma_i(n)$ is the set of light paths entering node $n \in V$. Similarly, we denote by $\Theta_o(n)$ the set of physical links outgoing from node $n \in V$ and $\Theta_i(n)$ is the set of physical links entering node $n \in V$. The numbers of transmitters and receivers at node n are denoted by T_n and R_n, respectively. The origin and destination nodes of logical link u are denoted by $O(u)$ and $D(u)$. Given the above three sets of variables, the constraints of topology design in WDM networks are given as follows:

- Network connectivity: First the logical topology should guarantee the connectivity; i.e., every node can reach another through the links in the logical topology. This can be represented in the following sets of inequalities:

$$\sum_{u\in\Gamma_o(n)} X_u \le T_n, \quad \forall n \in V, \tag{6.4}$$

and

$$\sum_{u\in\Gamma_i(n)} X_u \le R_n, \quad \forall n \in V, \tag{6.5}$$

which means that the numbers of outgoing and incoming light paths cannot be greater than the numbers of transmitters and receivers.
- Routing paths: The constraints are given by

$$\sum_{u\in\Gamma_o(n)} t_u^{sd}(S_v) - \sum_{u\in\Gamma_i(n)} t_u^{sd}(S_v) = \begin{cases} \lambda_{sd}, & \text{if } s = i, \\ -\lambda_{sd}, & \text{if } d = i, \\ 0, & \text{otherwise,} \end{cases} \tag{6.6}$$

for all $s, d, n \in V$ and $S_v \in \mathcal{S}$, which means that the routing is continuous in the network, and

$$t_u^{sd}(S_v) \leq X_u \lambda_{sd}, \quad \forall s,d \in V, S_v \in S, u \in \mathcal{E}, \tag{6.7}$$

which means that the routing can be carried out only over the logical links.

- Mapping from logical links to physical links: The constraints include

$$Y_{uv} \leq X_u, \quad \forall u \in \mathcal{E}, v \in E_O, \tag{6.8}$$

which reveals the relationship between the existence of a logical link and the corresponding routing through the optical fiber, and

$$\sum_{v \in \Theta_o(n)} Y_{uv} - \sum_{v \in \Theta_i(n)} Y_{uv} = \begin{cases} 1, & \text{if } O(u) = i, \\ -1, & \text{if } D(u = i), \\ 0, & \text{otherwise}, \end{cases} \tag{6.9}$$

for all $u \in \mathcal{E}$ and $i \in V$, which denotes the continuity of the logical links over the physical links, and

$$\sum_{s,d \in V} t_u^{sd}(S_v) \leq \left(\sum_{s,d \in V} \lambda_{sd} \right) (1 - Y_{uv}), \tag{6.10}$$

for all $u \in \mathcal{E}$, $v \in E_O$, $S_v \in S/\{S_O\}$.

- Number of wavelengths: Since the number of wavelengths on each physical link is limited, we have

$$\sum_{u \in \mathcal{E}} Y_{uv} \leq W_v, \quad \forall v \in E_O, \tag{6.11}$$

where W_v is the upper bound for the number of wavelengths on one optical fiber.

In traditional data communication networks, the optimization of the network topology is related to the data traffic requirements. For example, the objective function can be set as the maximum number of logical links over one physical link:

$$\min_{\{X_u\}_u, \{Y_{uv}\}_{u,v}, \{t_u^{sd}(S_v)\}_{u,s,d,S_v}} \quad O, \tag{6.12}$$

where

$$O \geq \sum_{s,d \in V} t_u^{sd}(S_v), \quad \forall S_v \in S, \quad \forall u \in \mathcal{E}. \tag{6.13}$$

As we will see, for communication network design in a CPS, the corresponding objective function may be dependent on the physical dynamics or a combination of the metrics of physical dynamics (e.g., metrics measuring the system stability) and data traffic (e.g., the congestion level).

6.2.3 OPTIMIZATION PROCEDURE

Since the variables are discrete, the optimization is usually an NP-hard problem except for very special cases. Hence heuristic algorithms are needed for the solution in practice. Ref. [43] provided the following two steps for the optimization:

- Step 1: Using the tabu search for topology design.
- Step 2: Using the Greedy Disjoint Alternate Path (GDAP) algorithm for routing and wavelength assignment.

Tabu search

We first provide a brief introduction to the tabu search. The basic idea of tabu search is to perturb an original topology design by a "move," which results in a "neighborhood" solution. For each current topology, we evaluate all neighboring solutions by optimizing the objective function in the routing and wavelength assignment, and choose the optimal one. The search procedure stops after a predetermined number of steps.

The key component in the tabu search is the definition of "move." An effective approach is given in Ref. [43], whose rationale can be found therein. Suppose that we have obtained a topology design (an initial design can be obtained from a simple heuristic approach). Then we find an l-hop cycle where $l \leq 8$. The selection of the next hop, if the current node in the cycle is the ith one, follows the subsequent rule:

- If i is odd, select an incoming logical link and hop to the node on the opposite side.
- If i is even, select any node that has not been in the cycle yet.

When a cycle is found, evaluate the degree of each node. If the degree of a node is increased after the cycle selection, remove superfluous edges in the graph. An example is illustrated in Fig. 6.3.

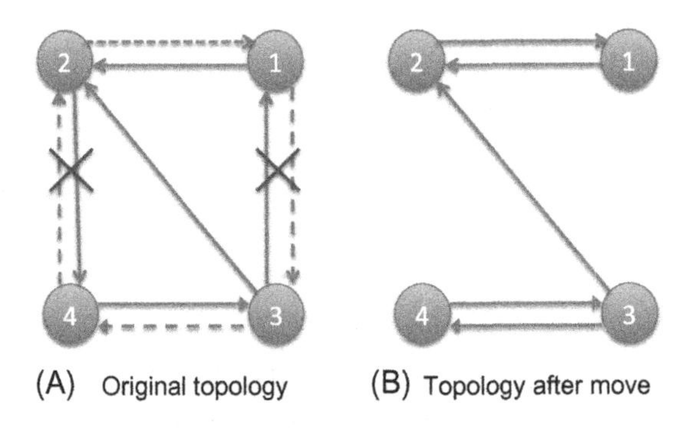

FIG. 6.3

An example of tabu search. The selected cycle is $1 \rightarrow 3 \rightarrow 4 \rightarrow 2 \rightarrow 1$.

GDAP algorithm

The GDAP algorithm is essentially a greedy algorithm. A detailed introduction can be found in Ref. [43]. Here we provide a brief introduction. Below is the notation used in the algorithm:

- $OR(i)$: the set of already routed logical links outgoing from node i.
- $IR(i)$: the set of already routed logical links entering node i.
- $ON(i)$: the set of not routed logical links outgoing from node i.
- $IN(i)$: the set of not routed logical links entering node i.
- V_{ij}: the logical link between nodes i and j.
- O: the set of nodes, which is initialized to the set of all nodes.

The GDAP algorithm is carried out in the following steps:

- Step 0: For the logical links between nodes with direct physical links, route them through the direct link; update $OR(i)$, $IR(i)$, $ON(i)$, and $IN(i)$.
- Step 1: If O is empty, stop the procedure. Otherwise, randomly choose a node i from O and then remove it from O.
- Step 2: Check the set $ON(i)$. If it is empty, jump to Step 3. Otherwise, choose one logical link from $ON(i)$. Carry out the shortest path routing such that the route does not intersect any route of logical links in $OR(i)$ and $IR(i)$. If no such path exists, simply carry out the shortest path routing, regardless of the path intersection with existing ones. If no such feasible path exists, give up the routing for this logical link.
- Step 3: Check the set $IN(i)$. If it is empty, jump to Step 1. Otherwise, choose one logical link from $IN(i)$. Carry out the shortest path routing such that the route does not intersect any route of logical links in $IR(i)$ and $OR(i)$. If no such path exists, simply carry out the shortest path routing, regardless of the path intersection with existing ones. If no such feasible path exists, give up the routing for this logical link.

It is easy to see that the GDAP algorithm tries to route the logical links using physical links that have not been used, so that the load over one physical link can be reduced.

6.3 OPTIMIZATION BASED ON TOPOLOGY DESIGN

In this section we consider topology design for guaranteeing the stability of the physical dynamics. We formulate the topology design as an optimization problem. The constraints on the communication network were explained in the previous section. Hence in this section, we discuss the objective function and the optimization procedure.

6.3.1 OBJECTIVE FUNCTION

A major challenge is the definition of the objective function, which should be directly related to the stability of the physical system. Here we introduce two types of objective functions; one is related to matrix eigenvalues and the other one is directly related to the cost of physical dynamics.

Eigenvalue-based objective function

In many situations, the system stability is highly dependent on the eigenvalues of matrices related to the network topology. Here we follow Ref. [100] to derive the objective function of topology design of communication networks for smart grids.

First we represent the power network using a graph, in which each vertex is a generator while each edge is a transmission line linking two generators. When two nodes i and j are directly connected, we denote this by $i \sim j$. The phase and frequency of node (generator) i are denoted by δ_i and f_i, respectively.

We assume that the system is close to an equilibrium point. We denote the standard frequency by f_0 and the frequency deviation of generator i by Δf_i. The angle deviation $\delta_i - f_0 t - \theta_i$ (θ_i is the initial phase of generator i) is denoted by $\Delta\delta_i$. Then when Δf_i and $\Delta\delta_i$, $i = 1, \ldots, N$, are both sufficiently small, it is easy to verify that the dynamics can be linearized to the following form:

$$\begin{cases} \Delta\dot{\delta}_i(t) = \Delta f_i(t), \\ M_i \Delta\dot{f}_i(t) + D_i \Delta f_i(t) = \Delta P^i_{\mathrm{m}}(t) - \sum_{k\sim i} c_{ik}(\Delta\delta_i - \Delta\delta_k), \end{cases} \tag{6.14}$$

where ΔP^i_{m} is the difference between the mechanical power and the stable one, and

$$c_{ik} = \frac{V^2 R_{ik}}{|Z_{ik}|^2} \sin\delta_{ik} - \frac{V^2 X_{ik}}{|Z_{ik}|^2} \cos\delta_{ik}, \tag{6.15}$$

where δ_{ik} is the stable phase difference between generators i and k in the equilibrium state, $Z_{ij} = R_{ij} + jX_{ij}$ is the impedance of the transmission line between generators i and j, E_i is the voltage of generator i, and Y_i is the shunt admittance. The derivation procedure can be found in Ref. [100]. The power network node is illustrated in Fig. 6.4.

To facilitate the analysis, we make the following assumptions, which further simplify the dynamics of the power grid:

- The damping constant is identical for all generators, and is denoted by D.
- The phase difference δ_{ik} is relatively small. Hence we can assume

$$c_{ik} = -\frac{V^2 X_{ik}}{|Z_{ik}|^2}, \tag{6.16}$$

  which implies $c_{ik} = c_{ki}$.
- All the rotor inertias are the same and equal to 1, i.e., $M_n = 1$. It is easy to extend this to the general case.

For notational simplicity, we have $c_{ik} = 0$ if generators i and k are not adjacent.

We assume that there exist communications between some adjacent generators. We denote by $i \leftrightarrow j$ that generators i and j can communicate with each other.

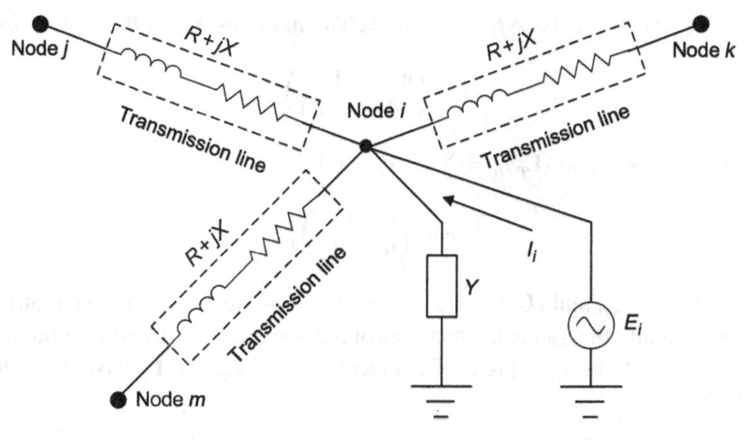

FIG. 6.4

Illustration of the generator and transmission line.

A necessary condition for $i \leftrightarrow j$ is $i \sim j$. We assume that the mechanical power is adjusted according to the feedbacks of frequencies from the communication channels, i.e.,

$$\Delta P_{\mathrm{m}}^{i}(t) = g(\Delta\delta_i(t), \{\Delta f_k(t-d)\}_{k\leftrightarrow i}), \tag{6.17}$$

where d is the communication delay of the communication link, which is assumed to be a constant.[1] Note that the change of mechanical power can be realized through the governor, fast valving, the voltage control of the Power System Stabilizer, or other devices. Here we do not explicitly model such approaches.

For simplicity, we assume that the control of mechanical power is a linear function of the observations without any communication delay, i.e.,

$$\Delta P_{\mathrm{m}}^{i}(t) = \sum_{k\leftrightarrow i} g_{ik}(\Delta f_k(t) - \Delta f_i(t)). \tag{6.18}$$

We assume $g_{ik} = g_{ki}$; i.e., the control gains are symmetric. Similarly, we define $g_{ik} = 0$ if generators i and k cannot communicate with each other. We also assume that $g_{ik} \geq 0$ since it is desirable to decrease the mechanical power when Δf_i is large while the neighboring frequency deviations are small.

Incorporating the control on the mechanical power, the linearized system dynamics can be written in the following vector form:

$$\dot{\mathbf{x}}(t) = \mathbf{A}\mathbf{x}(t) + \mathbf{B}\mathbf{x}(t), \tag{6.19}$$

[1] In practice, the communication delay can be different for different links or can even be random. The assumption of constant delay can substantially simplify the analysis and provide insights for future study on the generic case of communication delay.

where $\mathbf{x} = (\Delta\delta_1, \ldots, \Delta\delta_N, \Delta f_1, \ldots, \Delta f_N)^T$. The matrices $\mathbf{A}$ and $\mathbf{B}$ are given by

$$\mathbf{A} = \begin{pmatrix} \mathbf{0} & \mathbf{I} \\ -\mathbf{L}_p & -D\mathbf{I} \end{pmatrix}, \tag{6.20}$$

where $(\mathbf{L}_p)_{ij} = -c_{ij}$ and $(\mathbf{L}_p)_{ii} = \sum_{k\sim i} c_{ik}$, and

$$\mathbf{B} = \begin{pmatrix} \mathbf{0} & \mathbf{0} \\ \mathbf{0} & -\mathbf{L}_c \end{pmatrix}, \tag{6.21}$$

where $(\mathbf{L}_c)_{ij} = -g_{ij}$ and $(\mathbf{L}_c)_{ii} = \sum_{k\sim i} g_{ik}$. It is easy to verify that the matrices $\mathbf{L}_p$ and $\mathbf{L}_c$ are actually the Laplacian matrices of the power network and communications with weights $\{c_{ik}\}$ and $\{g_{ik}\}$ [101]. The eigenvalues of $\mathbf{L}_p$ and $\mathbf{L}_c$ have the following characteristics:

- All eigenvalues of $\mathbf{L}_c$ are nonnegative since all the weights are positive. The smallest eigenvalue is 0.
- There is one zero eigenvalue for $\mathbf{L}_p$. The other eigenvalues may be positive or negative since the weights can be positive or negative.

Obviously, the point of origin $\Delta\delta_i = 0$ and $\Delta f_i = 0$ (i.e., the frequencies equal the standard one) is a stationary point; however, its stability is unknown and it is of key importance to the power grid. Following the definition of stability in Ref. [102], we say that the point of origin $\Delta\delta_i = 0$ and $\Delta f_i = 0$ is stable, if for all $t \geq 0$, for any $\epsilon > 0$, there exists a $\theta > 0$ such that

$$\|\mathbf{x}(0)\| \leq \theta \Rightarrow \|\mathbf{x}(t)\| \leq \epsilon. \tag{6.22}$$

Then the following theorem discloses the sufficient and necessary condition for the stability of the equilibrium point of the power grid:

Theorem 33. *The origin of the physical dynamics in Eq. (6.14) is stable if and only if the eigenvalues of the following matrix have negative real parts:*

$$\mathbf{D} = \begin{pmatrix} \mathbf{0} & \mathbf{I} \\ -\mathbf{L}_p & -\mathbf{L}_c \end{pmatrix}. \tag{6.23}$$

The proof is very similar to that of Theorem 4 in Ref. [102], which discusses bird flocking.

The following corollary provides a condition for the stability of the point of origin in a special case:

Corollary 10. *Suppose that the Laplacian matrices $\mathbf{L}_c$ and $\mathbf{L}_p$ have the same eigenstructure and real eigenvalues. The point of origin is stable if and only if the following inequality holds:*

$$\max_{0\leq\rho\leq 1} \max_{\|\mathbf{t}\|=1} -\rho \sum_{i=1}^{N} \lambda_i^c t_i^2 - \sqrt{\rho(1-\rho)} \min_i t_i(\lambda_i^p + 1) < 0, \tag{6.24}$$

where λ_i^c and λ_i^p are the eigenvalues of $\mathbf{L}_c$ and $\mathbf{L}_p$, respectively.

Physical dynamics cost-based objective function

The previous formulation of the objective function can only guarantee whether the physical dynamics is stable. It does not measure the system performance given the corresponding stability. Hence a more detailed approach is to define the objective function as the system operation cost of the physical dynamics, which has been adopted in Ref. [25]. In this study, a dynamical system consisting of N linear subsystems is considered. For subsystem i, the dynamics is given by

$$\mathbf{x}_i(k+1) = \mathbf{A}_i\mathbf{x}_i(k) + \mathbf{B}_i\mathbf{x}_i(k), \tag{6.25}$$

where k is the discrete-time index, and $(\mathbf{A}_i, \mathbf{B}_i)$ are parameters. The set of all undirected graphs having N vertices is denoted by $\mathcal{G}$.

The edge set of graph $g \in \mathcal{G}$ is denoted by $E(g)$. The corresponding adjacency matrix $\mathcal{A}_{ij}(g)$ is defined as

$$\mathcal{A}_{ij}(g) = \begin{cases} 1, & \text{if } (i,j) \in E(g), \\ 0, & \text{otherwise.} \end{cases} \tag{6.26}$$

It is assumed that the control action is linear; i.e.,

$$\mathbf{u}(k) = \mathbf{K}(g)\mathbf{x}(k), \tag{6.27}$$

where the control and state vectors $\mathbf{u}$ and $\mathbf{x}$ are obtained by stacking the vectors $\{\mathbf{u}_i\}_{i=1,\ldots,N}$, and the feedback gain matrix $\mathbf{K}$ is determined by the graph topology g.

We then define the cost function to be

$$J(\mathbf{Q}, \mathbf{x}_0; \mathbf{u}) = \sum_{k=0}^{\infty} \mathbf{x}^T(k)\mathbf{Q}\mathbf{x}(k) + \|\mathbf{u}(k)\|^2, \tag{6.28}$$

where $\mathbf{Q}$ is a predetermined positive definite matrix and $\mathbf{x}_0$ is the initial state. We further define

$$J(\mathbf{Q}, \mathbf{u}) = \sup_{\|\mathbf{x}_0\| \leq 1} J(\mathbf{Q}, \mathbf{x}_0; \mathbf{u}) \tag{6.29}$$

and

$$J^*(\mathbf{Q}) = \inf_{\mathbf{u}} J(\mathbf{Q}, \mathbf{u}). \tag{6.30}$$

Since the optimal control law can be obtained from the linear quadratic regulation (LQR) strategy, the optimal cost function is also given by

$$J^*(\mathbf{Q}) = \lambda_{\max}(\mathbf{P}), \tag{6.31}$$

where $\lambda_{\max}$ denotes the maximum eigenvalue, and the matrix $\mathbf{P}$ is determined from the Riccati equation:

$$\mathbf{P} = \mathbf{A}^*\mathbf{P}\mathbf{A} + \mathbf{Q} - \mathbf{A}^T\mathbf{P}\mathbf{B}(\mathbf{B}^T\mathbf{P}\mathbf{B} + \mathbf{I})^{-1}\mathbf{B}^T\mathbf{P}\mathbf{A}, \tag{6.32}$$

where $\mathbf{A} = \text{diag}(\mathbf{A}_1, \ldots, \mathbf{A}_N)$ and $\mathbf{B} = \text{diag}(\mathbf{B}_1, \ldots, \mathbf{B}_N)$.

The formulation has the following pros and cons:

- The cost function is directly related to the system operation cost, instead of only the stability.
- These subsystems have no coupling in their dynamics, which could limit their application.
- Possible random perturbations to the system dynamics are not considered.

Sum cost of control- and communication-based objective function

In Ref. [103], the objective function for topology design is the sum of costs of the physical dynamics and communication networks. The following distributed control system is considered:

$$\mathbf{x}^{(i)}(k+1) = \mathbf{A}_{i,i}\mathbf{x}^{(i)}(k) + \mathbf{B}_i\mathbf{u}^{(i)}(k) + \sum_{j\neq i}\mathbf{A}_{i,j}\mathbf{x}^{(j)}(k), \tag{6.33}$$

where $\mathbf{x}^i$ and $\mathbf{u}^{(i)}$ are the system state and control action of the ith subsystem.

The topology of the communication network is specified by the following indicator:

$$\delta_{i,j} = \begin{cases} 1, & \text{if nodes } i \text{ and } j \text{ have communications,} \\ 0, & \text{otherwise.} \end{cases} \tag{6.34}$$

The control action $\mathbf{u}^{(i)}$ of subsystem i is obtained from the feedback through the communication network; i.e.,

$$\mathbf{u}^{(i)}(k) = \sum_j \mathbf{K}_{i,j}\mathbf{x}^{(j)}(k). \tag{6.35}$$

Obviously, we have

$$\delta_{i,j} = 0 \leftrightarrow \mathbf{K}_{i,j} = 0, \tag{6.36}$$

namely the feedback gain matrix $\mathbf{K}_{i,j}$ is zero if there is no communication connection between nodes i and j.

The cost function for communication network design is defined as

$$C = \sum_{l=0}^{\infty}\mathbf{x}^T(k+l)\mathbf{Q}\mathbf{x}(k+l) + \mathbf{u}^T(k+l)\mathbf{R}\mathbf{u}(k+l) + \sum_{i\neq j}c_{ij}\delta_{ij}, \tag{6.37}$$

where $\mathbf{x}$ and $\mathbf{u}$ are the vectors of the overall system state and control action, $\mathbf{Q}$ and $\mathbf{R}$ are nonnegative definite matrices, and c_{ij} is the cost of building a communication link between nodes i and j. Again the objective function can be optimized using discrete optimization tools.

Synchronization-based objective function

Another type of objective function can be established for the special type of dynamics called oscillator networks [104]. As illustrated in Fig. 6.5, we consider a network of oscillators, each of which consists of an inductor, a capacitor, and a current source. We denote by L_i, C_i, and d_i the inductance, capacitance, and current of node i. When the oscillators are isolated, the oscillation frequency of oscillator i is given by

$$w_i = \frac{1}{\sqrt{L_i C_i}}. \tag{6.38}$$

Now the oscillators are connected with links, each of which has resistance R_i. Our goal is to study the synchronization of the oscillators, i.e., whether the frequencies of the oscillators will converge to the same value.

When the nodes are connected, we consider the voltage $v_i(t)$ and the corresponding integral $\int_0^t v_i(t)dt$ as the state of node i. Then the system state is given by

$$\mathbf{x}(t) = \left(\int_0^t v_1(t)dt, \ldots, \int_0^t v_N(t)dt, v_1, \ldots, v_N \right)^T. \tag{6.39}$$

The system dynamics is then given by

$$\dot{\mathbf{x}}(t) = \begin{pmatrix} 0 & \mathbf{I} \\ -\mathbf{H} & -\mathbf{K} \end{pmatrix} \mathbf{x}(t) + \begin{pmatrix} 0 \\ \mathbf{d} \end{pmatrix}, \tag{6.40}$$

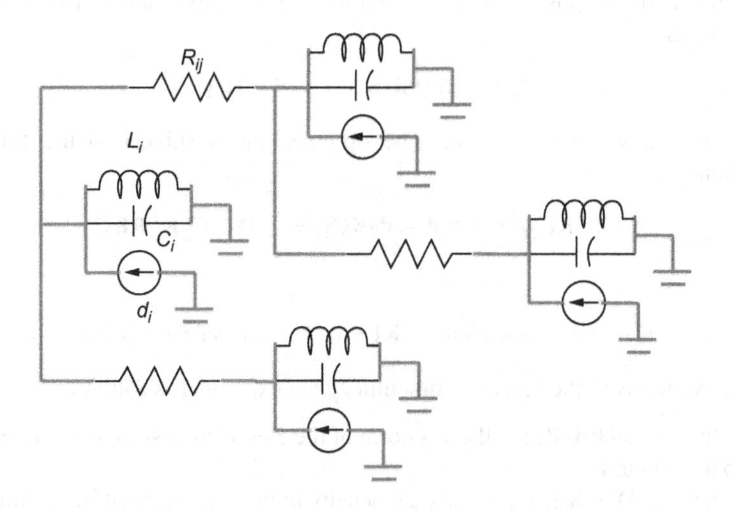

FIG. 6.5

Illustration of the oscillator network.

where $\mathbf{d} = (d_1, \ldots, d_N)$ is the vector containing all the currents of nodes, $\mathbf{H} = \text{diag}(1/L_iC_i)$, and $\mathbf{K}$ is the conductance matrix of the network interconnection.

The system dynamics in (6.39) can be rewritten in the following standard form of optimal control:

$$\begin{cases} \dot{\mathbf{x}}(t) = \mathbf{A}\mathbf{x}(t) + \mathbf{B}_1\mathbf{d} + \mathbf{B}_2\mathbf{u}(t), \\ \mathbf{z}(t) = \mathbf{C}_1\mathbf{x}(t) + \mathbf{D}\mathbf{u}(t), \\ \mathbf{y}(t) = \mathbf{C}_2\mathbf{x}(t), \\ \mathbf{z}(t) = -\mathbf{K}\mathbf{x}, \end{cases} \tag{6.41}$$

where

$$\mathbf{A} = \begin{pmatrix} 0 & \mathbf{I} \\ -\mathbf{H} & 0 \end{pmatrix}, \quad \mathbf{B}_1 = \mathbf{B}_2 = \begin{pmatrix} 0 \\ \mathbf{I} \end{pmatrix}, \tag{6.42}$$

and

$$\mathbf{C}_1 = \begin{pmatrix} \mathbf{Q}^{\frac{1}{2}} \\ 0 \end{pmatrix}, \quad \mathbf{C}_2 = (0, \mathbf{I}), \quad \mathbf{D} = \begin{pmatrix} 0 \\ \mathbf{R}^{\frac{1}{2}} \end{pmatrix}. \tag{6.43}$$

Considering the feedback law, the closed-loop system can be written as

$$\begin{cases} \dot{\mathbf{x}}(t) = (\mathbf{A} - \mathbf{B}_2\mathbf{K}\mathbf{C}_2)\mathbf{x}(t) + \mathbf{B}_1\mathbf{d}, \\ \mathbf{z}(t) = \mathbf{C}_1\mathbf{x}(t) + \mathbf{D}\mathbf{u}(t), \\ \mathbf{z}(t) = \begin{pmatrix} \mathbf{Q}^{\frac{1}{2}} \\ -\mathbf{R}^{\frac{1}{2}}\mathbf{K}\mathbf{C}_2 \end{pmatrix} \mathbf{x}(t). \end{cases} \tag{6.44}$$

Then for the purpose of synchronization, the objective function is defined in Ref. [104] as

$$J_\gamma = \text{trace}(\mathbf{P}\mathbf{B}_1\mathbf{B}_1^T) + \gamma\|\mathbf{W} \circ \mathbf{K}\|_1, \tag{6.45}$$

where $\mathbf{W}$ is a weighting matrix. The optimization is subject to the following constraints:

$$(\mathbf{A} - \mathbf{B}_2\mathbf{K}\mathbf{C}_2)^T\mathbf{P} + \mathbf{P}(\mathbf{A} - \mathbf{B}_2\mathbf{K}\mathbf{C}_2) = -(\mathbf{Q} + \mathbf{C}_2^T\mathbf{K}^T\mathbf{R}\mathbf{K}\mathbf{C}_2) \tag{6.46}$$

and

$$\mathbf{K} \text{ is semidefinite positive}, \ \mathbf{K}\mathbf{1} = 0, \ \mathbf{P} \text{ is semidefinite positive}. \tag{6.47}$$

The two terms in the objective function J_γ are explained as follows:

- The term $\text{trace}(\mathbf{P}\mathbf{B}_1\mathbf{B}_1^T)$ is the $\mathcal{H}_2$ norm of the closed-loop system, from the input $\mathbf{d}$ to the output $\mathbf{z}$.
- The term $\gamma\|\mathbf{W} \circ \mathbf{K}\|_1$ represents the penalty to the cost incurred by building the coupling links.

6.3.2 OPTIMIZATION OF TOPOLOGY

We now begin to study the approaches to optimize the different objective functions. The main challenge of the optimization is the discreteness of the variables, which makes the optimization an NP-hard problem except for trivial cases. It is possible to apply existing software packages to discrete optimizations. However, in the context of topology design for physical dynamics in a CPS, we can use some special approaches to address this challenge.

Greedy algorithm

A heuristic approach that is often quite effective is the greedy algorithm; i.e., choose one resource (e.g., a communication link in the topology design) that can optimize the instantaneous performance gain and add to the pool. It has been shown that if the problem has a matroid structure, the greedy algorithm is optimal. However, many problems unfortunately do not have such a structure.

We apply the principle of the greedy algorithm to the eigenvalue-based objective function. We assume that the feedback gain g_{ik} is a constant equal to g_0. It is more interesting to study the joint design of communication network and control law, which involves mixed integer programming and is thus more complicated.

As we have found, the stability of a power system subject to no communication delay is determined by the eigenvalues of matrix $\mathbf{D}$. Hence the objective is minimizing the maximum eigenvalue of $\mathbf{D}$ in order to make all eigenvalues negative. Meanwhile, the communication network is subject to costs since the communication links are not free. We denote by a_{ik} the cost of building the bidirectional communication link built between nodes i and k and assume that the total budget of the cost is a_{tot}. Then the design of communication topology can be formulated as the following optimization problem:

$$\begin{aligned} &\min_{\{z_{i,k}\}_{i\neq k}} \lambda^{\mathbf{D}}_{\max} \\ \text{s.t.} &\sum_{i\neq k} z_{i,k}a_{ik} \leq a_{\text{tot}} \\ &z_{i,k} \in \{0,1\}, \quad \forall i \neq k, \end{aligned} \tag{6.48}$$

where $z_{i,k}$ is the variable and denotes building (not building) the link between generators i and k when $z_{i,k} = 1(0)$.

Obviously the optimization problem in Eq. (6.48) is binary integer programming, which is generally NP-hard. In this book, we adopt the greedy algorithm in which the cost function is minimized in each step. The procedure is given in Procedure 2.

Relaxed optimization

First we can relax the variables to real numbers. For example, the binary variable indicating the existence of a communication link can be relaxed to a continuous weight of the corresponding link in a complete graph. There is still no existing study on applying this approach to the eigenvalue-based objective function in topology design. Here we introduce the optimization of eigenvalues in a graph by adjusting the edge weights. In Ref. [105], topology design is studied for the purpose of increasing the network connectivity. In summary, edges are added to the corresponding graph in order to increase the second smallest eigenvalue of the corresponding Laplacian matrix.

PROCEDURE 2 COMMUNICATION NETWORK TOPOLOGY DESIGN

1: Initialize the communication links as an empty set.
2: **while** The total cost is less than a_{tot} **do**
3:    **for** Each communication link between i and j with $z_{ij} = 0$ **do**
4:       Set $z_{ij} = 1$ temporarily.
5:       Compute the maximum eigenvalue of $\mathbf{D}$.
6:    **end for**
7:    Choose the link with the minimum cost function.
8:    Set the corresponding $z_{ij} = 1$.
9: **end while**

We consider a graph $G = (V, E)$, whose Laplacian is denoted by $\mathbf{L}$, where

$$\mathbf{L}_{ij} = \begin{cases} \text{the degree of node } i, & \text{if } i = j, \\ -1, & \text{if } i \neq j. \end{cases} \tag{6.49}$$

The second smallest eigenvalue of $\mathbf{L}$, denoted by $\lambda_2(\mathbf{L})$, has substantial effects on many properties:

- A large λ_2 implies higher network connectivity; i.e., it is harder to bisect a graph with a larger λ_2.
- A continuous-time Markov chain with unit rates has a faster mixing rate if the corresponding graph has a larger λ_2.
- λ_2 is also closely related to the stability of a networked control system [106].

Due to the importance of λ_2, we follow Ref. [105] to study the maximization of λ_2. We assume that there have existed a set of edges in the graph, thus forming a base graph denoted by $G_{\text{base}} = (V, E_{\text{base}})$, as illustrated in Fig. 6.6. We assume that the new edges can only be selected from a set E_{cand}. Then the goal is to select k edges from E_{cant} such that λ_2 is maximized in the new graph. Hence the topology design can be formulated as the following optimization problem:

$$\begin{aligned} &\max \lambda_2(\mathbf{L}(V, E_{\text{base}} \cup E)) \\ \text{s.t. } &|E| = k \\ &E \subset E_{\text{cand}}. \end{aligned} \tag{6.50}$$

To facilitate the mathematical analysis, we define the variable $\mathbf{x}$ as an $|E_{\text{cand}}|$-dimensional binary vector; i.e.,

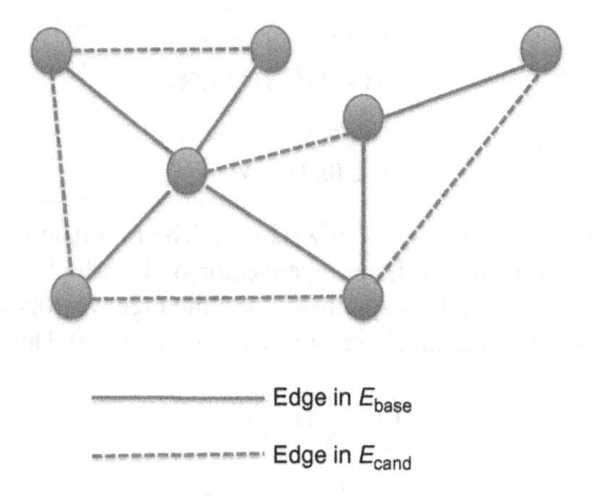

FIG. 6.6

Illustration of topology design.

$$x_i = \begin{cases} 1, & \text{if the } i\text{th edge in } E_{\text{cand}} \text{ is selected,} \\ 0, & \text{otherwise.} \end{cases} \tag{6.51}$$

Hence the optimization problem can be rewritten as

$$\begin{aligned} &\max \lambda_2 \left(\mathbf{L}_{\text{base}} + \sum_{l=1}^{|E_{\text{cand}}|} x_l \mathbf{a}_l \mathbf{a}_l^T \right) \\ &\text{s.t. } \mathbf{1}^T \mathbf{x} = k \\ &\quad x_i \in \{0,1\}, \quad \forall i, \end{aligned} \tag{6.52}$$

where $\mathbf{a}_l$ is a vector with all zero elements except for a unit element in the lth dimension.

Since the optimization is discrete, which is usually NP-hard, an effective approach is to relax the constraint of binary variables x_i. We define

$$\mathbf{L}(\mathbf{x}) = \mathbf{L}_{\text{base}} + \sum_{l=1}^{|E_{\text{cand}}|} x_l \mathbf{a}_l \mathbf{a}_l^T, \tag{6.53}$$

and relax the constraint $x_i \in \{0, 1\}$ to $x_i \in [0, 1]$. Then the optimization problem in Eq. (6.52) is changed to

$$\begin{aligned} &\max \lambda_2(\mathbf{L}(\mathbf{x})) \\ &\text{s.t. } \mathbf{1}^T \mathbf{x} = k \\ &\quad x_i \in [0,1], \quad \forall i. \end{aligned} \tag{6.54}$$

To facilitate the optimization, we can rewrite the optimization problem in Eq. (6.54) as

$$\begin{aligned}
&\max s \\
&\text{s.t. } s(\mathbf{I} - \mathbf{1}\mathbf{1}^T/n) \preceq \mathbf{L}(\mathbf{x}) \\
&\mathbf{1}^T\mathbf{x} = k \\
&x_i \in [0,1], \quad \forall i.
\end{aligned} \tag{6.55}$$

Note that the first constraint means $s \leq \lambda_2(\mathbf{L}(\mathbf{x}))$. The rationale is given as follows. First it is easy to verify that **1** is an eigenvector of **L**, which corresponds to the smallest eigenvalue 0. Since **L** is symmetric and the eigenvectors are orthogonal to each other, then for any other eigenvector **v**, we have $\mathbf{1}^T\mathbf{v} = 0$. Hence

$$\begin{aligned}
&\left(\mathbf{I} - \frac{1}{N}\mathbf{1}\mathbf{1}^T\right)\mathbf{v} \\
&= \mathbf{I}\mathbf{v} - \frac{1}{N}\mathbf{1}(\mathbf{1}^T\mathbf{v}) \\
&= \mathbf{v},
\end{aligned} \tag{6.56}$$

which means that **v** is also an eigenvector of $\mathbf{I} - \frac{1}{N}\mathbf{1}\mathbf{1}^T$ corresponding to eigenvalue 1. Meanwhile, we can also verify that **1** is an eigenvalue of $\mathbf{I} - \frac{1}{N}\mathbf{1}\mathbf{1}^T$, corresponding to eigenvalue 0. Hence we have

$$\begin{aligned}
&\mathbf{L}(\mathbf{x}) - s\left(\mathbf{I} - \frac{1}{N}\mathbf{1}\mathbf{1}^T\right) \\
&= \mathbf{U}\text{diag}(0, \lambda_2 - s, \ldots, \lambda_N - s)\mathbf{U}^T,
\end{aligned} \tag{6.57}$$

whose nonnegativity implies $\lambda_2 \geq s$.

The optimization problem in Eq. (6.55) can be efficiently solved using semidefinite programming (SDP). The solution provides an upper bound for the maximum λ_2. Moreover, the new edges to be added to the graph can be chosen as the k edges having the largest weights.

The following greedy algorithm was also tested in Ref. [105]. First the Fiedler vector **v** of the Laplacian is defined as the normalized eigenvector of the Laplacian. In the greedy algorithm, one edge is added at a time. The new edge is selected as the one between nodes i and j (edge (i,j) should be in the candidate set E_{cand}) that has the largest $(v_i - v_j)^2$. The rationale is due to the following fact:

$$\frac{\partial \lambda_2(\mathbf{L})}{\partial x_l} = (v_i - v_j)^2, \tag{6.58}$$

where x_l is the indicator for the edge between nodes i and j.

Decomposition-based optimization

In Ref. [103], it was proposed to decompose the overall system to subsystems using hierarchical lower block triangular (LBT) decomposition, which is illustrated in Fig. 6.7. Each hierarchy is a set of nodes that are strongly coupled. The nodes in one hierarchy are only affected by themselves and nodes in upper hierarchies.

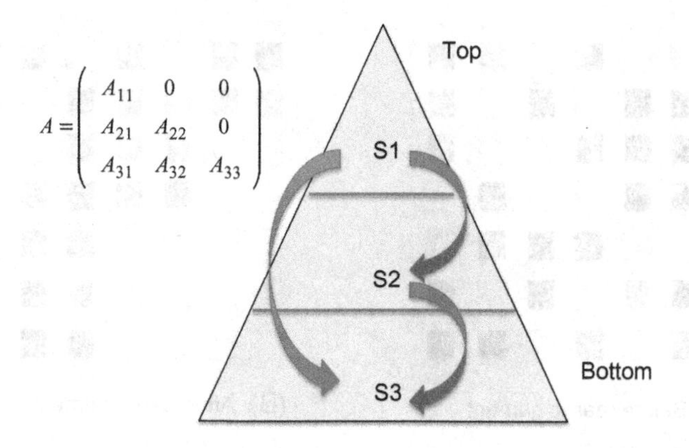

FIG. 6.7

Illustration of the hierarchical decomposition.

Based on the concept of hierarchical systems, the topology design is carried out in the following five steps:

- First the system is decomposed to clusters of subsystems having the above hierarchical LBT decomposition. The topology of the system is first encoded into a matrix Γ, where

$$\Gamma_{ij} = \text{the number of nonzero entries in matrix } A_{ij}. \tag{6.59}$$

  Then the Dulmage-Mendelsohn (DM) Decomposition is applied to permute the rows and columns of Γ such that the generated matrix Γ' has an LBT structure. The number of diagonal blocks in Γ' is denoted by $\hat{N}_c$. An index set $\hat{C}_i$ is created for every diagonal block. The indices of all subsystems belonging to the ith block are stored in $\hat{C}_i$. Select a minimal cluster size $c_{\min}$. For each index set $\hat{C}_i$, if its size is larger than $c_{\min}$, then label it as a fixed cluster; otherwise, label it as a free cluster. Then add every free cluster to a fixed cluster. The addition rule is: for a free cluster $\hat{C}_k$, it is added to the fixed cluster $\hat{C}_j$ having the largest number of connections to $\hat{C}_k$. Finally, this step outputs a set of node clusters $\{C_i\}$ having the hierarchal structure. This procedure is illustrated in Fig. 6.8.
- For the overall system, we design a preliminary controller $\hat{K}$ assuming a full communication network topology within the cluster. In the design of the preliminary controller, we ignore the communication cost and minimize only the cost of physical dynamics. Hence the controller design becomes an LQR one.
- When the preliminary controller has been designed, the cost function is then derived from the local control strategy of each cluster. Such an inverse optimal control problem can be solved using the LMI approach.
- When the local control cost function is obtained, the control and communication costs are optimized for each cluster. The corresponding control feedback gain matrix and topology matrix of cluster i are denoted by $\mathbf{K}^c_{i,i}$ and $\mathbf{D}_{i,i}$.

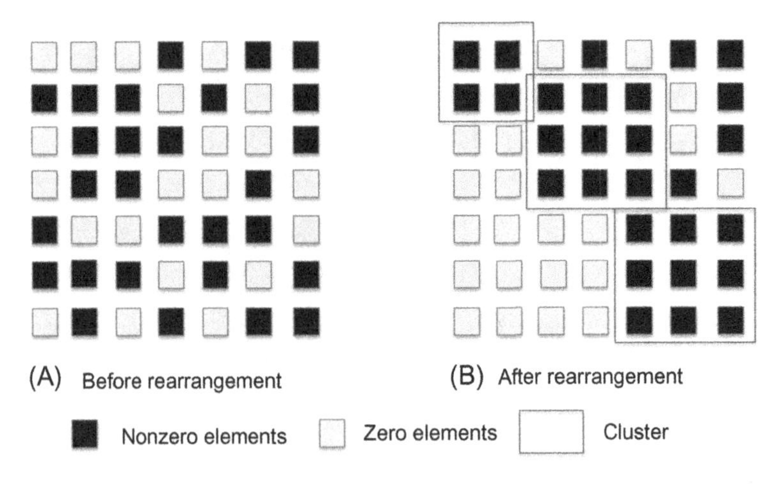

FIG. 6.8

Illustration of structure decomposition.

- Finally, the overall system control strategy and communication topology are obtained from the local ones, by assuming no communications among the clusters, namely the overall system control feedback gain matrix is given by

$$\mathbf{K} = \text{diag}(\mathbf{K}_{ii}^{c}) \tag{6.60}$$

and the communication network topology of the overall system is given by

$$\mathbf{D} = \text{diag}(\mathbf{D}_{ii}^{c}). \tag{6.61}$$

Structure-based optimization

In the previous approaches of topology design, the problem is usually formulated as an optimization problem and then solved using existing generic approaches such as integer programming or convex optimization. In the following discussion, we introduce the work in Ref. [25], which exploited the structure of the control system. Note that the system description and cost function have been given in Eq. (6.25) and Eq. (6.27), respectively.

Since different network topologies, or graphs denoted by g, result in different performances, we define the value of graph g as

$$V(g) = J_g^*(\mathcal{Q}(g)). \tag{6.62}$$

If there is no constraint on the communication network topology, the optimal g having the largest value is the full graph; namely each controller receives the information from all sensors.

In order to relate the cost function to the overhead incurred by the communications, Ref. [25] proposes to make the matrix $\mathbf{Q}$ in Eq. (6.28) dependent on the graph topology. Hence the following mapping $\mathcal{Q}$ is defined for this purpose:

Definition 18. A mapping $\mathcal{Q}$ maps from graphs having the fixed vertex set to the set of nonnegative definite matrices. We say $\mathcal{Q}$ is

- Edge separable without interference if

$$\mathcal{Q}(g) = \mathcal{Q}_0 + \sum_{i \sim j} \mathbf{P}_{ij}, \tag{6.63}$$

  where $\mathcal{Q}_0$ is a predetermined nonnegative definition matrix, $i \sim j$ means that vertices i and j are adjacent in the graph, and each matrix $\mathbf{P}_{ij}$ is nonnegative definite, in which all blocks equal zero except for the (i, i)th, $(j, j)th$, $(i, j)th$, and (j, i)th. In this case, each edge ij incurs a cost to the overall cost function.
- Edge separable with externalities if all the diagonal blocks in each $\mathbf{P}_{ij}$ could be nonzero. This model is useful when the actions of vertices may have an influence on each other.
- Nonseparable or with interference if $\mathcal{Q}(g)$ has no special structure.

To make the cost function reasonable, we also require the following two regulating properties:

- Nondecreasing: if for any two graphs g and g' having the same set of vertices, $g \preceq g'$ implies $\mathcal{Q}(g) \leq \mathcal{Q}(g')$.
- Structure compatible: for any g, if two vertices are not connected, the corresponding block in $\mathcal{Q}(g)$ is zero.

Ref. [25] explores the relationship between the graph g and the cost function. We can introduce a partial order $\preceq$ of all graphs having the same set of vertices: we say $g' \preceq g$ if all the edges in g' are also edges in g. It is easy to prove the following theorem:

Theorem 34. *The following statements are correct:*

- *If $g' \preceq g$, then we have $J_g^*(\mathbf{Q}) \leq J_{g'}^*(\mathbf{Q})$ for any positive definite $\mathbf{Q}$.*
- *If $\mathbf{Q} \leq \mathbf{Q}'$, then we have $J_g^*(\mathbf{Q}) \leq J_g^*(\mathbf{Q}')$ for all graphs g with the same set of vertices.*

For further discussion, we need the following definition, which is also illustrated in Fig. 6.9:

Definition 19. We say that a graph g is a clique graph if every connected component of it is a complete subgraph (i.e., each pair of vertices within the same component is connected).

Then the following theorem shows the value function of a clique graph:

Theorem 35. *Suppose that g is a clique graph and the mapping $\mathcal{Q}$ is structure compatible. We then have*

$$V(g) = \lambda_{\max}(\mathbf{P}(g)), \tag{6.64}$$

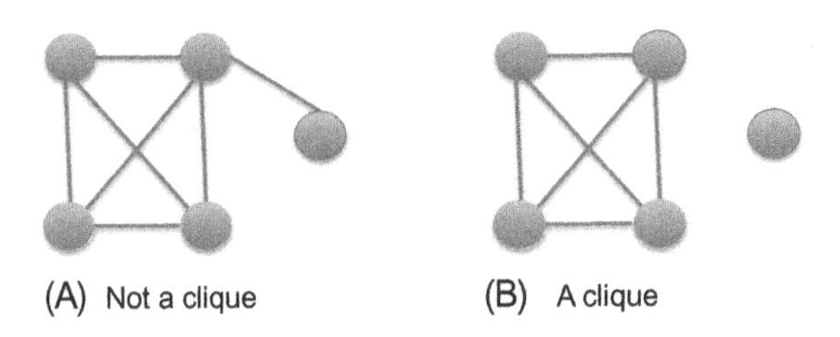

FIG. 6.9

Illustration of a clique.

where the matrix $\mathbf{P}(g)$ *satisfies*

$$\mathbf{P}(g) = \mathbf{A}^*\mathbf{P}(g)\mathbf{A} + \mathcal{Q}(g) - \mathbf{A}^*\mathbf{P}(g)\mathbf{B}(\mathbf{B}^T\mathbf{P}(g)\mathbf{B} + \mathbf{I})^{-1}\mathbf{B}^T\mathbf{P}(g)\mathbf{A}. \tag{6.65}$$

The following theorem shows that if the mapping $\mathcal{Q}$ satisfies certain conditions, it is more desirable not to carry out communications and the optimal strategy is to carry out completely decentralized control:

Theorem 36. *Suppose that the mapping* $\mathcal{Q}$ *is nondecreasing and structure compatible, and* g *is a clique graph. Then we have*

$$g \preceq g' \Rightarrow V(g) \leq V(g'). \tag{6.66}$$

In particular, the optimal control strategy for minimizing the overall cost is fully decentralized control.

Since the optimal communication topology is trivial if the mapping $\mathcal{Q}$ is nondecreasing, we drop the nondecreasing assumption such that certain communications can help to minimize the overall cost.

Assumption 2. We assume that the graph g and mapping $\mathcal{Q}$ satisfy either of the following conditions:

- g is a clique graph and the mapping $\mathcal{Q}$ is structure compatible.
- g is a complete graph and $\mathcal{Q}$ is an arbitrary mapping.

The following theorem shows the conditions (both necessary and sufficient) for a graph g' to be better than the fixed graph g.

Theorem 37. *Suppose that* g *and* $\mathcal{Q}$ *satisfy Assumption 2. A graph* g' *having the same set of vertices as* g *satisfies* $V(g') \leq V(g)$ *if and only if* $\lambda^* \leq \lambda_{\max}(\mathbf{P}(g))$, *where* λ^* *is the minimum value of the following optimization problem:*

$$\begin{aligned} &\min_{\mathbf{P}',\mathbf{K}'} \lambda_{\max}(\mathbf{P}') \\ &\text{s.t. } \mathbf{P}' \geq (\mathbf{A} + \mathbf{B}\mathbf{K}')^T\mathbf{P}'(\mathbf{A} + \mathbf{B}\mathbf{K}') + \mathcal{Q}(g') + (\mathbf{K}')^T\mathbf{K}' \\ &\mathbf{K}' \in \mathcal{K}(g'). \end{aligned} \tag{6.67}$$

Remark 14. Although this theorem provides the sufficient and necessary condition for a topology better than a certain clique topology, the condition is not easy to verify since the solution to the optimization problem Eq. (6.67) is not straightforward. In Ref. [25], numerically efficient approaches using the LMI tools are proposed for both conditions.

In the above discussion, we compare an arbitrary graph with a clique graph. The following theorem shows how to compare two arbitrary graphs:

Theorem 38. *For two graphs g and g', we have $V(g') < V(g)$ if there exist $\mathbf{K} \in \mathcal{K}(g')$ and a positive definite matrix* **P** *such that*

$$\mathbf{P} = (\mathbf{A} + \mathbf{BK})^T \mathbf{P}'(\mathbf{A} + \mathbf{BK}) + \mathcal{Q}(g') + (\mathbf{K})^T \mathbf{K} \tag{6.68}$$

and

$$\mathcal{Q}(g) \geq \mathcal{Q}(g') + (\mathbf{K} + \mathbf{S}^{-1}\mathbf{B}^T\mathbf{PA})^T \mathbf{S}(\mathbf{K} + \mathbf{S}^{-1}\mathbf{B}^T\mathbf{PA}), \tag{6.69}$$

where

$$\mathbf{S} = \mathbf{B}^T\mathbf{PB} + \mathbf{I}. \tag{6.70}$$

6.4 TEAM DECISION THEORY

An alternative approach to study communication network topology design is to consider the physical dynamics nodes as members of a team and study how they cooperate in the team to achieve the optimal reward. Interestingly, this topic has been studied in economics under the name of economic theory of teams [107]. Essentially, it considers how an organization, such as a commercial company, builds the information structure in order to benefit the team decision making. For example, it has been found that, under certain assumptions, the information structure of "emergency conference," in which all decisions are taken jointly when any corresponding information variable assumes an exceptional value, is more effective than the structure of dividing the agents into equal groups, within each of which the agents share all information [107, Chapter 6.13, p. 223].

In this section, we first introduce some basic results on the theory of team decision and then an application in optimal control. Although they do not provide systematic algorithms for optimizing the communication network topology (or equivalently the information structure), they do provide guidelines and insights for selecting the topology.

6.4.1 TEAM DECISION THEORY

We first introduce the system model in team decision theory and then provide some results related to the selection of network topology (information structure).

System model

We assume that there are a total of N members in a team. The action taken by member i is a scalar denoted by a_i. There is also a variable representing the environment, which is denoted by x, belonging to a certain set X. The payoff to the team is defined as

$$w(x, \mathbf{a}) = \mu_0 + 2\mathbf{a}^T \mu(x) - \mathbf{a}^T \mathbf{Q}\mathbf{a}, \tag{6.71}$$

where μ_0 is a constant which could be selected as zero, $\mathbf{a} = (a_1, \ldots, a_N)^T$, μ is a vector function on X, and $\mathbf{Q}$ is a fixed positive definite matrix.

Each member receives an observation on x, which is denoted by ξ_i for member i. Meanwhile, each member receives some information about the environment variable x. For member i, its information space is denoted by Y_i and the received information (or signal) is given by

$$y_i = \eta_i(x). \tag{6.72}$$

Note that y_i can be greater than ξ_i since member i may share information from other members. The N-tuple $\eta = (\eta_1, \ldots, \eta_N)$ is called the information structure of the team. The action of each member will be dependent on the information:

$$a_i = \alpha_i(y_i), \quad i = 1, \ldots, N. \tag{6.73}$$

Given the information structure η, the team will endeavor to maximize the expected payoff:

$$\Omega(\eta, \alpha) = E[w(x, \alpha_1(\eta_1(x)), \ldots, \alpha_N(\eta_N(x)))]. \tag{6.74}$$

The value of the information structure is defined as

$$V(\eta) = \max_{\alpha} \Omega(\eta, \alpha) - \max_{\mathbf{a}} E[w(x, \mathbf{a})], \tag{6.75}$$

where the second term denotes the optimal payoff when there is no observation on state x (thus being open loop).

Values of multiple information structures

In Ref. [107], the following information structures have been discussed in detail in order to compare their values:

- Complete communications: In this case, all the members share all information; i.e.,

$$\eta_i(x) = \xi(x) = \{\xi_j, j = 1, \ldots, N\}. \tag{6.76}$$

It is shown in Ref. [107] that the optimal decision is given by

$$\alpha = \mathbf{Q}^{-1} E[\mu | \xi]. \tag{6.77}$$

A more special case is complete information, in which we assume $\xi(x) = x$, i.e., the random state x is revealed to all members. In this case, the value of the information structure is given by

$$V_{ci} = E[\mu^T \mathbf{Q}^{-1} \mu]. \tag{6.78}$$

- No communications: In this case, the members do not share any information, namely

$$\eta_i(x) = \xi_i(x). \tag{6.79}$$

It has been shown that the value of this information structure is given by

$$V_{no} = \sum_{i=1}^{N} \frac{1}{Q_{ii}} E[E^2[\mu_i|\xi_i]], \tag{6.80}$$

when the observations of different members are mutually independent.

- Partitioned communications: We assume that the team members are partitioned into groups $\{I_k\}$, within each of which all members share the same information (thus forming a complete graph); i.e.,

$$\xi^k = \{\xi_i\}_{i \in I_k} \tag{6.81}$$

and

$$\eta(x_i) = \xi^k, \quad i \in I_k. \tag{6.82}$$

It has been shown that the optimal actions taken by the members in the kth group are given by

$$\alpha^k = \mathbf{k}^{-1} E[\mu^k | \xi^k], \tag{6.83}$$

where the superscript k denotes the terms corresponding to the members in group k. The value of the information structure of partitioned communications is given by

$$V_{pc} = \sum_k E[E^T[\mu^k|\xi^k] \mathbf{Q}_k E[\mu^k|\xi^k]]. \tag{6.84}$$

- Reporting exceptions: We assume that there are two types of values for $\mu_i(x)$, i.e., ordinary values and exceptional values. We denote by R_i the set of exceptional values for member i. If $\mu_i(x)$ is not in R_i, then member i will make a decision based only on its own observation; if $\mu_i \in R_i$, then it will inform a centralized agent, which will compute and send the decision. It is based on all the observations reported by the members with exceptional observations, to every member, as illustrated in Fig. 6.10.

Formally, we define the information structure of reporting exceptions as follows. The set of exceptional values R_i of member i is a given subset of the real

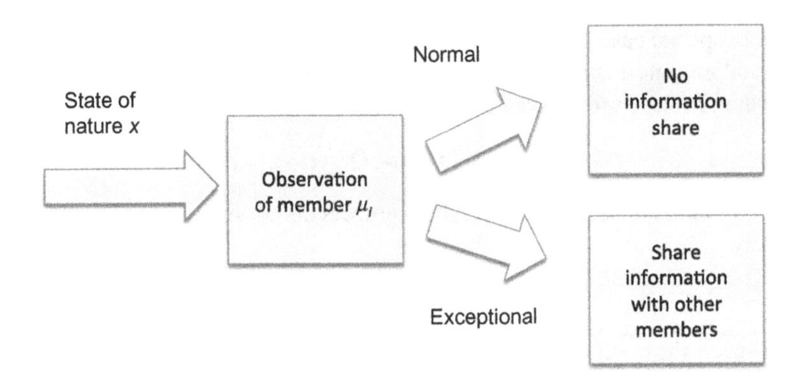

FIG. 6.10

Illustration of the information structure of reporting exceptions.

line. For each possible state of the environment x, $J(x)$ is the set of all members in the exceptional state. The information structure is defined as

$$\eta_i(x) = \begin{cases} \mu_i(x), & \text{if } \mu_i(x) \neq R_i, \\ \{\mu_j(x)\}_{j \in J(x)}, & \text{if } \mu_i(x) \in R_i. \end{cases} \tag{6.85}$$

It is found that, in the information structure of reporting exceptions, the optimal decision rule is given by

$$\alpha(y_i) = \begin{cases} \mu_i(x)/q_{ii}, & \text{if } i \notin J(x), \\ (\mathbf{Q}_{J(x)}^{-1}\mu^J)_i(x), & \text{if } \mu_i(x) \in R_i. \end{cases} \tag{6.86}$$

The value of the information structure is given by

$$V_{re} = \sum_{j=1}^{N} \left(s_{Rj}^2 E\left(q_{J(x)}^{jj}\right) + s_{Oj}^2 (1 - p_j) \left(\frac{1}{q_{jj}}\right) \right), \tag{6.87}$$

where $p_j = P(\mu_j(x) \in R_j)$, and

$$s_{Oi}^2 = E[\mu_i^2 | \mu_i(x) \neq R_i] \tag{6.88}$$

and

$$s_{Ri}^2 = E[\mu_i^2 | \mu_i(x) \in R_i]. \tag{6.89}$$

- Emergency conference: The emergency conference structure is a special form of reporting exceptions; i.e., all the decisions are made jointly if any member observes an exceptional value. Hence the information structure is given by

$$\eta_i(x) = \begin{cases} \mu_i(x), & \text{if } \forall j\ \mu_j(x) \neq R_j, \\ \mu(x), & \text{if } \exists\, \text{j}\ \mu_j(x) \in R_j. \end{cases} \tag{6.90}$$

It is found that the optimal decision rule is

$$\alpha_i(y_i) = \begin{cases} a_i^0 + \frac{\mu_i(x) - m_i^0}{q_{ii}}, & \text{if } x \notin \bar{R}, \\ \mathbf{Q}^{-1}\mu(x), & \text{if } x \in \bar{R}, \end{cases} \tag{6.91}$$

where

$$\bar{R} = \{x | \exists j, \text{ s.t. } \mu_j(x) \in R_j\}, \tag{6.92}$$

$$m_i^0 = E[\mu_i | \mu_i(x) \notin R_i], \tag{6.93}$$

and

$$\mathbf{a}^0 = \mathbf{Q}^{-1}\mathbf{m}^0. \tag{6.94}$$

The value of the information structure is shown to be

$$V_{ec} = \sum_i q^{ii} s_i^2 - \sum_i \left(q^{ii} - \frac{1}{q_{ii}} \right) s_{Oi}^2 P(x \in \bar{R}), \tag{6.95}$$

where

$$s_i^2 = \text{Var}(\mu_i) \tag{6.96}$$

and

$$s_{Oi} = \text{Var}(\mu_i | \mu_i(x) \notin \bar{R}). \tag{6.97}$$

In the above information structures, obviously the complete communication case has the best performance since it fully utilizes all the information. However, its cost for the communication structure is the highest. For the cases of partitioned communications, reporting exceptions, and emergency conference, we have the following conclusion:

Theorem 39. *With the same expected size of groups (i.e., the same expected amount of communications), we have*

$$V_{pc} \leq V_{re} \leq V_{ec}. \tag{6.98}$$

This conclusion shows that it is more desirable to have a dynamic information structure which is based on current observations.

Information structures in networks

In the previous discussion, all the communication is via a single hop. However, in the case of a networked information structure, it is desirable to study the impact of network topology in team decision making. Chapter 8 of Ref. [107] considered a networked information structure. Here we provide a simple discussion.

In the network, each member has output $b = \beta(z', e, \bar{b})$, where $z' = \xi'(x)$ is the message from the outside world and $e = \epsilon(x)$ is the noise (both depending on the state of nature x). We define the following two sets for each member i:

- Z_i: the set of possible messages z_i from the outside world.
- E_i: the set of possible values of noise e_i.

We also denote the set of messages that could be sent from member i to member j by B_{ij}. If there is no communication from i to j, then B_{ij} is an empty set. We also define the sets of all incoming and outgoing messages at member i as

$$\begin{cases} B_i = \prod_{j=0}^{N} B_{ji}, \\ \tilde{B}_i = \prod_{j=0}^{N} B_{ij}, \end{cases} \tag{6.99}$$

where the index 0 denotes special member—the nature.

When a member i receives messages $b_{0i}, \ldots, b_{Ni}$, its output messages are determined by

$$(b_{i0}, \ldots, b_{iN}) = \beta_i(b_{0i}, \ldots, b_{Ni}), \tag{6.100}$$

which is called the task function of member i. Here b_{0i} and b_{i0} denote the message received from the nature and the action taken by member i on the nature, respectively.

An effective approach to represent a network is to use a graph. Two types of edges are used in the graph:

- A single arrow means the transmission of a message from one member to another.
- A double arrow into member i means i receives information from the nature; a double arrow from member i means i takes an action on the nature.

Some examples of networked information structure are given in Fig. 6.11.

In Ref. [107], it is assumed that the message sets $\{B_{ij}\}$ have been predetermined. Hence the network topology of the information structure has been determined. There has been some qualitative discussion on how to optimize the network topology in the context of economic organizations; however, there is no quantitative discussion on the optimization of the network structure. Many typical topologies have been discussed in Ref. [107], which can provide much insight into the design of communication network topology.

Take Fig. 6.11A, for example. There are two members, 1 and 2. Member 1 receives observations $\mu = (\mu_1, \mu_2)$. It sends a message $\gamma(\mu)$ to member 2. The message sets are defined as

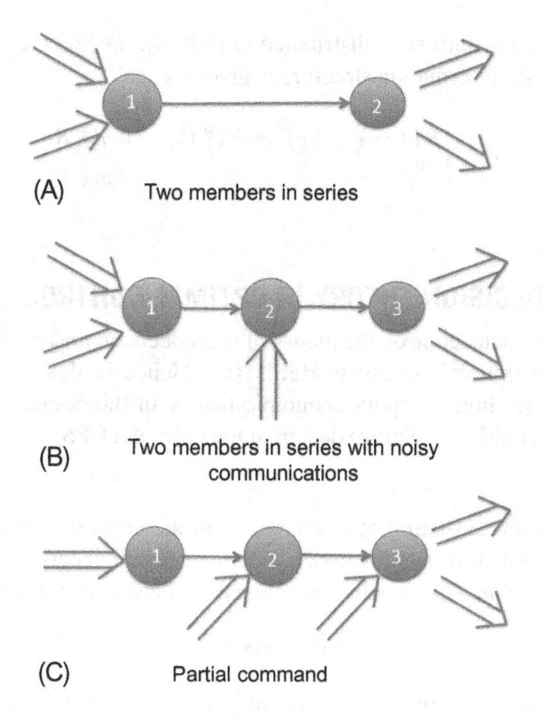

FIG. 6.11

Examples of networked information structures.

$$\begin{cases} B_{01} : \text{the space of real 2-tuples,} \\ B_{12} : \{(1,1),(-1,1),(1,-1),(-1,-1)\}, \\ B_{20} : =B_{12}, \\ B_{10} : \text{empty,} \\ B_{01} : \text{empty.} \end{cases} \tag{6.101}$$

We assume that the payoff function is given by

$$w(x,a) = \mu_1(x)a_1 + \mu_2(x)a_2 - qa_1a_2, \tag{6.102}$$

where a_1 and a_2 are binary. Obviously, member 1 should inform member 2 of the optimal decision, given the observations. Therefore the optimal message sent from member 1 to member 2 is given by

$$\gamma(\mu) = \begin{Bmatrix} (1,1) \\ (-1,1) \\ (1,-1) \\ (-1,-1) \end{Bmatrix} \text{ if } \begin{Bmatrix} \mu_1 + \mu_2 - q \\ -\mu_1 + \mu_2 + q \\ \mu_1 - \mu_2 - q \\ -\mu_1 - \mu_2 - q \end{Bmatrix} \text{ is the largest.} \tag{6.103}$$

Suppose that μ_i is uniformly distributed in $[-R, R]$, for $i = 1, 2$. Then it is found that the value of the information structure is given by

$$V = \begin{cases} R\left(1 - \frac{q}{R} + \left(\frac{q}{R}\right)^2 - \frac{1}{3}\left(\frac{q}{R}\right)^3\right), & \text{if } q \leq R, \\ \frac{2R}{3}, & \text{if } q \geq R. \end{cases} \tag{6.104}$$

6.4.2 TEAM DECISION THEORY IN OPTIMAL CONTROL

In the field of automatic control, the theory of team decision making has been applied in the context of optimal control in Ref. [108]. Hence in this section, we follow Ref. [108] to study how to apply economic theory in the context of topology (or equivalently information structure) design in the case of a CPS.

System model

We consider a team consisting of N members; in the context of a CPS, the system consists of N physical dynamics nodes. Each member i receives observations (or information) $\mathbf{y}_i$ and then takes action $\mathbf{u}_i$. The control law is denoted by γ_i; i.e.,

$$\mathbf{u}_i = \gamma(\mathbf{x}_i). \tag{6.105}$$

The payoff function of the team is denoted by J, which is given by

$$J = J(\gamma_1, \gamma_2, \ldots, \gamma_N), \tag{6.106}$$

namely the team payoff is a function of the individual control strategies. Now the problem is how to maximize the payoff function when the system is dynamical in time.

We further assume that the observation $\mathbf{y}_i$ is given by

$$\mathbf{y}_i = \mathbf{H}_i\mathbf{w} + \sum_j \mathbf{D}_{ij}\mathbf{u}_j, \tag{6.107}$$

where $\mathbf{H}_i$ and $\mathbf{D}_{ij}$ are matrices known to all team members, and $\mathbf{w}$ is a random vector representing the random perturbation in the environment.

We assume that the impact on the observations/information is one-directional, namely

$$\mathbf{D}_{ij} \neq 0 \Rightarrow \mathbf{D}_{ji} = 0. \tag{6.108}$$

This means that if a member i can affect the observation of member j, then the action of j does not affect that of i. This assumption guarantees the causality of the system.

Obviously, the information structure forms a tree, since the requirement in Eq. (6.108) eliminates the possibility of cycles in the information structure. We say that member j is related to i if $\mathbf{D}_{ij} \neq 0$, and denote it by $i \rightsquigarrow j$. If we can find a series of members $n_1, \ldots, n_k$ such that $j \rightsquigarrow n_1 \rightsquigarrow n_2 \ldots \rightsquigarrow n_k \rightsquigarrow i$, then we say that j is a precedent of i and denote it by $i \preceq j$. This structure is similar to Bayesian networks, although the concept of a Bayesian network had not been proposed at the time Ref. [108] was published.

We assume that the cost function of the team is given by

$$J(\gamma_1,\ldots,\gamma_N) = E\left(\frac{1}{2}\mathbf{u}^T\mathbf{Q}\mathbf{u} + \mathbf{u}^T\mathbf{S}\mathbf{w} + \mathbf{u}^T\mathbf{c}\right), \tag{6.109}$$

where $\mathbf{Q}$ is a positive definite matrix, $\mathbf{S}$ is a matrix with appropriate dimensions, and $\mathbf{c}$ is a vector with appropriate dimension.

State teams

The simplest case is for static teams; i.e., the observation of each member in the team is a linear transformation of the random vector $\mathbf{w}$:

$$\mathbf{y}_i = \mathbf{H}_i\mathbf{w}, \quad \forall i. \tag{6.110}$$

Radner [109] proposed the following theorem:

Theorem 40. *The following control law is optimal for the case of static teams:*

$$\mathbf{u}_i = \mathbf{A}_i\mathbf{y}_i + \mathbf{b}_i, \quad \forall i, \tag{6.111}$$

where $\mathbf{A}_i$ *and* $\mathbf{b}_i$ *are matrix and vector having the appropriate dimensions.*

Note that the computation of the optimal $\mathbf{A}_i$ and $\mathbf{b}_i$ can be found in Ref. [108].

Dynamic teams

For the case of dynamic teams, we focus on the special case of partially nested information structures, which is defined as follows:

Definition 20. We say that an information structure is partially nested if

$$j \preceq i \Rightarrow \mathcal{Y}_j \subset \mathcal{Y}_i, \tag{6.112}$$

for all i and j, where $\mathbf{Y}_i$ is the sub-σ-algebra over which $\mathbf{y}_i$ is defined.

Intuitively speaking, a team has a partially nested structure if the information known to a member is always known to all its followers. Then the following theorem shows the form of optimal control strategy when the information structure is partially nested:

Theorem 41. *For a dynamic team with partially nested information structure, the optimal control of each member i is unique and is linear in* $\mathbf{y}_i$.

6.5 CONCLUSIONS

In this chapter, we have discussed communication network topology design for cyber physical systems. In summary, topology design can be cast into an optimization problem:

- The objective function should take into consideration the operation of the physical dynamics, e.g., its operation cost, instead of metrics characterizing the communications (e.g., throughput or delay).
- The constraints should be related to the limited resources of communication networks, such as available links or the number of wavelengths. The hardware characteristics, e.g., the mapping between logical links and physical links, should also be incorporated.

The theory of team decisions can be applied in the network topology design, since the network topology is equivalent to the information structure in an economic system. However, it is still not quite clear how to apply the theory of team decisions in topology design.

CHAPTER

Communication network operation for CPSs

7

7.1 INTRODUCTION

It is important to optimize the operation of the communication network to control the physical dynamics in a cyber physical system (CPS). The networking mimics the nervous system in the human body. For example, in smart grids, a wide area monitoring system (WAMS) [110,111] can send various measurements such as phase, voltage, and frequency from sensors (often called phasor measurement units (PMUs) [1,112]) to control center(s). There have been various communication networks specifically for CPSs (particularly for industrial control systems), such as the supervisory control and data acquisition (SCADA) system, P-Net, and Foundation Fieldbus [27]. There are also many industrial standards such as IEC61850 (for electrical substations) [113] and EIA RS-485 (for industrial automation) [114]. Meanwhile, the impact of communication imperfections, such as delay and packet drop, on system stability has been intensively studied in automatic controls [19,115,116], and in power systems [117–119].

However, the current communication standards and infrastructures, either dedicated systems for CPSs (such as SCADA) or generic systems (such as 4G cellular systems), may not be able to fulfill the rapidly increasing demand of CPSs or may not be efficient for resource utilization. For example, for real-time control in a smart grid, the communication delay may need to be of the order of milliseconds [120], particularly to protect power grids in contingencies such as cascading failure. However, the communication delay in the current SCADA system could be a few seconds or even several minutes, which is far too long for real-time control in a smart grid. Moreover, traditional SCADA may provide a data rate of the order of kbps, while smart grids may need tens or even hundreds of Mbps [121]. Meanwhile, most studies in the field of communications and networking are focused on generic data transmissions (such as file downloading or video streaming), which are measured by traditional metrics such as throughput or delay, instead of quantities directly related to the physical dynamics in CPSs (e.g., the oscillation of voltages in power grids). These key differences are summarized in Table 7.1.

Although one may use a large amount of communication resources, such as a wideband 4G system or optical fiber network, for monitoring and controlling CPSs, this may incur a significant waste since the characteristics of physical dynamics of

Communications for Control in Cyber Physical Systems. http://dx.doi.org/10.1016/B978-0-12-801950-4.00007-X

Table 7.1 Comparison Between Communication Networks in CPSs and Traditional Data Networks

	Purpose	Real-Time or Not?	Performance Metrics
CPS	Control the system dynamics	Often highly real-time	System stability or cost functions of system state
Traditional	Convey the data, unaware of the data contents	Many are elastic, such as the Internet	Throughput or delay

CPSs are not incorporated to optimize the communication protocols. There have been some studies on the communication networking dedicated to controlling a CPS [23, 24,122]; however, there is still a lack of systematic studies and unified frameworks for designing communication networks for a CPS, which considerably retards the future development of CPSs and corresponding applications. In this chapter, we focus on the real-time operation of the communication networks in CPSs, particularly resource allocation and routing.

7.1.1 MAIN CHALLENGES

The main challenges of network operation in the context of a CPS include the following:

- The networking should be aware of the current system state of the physical dynamics. The eventual goal of networking in CPS is to optimize the operation of the physical dynamics, instead of communicating the messages. Delays and packet losses can be tolerated, as long as the physical dynamics runs smoothly. Hence, many existing networking algorithms in communication networks may not apply. Actually, as will be seen, some traditional metrics such as expected communication delay (which is important in the context of routing) may be misleading in a CPS.
- The networking algorithm requires joint optimization for both communications and controls. However, the system states of many physical dynamics are continuously valued (many are also continuous in time), while the operations in most communication networks are discrete in both time and values. It is challenging to handle both discrete and continuous subsystems simultaneously.
- As will be demonstrated in subsequent sections, in certain situations a single networking mode (e.g., a single routing scheme) may not stabilize the physical dynamics. Hence, it may be desirable to prepare multiple networking modes (e.g., multiple routing paths), in order to stabilize the dynamics. Then, it is challenging to figure out how to select these modes and how to switch among these modes.

7.1.2 MAIN APPROACHES

There have been no widely recognized frameworks for networking in CPSs. In this book, we introduce several potentially useful approaches for networking algorithm design in CPSs:

- Hybrid systems: Since one of the major challenges in networking in CPSs is the existence of both continuous and discrete system states, it is natural to introduce the theory of hybrid systems [123], which is studied to handle the case of hybrid system states.
- Optimizations: The networking for communication and control can be converted into the framework of optimizations, which can be solved by using efficient algorithms such as convex optimizations.
- Information interface: Besides the joint consideration of networking and control, we can also hold a different but more practical philosophy; i.e., designing the communication and control subsystems separately, thus facilitating the exploitation of existing approaches in both areas. Meanwhile, an interface will be designed to make the communication network aware of the control subsystems and the physical dynamics. The responsibility of the interface is to "translate" the requirement of control to quantities widely used in communication networks; or equivalently, the interface transforms the communication requirement in a CPS to traditional problems in communication networks. One such information interface, namely effective information and virtual queues, will be introduced in this chapter.

7.2 HYBRID SYSTEM MODELING FOR CPSs

In this section, we fit the communication system design in CPSs into the framework of hybrid systems. We first introduce the theory of hybrid systems. Then, we define communication/dynamics modes and fit the CPS into the framework of hybrid systems. Note that hybrid system modeling for the communications and physical dynamics in a CPS is proposed in Ref. [124]. In this book, we follow Ref. [124] to provide a brief introduction to this modeling.

7.2.1 HYBRID SYSTEMS

Originally proposed in 1966 [125], hybrid systems have been studied intensively, in terms of system stability, controllability and observability, and have been used in many applications such as energy management [126], industrial controls [127], and automotive control [128]. The key feature of a hybrid system is that it has both discrete and continuous system states.

Linear switching system

Here, we use linear switching systems to model the CPS, for simplicity, as illustrated in Fig. 7.1. The dynamics is described by the following difference equations[1]:

$$\begin{cases} \mathbf{x}(t+1) = \mathbf{A}_{k(t)}\mathbf{x}(t) + \mathbf{B}_{k(t)}\mathbf{u}(t) + \mathbf{n}(t), \\ \mathbf{y}(t) = \mathbf{C}_{k(t)}\mathbf{x}(t) + \mathbf{w}(t), \end{cases}$$

where $\mathbf{x}$ is the continuous system state, k is the discrete system state, $\mathbf{u}$ is the control action, $\mathbf{y}$ is the observation vector, $\mathbf{n}$ and $\mathbf{w}$ are both random noise, and $\mathbf{A}_k$, $\mathbf{B}_k$ and $\mathbf{C}_k$ are system parameters, which are dependent on the discrete system state. The set of the discrete system state is denoted by $\mathcal{K}$. Obviously, when the discrete system state k changes, the mode of the dynamics of the continuous system state $\mathbf{x}$ also changes. Hence the continuous system dynamics seems to be switched among multiple modes determined by the discrete system state. The observability, controllability, stability, and control strategies (for both $\mathbf{x}$ and k) have been intensively studied, which can be found in the survey book [123].

Control of linear switching systems

In this section we provide an introduction to an effective approach for controlling linear switching systems using linear quadratic regulator (LQR) theory. For simplicity,

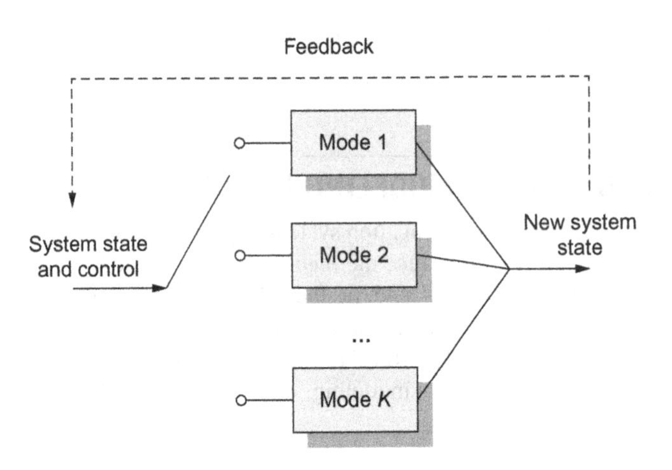

FIG. 7.1

Illustration of switching-based hybrid systems.

[1]In this book, we consider only discrete-time dynamics, since the communication system is digital and hence the switch of system dynamics mode changes only at discrete-time snapshots. It is not straightforward to extend to the purely continuous-time case, which is not very useful in practice.

we consider the following dynamical system with the continuous system state $\mathbf{x}$ and discrete system state k:

$$\mathbf{x}(t+1) = \mathbf{A}_{k(t)}\mathbf{x}(t) + \mathbf{B}_{k(t)}\mathbf{u}(t), \tag{7.1}$$

which is a special case of the generic linear switching system in Eq. (7.1) with $\mathbf{n} = \mathbf{w} = 0$ and $\mathbf{C} = \mathbf{I}$ (i.e., there is no noise and the system state is observed directly).

In the framework of the LQR, we define the cost function as

$$L(\mathbf{x}, \mathbf{u}, k) = \mathbf{x}^T \mathbf{Q}_k \mathbf{x} + \mathbf{u}^T \mathbf{R}_k \mathbf{u}, \tag{7.2}$$

where $\{\mathbf{Q}_k\}_k$ and $\{\mathbf{R}_k\}_k$ are positive definite matrices. Then for a time period T, the cost of system operation is defined as

$$J(\mathbf{x}(0), \pi_T) = \psi(\mathbf{x}(T)) + \sum_{t=0}^{T-1} L(\mathbf{x}(t), \mathbf{u}(\mathbf{x}(t)), k(\mathbf{x}(t))), \tag{7.3}$$

where ψ is the cost of the final time slot, $\mathbf{x}(0)$ is the initial system state, and π_T is the feedback control law. We can also let $T \to \infty$ such that

$$J(\mathbf{x}(0), \pi_\infty) = \sum_{t=0}^{\infty} L(\mathbf{x}(t), \mathbf{u}(\mathbf{x}(t)), k(\mathbf{x}(t))). \tag{7.4}$$

The goal of switched LQR control is to obtain the following value functions:

$$\begin{cases} V_T(\mathbf{z}) = \inf_{\pi_T} J(\mathbf{z}, \pi_T), \\ V_\infty(\mathbf{z}) = \inf_{\pi_\infty} J(\mathbf{z}, \pi_i nfty), \end{cases} \tag{7.5}$$

A standard approach for solving multistage optimization problems is dynamic programming (DP). We consider a control law $\xi = (\mathbf{u}, k)$ which maps from the system state $\mathbf{x}$ to the continuous action $\mathbf{u}$ and the discrete state selection k. We define an operator T_ξ as follows:

$$T_\xi[g](\mathbf{z}) = L(\mathbf{z}, \mathbf{u}(\mathbf{z}), k(\mathbf{z})) + g(\mathbf{A}_{k(\mathbf{z})}\mathbf{z} + \mathbf{B}_{k(\mathbf{z})}\mathbf{u}), \tag{7.6}$$

for any function g mapping from R^n to positive numbers. Then, the one-stage value iteration operation T, which is now independent of the control law, is given by

$$T[g](\mathbf{z}) = \inf_{\mathbf{u},k}\{L(\mathbf{z}, \mathbf{u}, k) + g(\mathbf{A}_{k(\mathbf{z})}\mathbf{z} + \mathbf{B}_k)\mathbf{u})\}. \tag{7.7}$$

The tth repetition of the value iteration is denoted by T^t. For the finite duration case, the optimal action taken at time slot t is given by

$$(\mathbf{u}, k)_{\text{opt}} = \arg\min_{\mathbf{u},k}\{L(\mathbf{z}, \mathbf{u}, k) + T^{T-t}[\psi](\mathbf{A}_{k(\mathbf{z})}\mathbf{z} + \mathbf{B}_k)\mathbf{u})\}. \tag{7.8}$$

Although the DP framework is theoretically simple, it is difficult in practice since the argument $\mathbf{z}$ (namely the initial value) in the function $T[g]$ is continuous. For the generic case, it takes infinitely many bits to perfectly describe the function $T[g]$. An approach to handle the continuous argument is to discretize it by dividing the real space into many bins. However, such an approach is impractical when the dimension of $\mathbf{z}$ is high, due to the curse of dimensions.

Fortunately, the special structure in the switched LQR problem can be used to simplify the DP. This is accomplished in Ref. [129]. To that goal, we define the Riccati mapping for each dynamics mode k:

$$\rho_k(\mathbf{P}) = \mathbf{Q}_k + \mathbf{A}_k^T\mathbf{P}\mathbf{A}_k - \mathbf{A}_k\mathbf{P}\mathbf{B}_k(\mathbf{R}_k + \mathbf{B}_k^T\mathbf{P}\mathbf{B}_k)^{-1}\mathbf{B}_k^T\mathbf{P}\mathbf{A}_l. \tag{7.9}$$

Then, we define the switched Riccati mapping as

$$\rho_{\mathcal{K}}(\mathcal{H}) = \{\rho_k(\mathbf{P})|k \in \mathcal{K}, \mathbf{P} \in \mathcal{H}\}, \qquad \mathcal{H} \in \mathcal{F}, \tag{7.10}$$

where $\mathcal{F}$ is the set of all subsets of semidefinite matrices with finitely many elements. The switched Riccati sets are defined as the sequence of sets $\{\mathcal{H}_t\}_{t=0}^{\infty}$ generated iteratively by

$$\mathcal{H}_{t+1} = \rho_{\mathcal{K}}(\mathcal{H}_t), \tag{7.11}$$

with the initial condition $\mathcal{H}_0 = \{\mathbf{Q}_f\}$. The analogy between the function value iteration and the Raccati sets value iteration is illustrated in Fig. 7.2.

The following theorem shows the relationship between the value functions and the switched Raccati sets, which provides an approach to calculate the value functions:

Theorem 42. *For any time slot t, the tth value function in Eq. (7.5) is given by*

$$V_t(\mathbf{z}) = \min_{\mathbf{P}\in\mathcal{H}_t} \mathbf{z}^T\mathbf{P}\mathbf{z}. \tag{7.12}$$

Remark 15. The key advantage of Eq. (7.12) is to manifest the explicit expression of the value function and thus avoid the discretization, by fully utilizing structure of the LQR control within each time slot. However, we still face the challenge of the scalability of the sets $\{\mathcal{H}_t\}$ since the cardinality of $\mathcal{H}_t$ satisfies

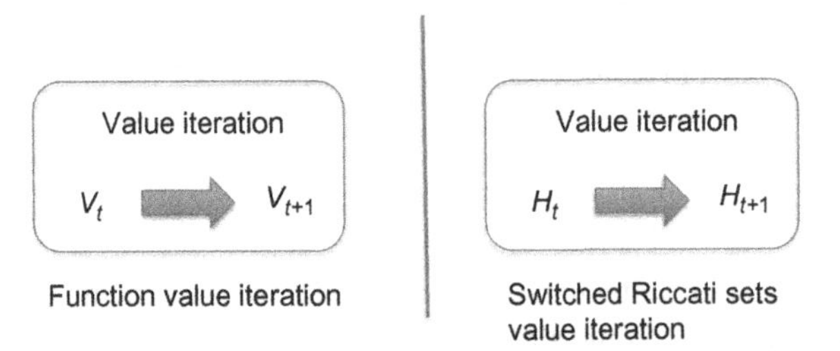

FIG. 7.2

Analogy between the function value iteration and the Raccati sets value iteration.

$$|\mathcal{H}_t| = |\mathcal{K}|^t. \tag{7.13}$$

If the set of discrete system states is sufficiently large, the size of $\mathcal{H}_t$ becomes prohibitively large very quickly.

As pointed out in the above remark, the key challenge here is the large number of matrices in each $\mathcal{H}_t$ when t becomes large. It was found in Ref. [129] that many matrices in $\mathcal{H}_t$ can actually be removed without affecting the optimal value functions. For a matrix $\mathbf{P}$ in $\mathcal{H}$, we say that $\mathbf{P}$ is algebraic redundant if we can find another $\mathbf{P}' \in \mathcal{H}$ such that

$$\mathbf{z}^T\mathbf{P}\mathbf{z} \geq \mathbf{z}^T\mathbf{P}'\mathbf{z}, \tag{7.14}$$

for any $\mathbf{z}$ with appropriate dimension. Intuitively speaking, $\mathbf{P}$ is algebraically redundant if it represents the value of a strategy that is suboptimal in all cases. Then it does not affect the optimality if we omit these algebraically redundant matrices. The following theorem provides a sufficient condition for the redundant matrices [129]:

Theorem 43. *$\mathbf{P} \in \mathcal{H}$ is algebraically redundant in $\mathcal{H}$ if there exist nonnegative constants $\{c_i\}_{i=1}^{|\mathcal{H}|-1}$ such that*

$$\begin{cases} \sum_{i=1}^{|\mathcal{H}|-1} c_i = 1, \\ \mathbf{P} \preceq \sum_{i=1}^{|\mathcal{H}|-1} c_i \mathbf{P}_i, \end{cases} \tag{7.15}$$

where $\{\mathbf{P}_i\}$ is an enumeration of all the matrices in $\mathcal{H}$ except $\mathbf{P}$.

Based on the above theorem, the procedure for removing algebraically redundant matrices and computing the sets $\{\mathcal{H}_t\}$ is given in Procedure 3.

Although Procedure 3 can substantially reduce the complexity of the DP procedure, its complexity is still too high for the case of large time durations. Hence, for practical systems, it is desirable to propose suboptimal but computationally efficient algorithms. One possible approach is to relax the condition of the redundancy in Eq. (7.14) to the following ϵ-redundancy:

PROCEDURE 3 REMOVING ALGEBRAICALLY REDUNDANT MATRICES

1: Set $\mathcal{H}_0 = \{\mathcal{Q}_f\}$.
2: **for** Each $t = 1, \ldots, T$ **do**
3: Compute $\mathcal{H}_t = \rho_{\mathcal{K}}(\hat{\mathcal{H}}_{t-1})$.
4: Set $\hat{\mathcal{H}}(t)$ to be an empty set.
5: **for** Each $\mathbf{P} \in \mathcal{H}(t)$ **do**
6: Check the conditions in Eq. (7.18).
7: **if** The conditions are not satisfied **then**
8: Add $\mathbf{P}$ to $\hat{\mathcal{H}}(t)$.
9: **end if**
10: **end for**
11: **end for**

$$\mathbf{z}^T(\mathbf{P}+\epsilon\mathbf{I})\mathbf{z} \geq \mathbf{z}^T\mathbf{P}'\mathbf{z}, \tag{7.16}$$

where ϵ is a predetermined positive number. The larger ϵ is, the more relaxed the concept of redundancy is and the more matrices can be removed from the sets $\mathcal{H}(t)$. Theorem 43 can be correspondingly changed to the following version:

Theorem 44. $\mathbf{P} \in \mathcal{H}$ *is* ϵ*-redundant in* $\mathcal{H}$ *if there exist nonnegative constants* $\{c_i\}_{i=1}^{|\mathcal{H}|-1}$ *such that*

$$\begin{cases} \sum_{i=1}^{|\mathcal{H}|-1} c_i = 1, \\ \mathbf{P}+\epsilon\mathbf{I} \preceq \sum_{i=1}^{|\mathcal{H}|-1} c_i\mathbf{P}_i, \end{cases} \tag{7.17}$$

where $\{\mathbf{P}_i\}$ *is an enumeration of all the matrices in* $\mathcal{H}$ *except* $\mathbf{P}$.

Then, the algorithm in Procedure 3 can be modified accordingly, and the constant ϵ can be set according to the tradeoff between the complexity and the optimization performance.

The modified algorithm using ϵ-redundancy can work for finite time horizons. However, the complexity increases at least linearly with respect to the time duration T, since the sets $\mathcal{H}(t)$ need to be stored for each t. Hence, the above approaches cannot work in the infinite horizon case, i.e., $T \rightarrow \infty$. A practical approach is to fix a sufficiently large T, obtain the optimal (or near-optimal) strategy π_T, and repeat π_T for each time slot in order to approximate the optimal strategy in the infinite time horizon. A key challenge for this approximate approach is whether the resulting strategy can yield stable system dynamics (in the finite time horizon, we do not have this concern). The following theorem provides a sufficient condition for stability (Lemma 7, [129]):

Lemma 7. *The system is exponentially stabilizing if there exist nonnegative constants, and if for each* $\mathbf{P} \in \mathcal{H}_t^{\epsilon}$ *there exist* $\{c_i\}_{i=1}^{|\mathcal{H}|-1}$ *such that*

$$\begin{cases} \sum_{i=1}^{|\mathcal{H}|-1} c_i = 1, \\ \mathbf{P} \preceq \sum_{i=1}^{|\mathcal{H}|-1} c_i(\mathbf{P}_i+(\kappa_3-\kappa_*)\mathbf{I}), \end{cases} \tag{7.18}$$

where $\{\mathbf{P}_i\}$ *is an enumeration of all the matrices in* $\mathcal{H}$ *except* $\mathbf{P}$, *and*

$$\kappa_* = \min \lambda_{\min}\{\mathbf{K}_i^T\mathbf{R}_i\mathbf{K}_i+\mathbf{Q}_i\}. \tag{7.19}$$

Based on this lemma, Ref. [129] provides an efficient algorithm for designing the control algorithm with the assurance of system stability.

7.2.2 HYBRID SYSTEM MODEL OF A CPS

Now our goal is to model the communication network and physical dynamics in a CPS using the hybrid system model. To fit a CPS with communication infrastructure into the framework of a hybrid system, we need to realize that the operation of the communication network can change the mode of the physical dynamics in the CPS, and the hybrid system model is a suitable tool to jointly describe the discrete states of the digital communication network and continuous system state of physical dynamics.

Communication mode

We first define *a communication mode as a configuration of the communication network in a CPS*, which is discrete. For example, if we consider only the physical layer, a communication mode may correspond to a combination of modulation and coding, e.g., QPSK + convolutional code with transmission rate 1/2; if we consider only the MAC layer, a communication mode may correspond to a set of active communication links scheduled for communications; in the network layer, a communication mode may correspond to a set of routes for different data traffic. A communication mode may be across different layers, e.g., it may specify the configurations of both physical and MAC layers. The communication mode may change with time due to the change of either the information source characteristics or the communication channel qualities. For example, in the physical layer, the mode may be changed if adaptive modulation/coding is employed; in the MAC layer, the active communication links may change with time, which can be determined by a scheduler.

Relationship between communication and dynamics modes

We have defined the communication mode. Now we map each communication mode to a mode of the physical dynamics, which means the law of system state evolution. When different communication modes are used, the controller in a CPS may receive different packets with different delays or different packet drop rates, thus resulting in different dynamics modes. Below we use examples in the physical layer and MAC layer to illustrate the mapping.

Example 2 (Physical Layer). Here we consider the physical layer communication from a sensor to a controller. Two possible modulations, QPSK and 8-PSK, can be used. Fig. 7.3 shows a comparison between QPSK and 8-PSK used by the sensor. We suppose that the same number of bits are used for quanizing each measurement at the sensor and the channel coding is fixed. Obviously, the delay when using QPSK is longer because each QPSK symbol carries fewer bits than the 8-PSK; meanwhile, since the constellation of QPSK is sparser, the demodulation error rate of QPSK is lower. The different communication delays and different error rates result in different actions of the controller and thus different dynamics modes. Then, the evolution of the system state can be written as (consider the discrete-time case)

$$\begin{aligned}\mathbf{x}(t+1) &= f(\mathbf{x}(t), g_{e,d}(\mathbf{x}), \mathbf{n}(t)) \\ &= \tilde{f}_{e,d}(\mathbf{x}(-\infty:t), \mathbf{n}(t)),\end{aligned} \tag{7.20}$$

where g is the control action law as a function of the observations, e and d represent the packet error probability level and delay, $\tilde{f}$ is the evolution law taking e, d, and g into account. Obviously, e and d denote the discrete state of a hybrid system, which is determined by the selection of modulation.

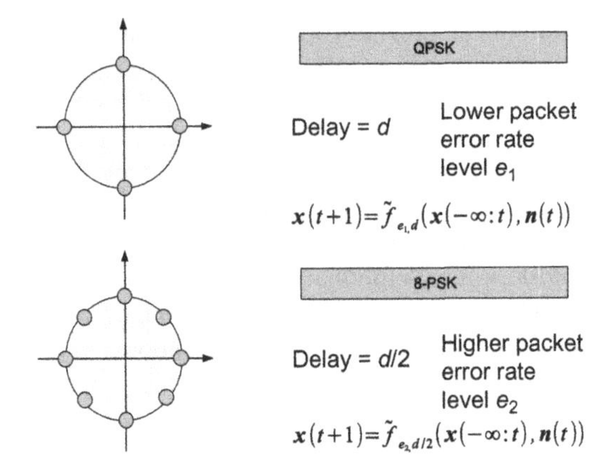

FIG. 7.3

Example of communication modes in the physical layer.

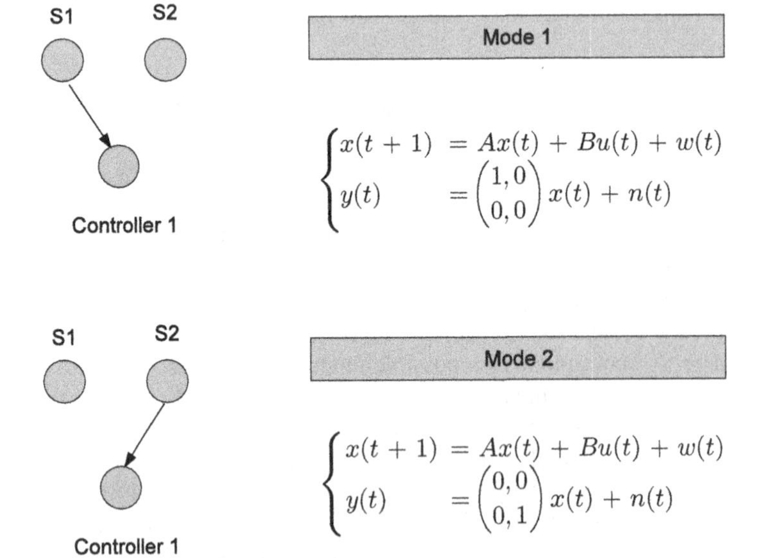

FIG. 7.4

Example of communication modes in the MAC layer.

Example 3 (MAC Layer). Fig. 7.4 shows a system with two sensors, each monitoring one dimension of the continuous system state $\mathbf{x}$, and one controller. We assume that only one sensor can transmit its measurement at a time, due to co-channel interference. If all other aspects of the communication infrastructure, e.g., modulation and code, are fixed, there are two communication modes in the CPS: one is sensor 1 being scheduled while the other is sensor 2 being scheduled. The equations describing the different dynamics corresponding to different communication modes are also given in Fig. 7.4. We observe that the system matrix **A** and control matrix **B** do not change, while the observation matrix **C** changes with the communication mode.

From the examples, we observe that each communication mode corresponds to a unique dynamics mode. Hence a communication mode is equivalent to a dynamics mode, if there are no other external factors affecting the dynamics mode. To fit a CPS into the hybrid system model, we can use the discrete state, i.e., $k(t)$ in (8.19), to indicate the current communication/dynamics mode, and use the continuous system state, i.e., $\mathbf{x}(t)$ in (8.19), to model the physical dynamics. For example, in Example 3, k equals either 1 or 2, indicating whether sensor 1 or 2 is scheduled; in each case, either matrix $\mathbf{C}_1 = (1, 0; 0, 0)$ or $\mathbf{C}_2 = (0, 0; 0, 1)$ is used in the dynamics equation. Then we can fully exploit the existing conclusions on the observability, controllability, or stability of hybrid systems to analyze or design the communication network in a CPS. Note that a rapid change of the MAC layer links is possible. For example, when the voltages in a microgrid are experiencing an oscillation, it is important to check the measurements of different sensors quickly, thus requiring a fast change of active links.

7.3 OPTIMIZATION OF SCHEDULING POLICY

In this section, by following Ref. [130], we study how to optimize the communication mode, as illustrated in Fig. 7.5, in the framework of hybrid systems. In particular, we focus on MAC layer scheduling in the communication network. We first decompose the problem into two stages and then study the optimization procedure.

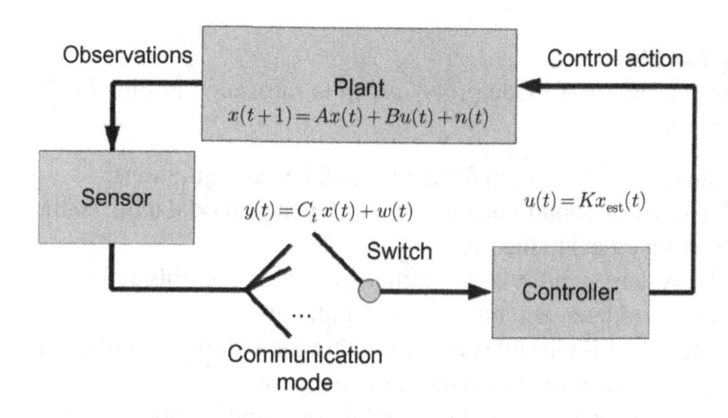

FIG. 7.5

Illustration of communication mode scheduling.

7.3.1 FUNDAMENTAL CHALLENGES

As we have seen, a CPS with communication infrastructure can be modeled as a hybrid system. Since our focus is on the operation of communication infrastructure, which can be abstracted as a communication mode, we have the following major challenges for the study within the framework of hybrid systems:

- Mode provisioning: Since the purpose of the communication network is to choose the communication mode, e.g., selecting the active communication links by the scheduler, we need to prepare a set of communication modes before the operation of a CPS. The preparation of communication modes is called mode provisioning. Obviously, this is an offline procedure.
- Mode scheduling: Given the set of communication modes obtained from the mode provisioning, the communication mode can be selected adaptively to the system state. We call this mechanism mode scheduling. Note that adaptive modulation/coding, traffic scheduling, and dynamic routing in traditional communication networks can be categorized as mode scheduling. Obviously, mode scheduling is an online procedure.

7.3.2 MODE PROVISIONING

As we have explained, the task of mode provisioning is to provide a set of communication modes for the communication network such that the hybrid systems of communication and control can be stabilized or more easily optimized. The principle of mode provisioning is that *there should exist a scheduling policy for the communication modes such that the system can be stabilized by intelligently switching among these modes*. Below, we introduce the generic procedure of provisioning and then use examples in the MAC and physical layers to illustrate the principle. We also provide criteria of system stability in the subsequent two propositions such that the mode provisioning can guarantee the stability of physical dynamics.

Generic procedure

The generic procedure of mode provisioning is illustrated in Fig. 7.6. It consists of the following steps:

1. Formulate the hybrid system for the given CPS configurations.
2. Add a basic set of communication modes (such as modulation, coding, or scheduling schemes) to the pool.
3. Check the system stability using certain criteria in controls and systems, which will be provided later, and the system complexity.
4. If the system is still unstable and still within the requirement of complexity, add proper unused communication modes to the pool.
5. Finally, the procedure either claims failure or outputs a pool of communication modes that can stabilize the physical dynamics.

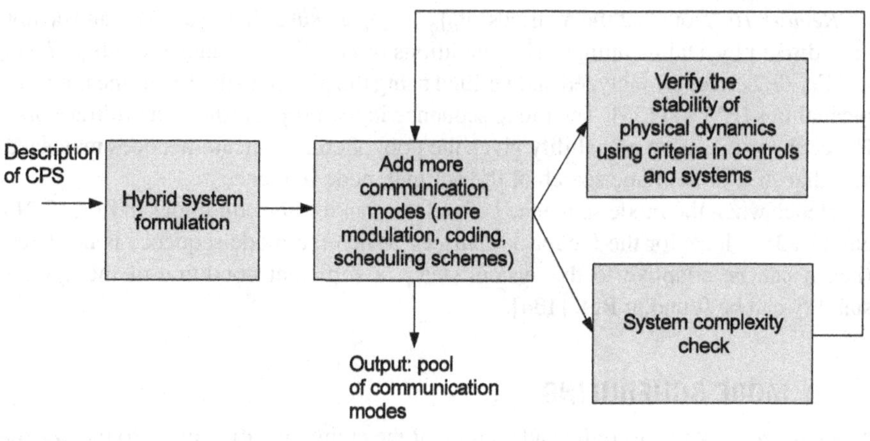

FIG. 7.6

Procedure of mode provisioning.

Illustration by examples

Example 4. [MAC Layer Mode Provisioning] In this example, we consider the selection of active communication links and follow the approach proposed in Ref. [131]. As we found in the previous discussion, scheduling in the MAC layer is equivalent to determining the observation pattern of the dynamical system. When a certain sensor is not selected, it is equivalent to setting the corresponding column in the feedback gain matrix $\mathbf{K}$ to zero. For example, consider a two-dimensional system in which two sensors sense each dimension of the state. When sensor 1 is selected while sensor 2 is not, we have $\mathbf{K} = \begin{pmatrix} K_{11}, 0 \\ K_{21}, 0 \end{pmatrix}$; on the contrary, when sensor 2 is selected while sensor 1 is not, $\mathbf{K} = \begin{pmatrix} 0, K_{12} \\ 0, K_{22} \end{pmatrix}$.

Therefore the mode provisioning in the MAC layer is equivalent to preparing a set of feedback gain matrices, each corresponding to a set of active communication links without interference and setting the corresponding columns to zero. When the mode sequence is fixed, the following proposition (Theorem 1 in Ref. [132]), which is obtained in the field of controls and systems, can be used to determine whether the system can be stabilized.

Proposition 3. *Suppose that the feedback gain matrix $\mathbf{K}_q$ and the mode sequence $1, 2, \ldots, Q$ have been fixed. If there exist Q positive definite matrices $\mathbf{P}_1, \ldots, \mathbf{P}_Q$ such that (τ is the time interval between two switches of routing modes)*

$$(e^{\tau \tilde{A}_q})^T \mathbf{P}_q e^{\tau \tilde{A}_q} - \mathbf{P}_{q-1} < 0, \quad \forall q = 2, \ldots, Q, \tag{7.21}$$

and

$$(e^{\tau \tilde{A}_1})^T \mathbf{P}_1 e^{\tau \tilde{A}_1} - \mathbf{P}_Q < 0, \tag{7.22}$$

where $\tilde{\mathbf{A}}_q \triangleq \mathbf{A} + \mathbf{B}\mathbf{K}_q\mathbf{C}$, the system is stable.

Remark 16. Note that the matrices $\{\mathbf{P}_q\}_{q=1,\ldots,Q}$ are subsidiary variables and do not have direct physical meanings. The conditions of negative definiteness in Eq. (7.21) and Eq. (7.22) can be analyzed and verified using the powerful theory of linear matrix inequalities (LMIs) [133]. The mode sequence in the proposition is an arbitrary one. Hence the verification of stability given the pool of communication modes may need an exhaustive or heuristic search of the optimal mode sequence.

Hence when the mode sequence is fixed, we can use the conditions in Eqs. (7.21) and (7.22) to look for the feedback matrices. When the mode sequence is not fixed (i.e., it can be adaptive to the system state), a sufficient condition of the system stability can be found in Ref. [134].

7.3.3 MODE SCHEDULING

Mode scheduling means online adaptation of the communication mode to the system state, given the pool of communication modes obtained in the offline procedure of mode provisioning. We will first introduce existing works on mode scheduling in generic hybrid systems. Then we introduce an optimization framework for the mode scheduling.

Centralized mode scheduling

For the generic case, mode scheduling is carried out in the following steps:

1. Define a cost function for the scheduling.
2. For each scheduling period, the scheduler receives certain observations on the current system state.
3. Based on the current system state, the scheduler chooses the communication mode that minimizes the expectation of the cost function using DP (or approximate dynamic programming (ADP) if needed).
4. If the scheduling is in a distributed manner, quantities characterizing the expected cost will be broadcast.

Note that the scheduler can be centralized or decentralized. The procedure will be discussed in the subsequent discussions.

There have been many studies on mode scheduling in generic hybrid systems. For example, Ref. [135] studied the optimization of both states by decomposing the optimization into master and slave problems. A special case of mode scheduling is to schedule the observations of sensors, which is called sensor selection (or sensor switching) in the field of control theory [136]. In the seminal work of Ref. [137], it has been shown that, for linear quadratic Gaussian (LQG) control, which is a very popular framework of controller synthesis due to the explicit and efficient computation of the control actions, the optimal sequence of selected sensors (it is assumed that only one sensor can be selected at a time) can be determined in advance and is independent of the observations. In Ref. [122], the communication delay is taken into account based on delayed Kalman filtering. The sensor selection problems have also been studied in many other publications [138].

We define the scheduling law as a mapping from the system state to the selection of communication mode. We want the scheduling law to minimize the expectation of the system cost, i.e.,

$$k_1^*(\mathbf{x}(0)) = \arg\min_{k_1} E[J|k_1, \mathbf{x}(0)], \tag{7.23}$$

where $\mathbf{x}(0)$ is the feedback of the system state, k_1 is the first mode selection when seeing the system state $\mathbf{x}(0)$, and J is a cost function. Intuitively, the scheduler determines the selection of communication mode to minimize the expected future cost J given the current system observation.

For simplicity, we assume that the cost function is the sum of quadratic functions, i.e.,

$$J = \sum_{t=1}^{T_f} \mathbf{x}^T(t)\mathbf{P}\mathbf{x}(t), \tag{7.24}$$

where $\mathbf{P}$ is a positive definite matrix and T_f is the total number of time slots under consideration. This cost function is reasonable if we desire a small norm of $\mathbf{x}(t)$ and different dimensions of $\mathbf{x}$ have different priorities. This is reasonable if $\mathbf{x}$ denotes the deviation of a system from a desired operation state. We assume that the communication mode can be changed every T_s time slots (which is called a scheduling period) and $t_f = T_f/T_s$ is an integer; i.e., within the window of optimization the scheduler can change the communication mode t_f times. For example, in practical cellular systems, T_s can be the period of each data frame (10 ms in 4G LTE systems).

Theoretically, the optimal scheduling law can be obtained from DP [139]. Carrying out the standard DP, we define the cost-to-go function $J_t^*(\mathbf{x}(t))$, $t = 1, \ldots, t_f$, as

$$\begin{aligned} &J_t^*(\mathbf{x}((t-1)T_s)) \\ &= \min_{k_t} E\left[\sum_{\tau=(t-1)T_s+1}^{T_f} \mathbf{x}^T(\tau)\mathbf{P}\mathbf{x}(\tau) \,\middle|\, \mathbf{x}((t-1)T_s)\right], \end{aligned} \tag{7.25}$$

which is the minimum expected cost in the future given the observation at the beginning of the scheduling period, i.e., $\mathbf{x}((t-1)T_s)$. Applying Bellman's equation in DP [139], we have

$$\begin{aligned} &J_t^*(\mathbf{x}((t-1)T_s) \\ &= \min_{k_t} \left\{ E\left[\sum_{\tau=(t-1)T_s+1}^{tT_s} \mathbf{x}^T(\tau)\mathbf{P}\mathbf{x}(\tau) \,\middle|\, \mathbf{x}((t-1)T_s), k_t\right] \right. \\ &\quad \left. + E\left[J_{t+1}^*(\mathbf{x}(tT_s)|\mathbf{x}((t-1)T_s), k_t)\right] \right\}, \end{aligned} \tag{7.26}$$

for $t = 1, \ldots, t_f - 1$, and the final condition is

$$J_{t_f}^*(\mathbf{x}((t_f-1)T_s) = \min_{k_{t_f}} E\left[\sum_{\tau=(t_f-1)T_s+1}^{T_f T_s} \mathbf{x}^T(\tau)\mathbf{P}\mathbf{x}(\tau) \,\middle|\, \mathbf{x}((t_f-1)T_s), k_t\right]. \quad (7.27)$$

Intuitively, $J_t^*(\mathbf{x}((t-1)T_s)$ is the minimal expected cost from the tth scheduling to the final time slot, given the final system state at the previous scheduling period. To obtain the cost-to-go functions, we can begin from the last scheduling period in Eq. (7.27) and compute the cost-to-go functions in a time-reversed order. The computation of $J_{t_f}^*(\mathbf{x}((t_f-1)T_s)$ can be an exhaustive search for every possible communication mode; for each communication mode, the cost within the T_s time slots can either be obtained from Monte-Carlo simulations or explicit expressions (if the noise is Gaussian). Once the cost-to-go functions are obtained, it is straightforward to obtain the optimal decision (i.e., the scheduling law):

$$k_1^*(\mathbf{x}(0)) = \arg\min_{k_t}\left\{E\left[\sum_{\tau=1}^{T_s} \mathbf{x}^T(\tau)\mathbf{P}\mathbf{x}(\tau)\middle| \mathbf{x}(0), k_t\right] + E\left[J_{t+1}^*(\mathbf{x}(T_s)|\mathbf{x}(0), k_1)\right]\right\}, \quad (7.28)$$

which can be computed using an exhaustive search among all possible selections of communication modes.

Although DP can theoretically solve the mode scheduling problem, the corresponding computation is infeasible in practice. The reason is that $\mathbf{x}$ is a continuous vector. To obtain the cost-to-go functions for every $\mathbf{x}$, we have to compute uncountably many functions, which is impossible in practical systems.

To alleviate the computational challenge, we can apply ADP [140]. We can assume an explicit expression for the cost-to-go function, which is usually chosen as a mathematically tractable form. For example, in Ref. [141], the cost-to-go function for communication mode scheduling is assumed to be a quadratic function, i.e.,

$$J_t(\mathbf{x}) = \mathbf{x}^T \Sigma_t \mathbf{x}, \quad (7.29)$$

where Σ_t is a positive definite matrix. Then the problem of estimating the cost-to-go functions is converted to the estimation of the matrices $\Sigma_1, \ldots, \Sigma_{t_f}$. An algorithm for estimating the matrices $\{\Sigma_t\}_t$ is proposed in Ref. [130], which is based on Monte-Carlo simulation and matrix equations. We also define the conditional cost

$$J_t(\mathbf{x}|\mathbf{z}) = \mathbf{x}^T \Sigma_{t,\mathbf{z}} \mathbf{x}, \quad (7.30)$$

where $\Sigma_{t,\mathbf{z}}$ is a symmetric matrix dependent on $\mathbf{z}$ and t, and $\mathbf{z} = (z_1, \ldots, z_N)$ is the transmission status of the sensors ($z_i = 1$ or 0 means that sensor i is active or not).

Distributed scheduling

In the previous discussion of the scheduling algorithm, an implicit assumption is the centralized scheduling; i.e., a centralized scheduler observes the system state $\mathbf{x}$

and then maps it to the decision of the communication mode selection. However, such centralized scheduling needs to convey the sensor observations to the scheduler, which may incur much communication overhead and is difficult in many practical applications. Hence it is necessary to study the decentralized scheduling of the communication mode; i.e., the sensors determine their communication modes based on their local observations. Below, we use MAC layer scheduling as an example to illustrate decentralized scheduling.

We can take MAC layer scheduling as an example of distributed scheduling. We assume that the scheduling follows the above optimization framework and the quadratic approximation of the cost-to-go functions in Eq. (7.29) has been obtained in offline computations. Then we assume that the positive definite matrix Σ_1 has dominant diagonal elements; i.e., the diagonal elements have much larger magnitude than the off-diagonal elements. This assumption has been demonstrated to be valid in numerical simulations of voltage control in microgrids [130].

Then the cost can be approximately decomposed to the sum of local costs, which is given by

$$J_1(\mathbf{x}, \mathbf{z}) \approx \sum_{n=1}^{N} \sigma_{n,z_n} x_n^2, \tag{7.31}$$

where

$$\sigma_{n,z_n}|_{z_n=z} = \frac{1}{\#(z_n = z)} \sum_{z_n=z} \left(\Sigma_{1,\mathbf{z}}\right)_{nn}, \tag{7.32}$$

where $\#(z_n = z)$ is the number of matrices having $z_n = z$. Recall that $\Sigma_{1,\mathbf{z}}$ is the symmetric matrix for approximating the conditional cost in Eq. (7.30). The cost can be further written as

$$\begin{aligned} J_1(\mathbf{x}, \mathbf{z}) &= \sum_{n=1}^{N} \left(\sigma_{n,1} - \sigma_{n,0}\right) x_n^2 I(z_n = 0) \\ &= \sum_{n=1}^{N} c_n I(z_n = 0), \end{aligned} \tag{7.33}$$

where $I(z_n = 0)$ is the characteristic function of the event $z_n = 0$ (i.e., node n is not scheduled) and

$$c_n = (\sigma_{n,1} - \sigma_{n,0}) x_n^2, \tag{7.34}$$

which is the cost contributed by node n if it is not scheduled. Then, the problem is converted to the following one: every node (say node n) is associated with a potential cost c_n; if node n is scheduled, the cost is zero; otherwise, the cost is c_n; then, we study how to arrange the transmission within the constraint of communication interference such that the sum cost is minimized. If we define the total reward as the negative of the total cost, then the problem is essentially the same as the problem of maximizing the total throughput of the network, or more abstractly the problem of maximizing the total weight of matching in a graph. Many distributed algorithms for the maximization of weighted matching [142], such as a greedy algorithm, can be applied to the distributed scheduling in CPS communications. Procedure 4 shows simple distributed scheduling by broadcasting a number of bits for scheduling before data transmission, in which it is assumed that a mechanism of broadcasting has been implemented (e.g., each transmitter is assigned a very short time slot).

PROCEDURE 4 DISTRIBUTED SCHEDULING BY BROADCASTING

1: The sensors compute the cost matrices $\Sigma_{t,\mathbf{z}}$.
2: **for** Each scheduling period **do**
3: **for** Every sensor **do**
4: Calculate σ_{n,z_n} in Eq. (7.32).
5: Calculate the cost c_n in Eq. (7.34) and quantize it using the assigned number of bits.
6: Broadcast the quantized c_n.
7: **end for**
8: **for** Every sensor **do**
9: Compare the local cost with received broadcast messages.
10: **if** Its cost is higher than those of the nodes it may interfere **then**
11: Transmit its packet.
12: **else**
13: Do not transmit.
14: **end if**
15: **end for**
16: **end for**

7.4 SCHEDULING: OTHER APPROACHES

There are also other approaches to handle MAC layer scheduling in the context of a CPS. In this section, we introduce two of them, namely optimization-based scheduling and effective information-based scheduling.

7.4.1 OPTIMIZATION-BASED SCHEDULING

Ref. [22] proposed a framework to optimize the scheduling issue in linear dynamical systems.

System model

The CPS system considered in Ref. [22] consists of a linear time-invariant (LTI) dynamical system and a communication network, which is illustrated in Fig. 7.7.

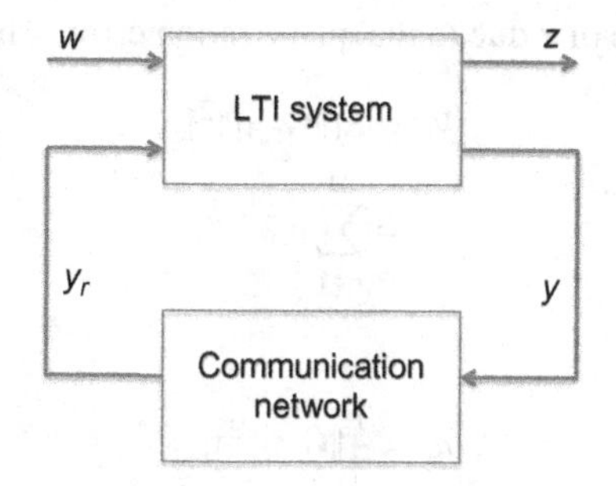

FIG. 7.7

System model of LTI system and communication network.

The system dynamics is given by

$$\begin{cases} \mathbf{z} = \mathbf{G}_{11}(\psi)\mathbf{w} + \mathbf{G}_{12}(\psi)\mathbf{y}_r, \\ \mathbf{y} = \mathbf{G}_{21}(\psi)\mathbf{w} + \mathbf{G}_{22}(\psi)\mathbf{y}_r, \end{cases} \tag{7.35}$$

where $\{\mathbf{G}_{ij}\}$ are the LTI operations that are dependent on the design parameter ψ. It is assumed that $\mathbf{y}(t)$ and $\mathbf{y}_r(t)$ are both M-dimensional real vectors. $\mathbf{y}$ consists of M scalar signals, $y_1, \ldots, y_M$, sent through the communication network at each time slot, while $\mathbf{y}_r$ is the received signals $y_{r1}, \ldots, y_{rM}$. $\mathbf{y}$ and $\mathbf{y}_r$ are related by a memoryless scalar quantizer.

A uniform quantizer is used, which partitions the interval $[-1, 1]$ into 2^b segments. It is assumed that each scalar signal y_i has an amplitude bound s_i. Hence the signal y_i is put into the quantizer after being normalized by s_i, namely

$$y_{ri}(t) = s_i Q_{b_i}\left(\frac{y_i(t)}{s_i}\right), \tag{7.36}$$

where Q_{b_i} denotes the uniform quantization function with b_i bits. The quantization error is given by

$$q_i(t) = y_{ri}(t) - y_i(t), \tag{7.37}$$

which is assumed to be uniformly distributed in the interval $s_i[-2^{-b_i}, 2^{-b_i}]$.

Then the system dynamics can be rewritten as

$$\begin{cases} \mathbf{z} = \mathbf{G}_{zw}(\psi)\mathbf{w} + \mathbf{G}_{zq}(\psi)\mathbf{q}, \\ \mathbf{y} = \mathbf{G}_{yw}(\psi)\mathbf{w} + \mathbf{G}_{yq}(\psi)\mathbf{q}, \end{cases} \tag{7.38}$$

where $\mathbf{G}_{zw}$, $\mathbf{G}_{zq}$, $\mathbf{G}_{yw}$, and $\mathbf{G}_{yq}$ are closed-loop transfer matrices. Since $\mathbf{w}$ is independent of $\mathbf{q}$, the variance of $\mathbf{z}$ is then given by

$$\begin{aligned} V &= E[\|\mathbf{z}\|^2] \\ &= V_q + E[\|\mathbf{G}_{zw}\mathbf{w}\|^2], \end{aligned} \tag{7.39}$$

where V_q is the variance of $\mathbf{z}$ due to the quantization error, which is given by

$$\begin{aligned} V_q &= E[\|\mathbf{G}_{zq}\mathbf{q}\|^2] \\ &= \sum_{i=1}^{M} a_i 2^{-2b_i}, \end{aligned} \tag{7.40}$$

with

$$a_i = \frac{1}{3}\|\mathbf{G}_{zqi}\|^2 s_i^2. \tag{7.41}$$

Communication constraints

The vector of bits allocated to every quantized signal is denoted by $\mathbf{b}$ while the corresponding communication rates are denoted by $\mathbf{r}$ (bits per second). It is easy to verify that

$$b_i = \frac{c_s r_i}{f_s}, \tag{7.42}$$

where f_s is the sampling frequency and c_s is the channel coding rate (information bits per coded bit). The following constraints are used for the bit allocations:

$$\begin{cases} f_i(\mathbf{b}, \theta) \le 0, \quad i = 1, \ldots, m_f, \\ \mathbf{h}_i^T \theta \le d_i, \quad i = 1, \ldots, m_h, \\ \theta_i \ge 0, \quad i = 1, \ldots, m_\theta, \\ b_{il} \le b_i \le b_{iu}, \quad i = 1, \ldots, M. \end{cases} \tag{7.43}$$

These constraints are explained as follows:

- We assume that functions $\{f_i\}$ are convex; they are monotonically increasing in b and monotonically decreasing in θ. They represent the constraints on the communication capacities. For example, in the Gaussian noise channel, we have

$$f(b, W, P) = b - \alpha W \log_2\left(1 + \frac{P}{NW}\right) \le 0, \tag{7.44}$$

  where W is the bandwidth, P is the transmission power, N is the noise power spectrum density, and α is the discount on the Shannon capacity due to the performance loss in practical coding schemes.
- The constraints $\mathbf{h}_i^T \theta \le d_i$ represent the limitations on the communication resources. Take the broadcast channel as an example. Suppose that the total available power is P and the total bandwidth is W. Then the constraints become

$$W_1 + W_2 + \cdots + W_n = W \tag{7.45}$$

  and

$$P_1 + P_2 + \cdots + P_n = P. \tag{7.46}$$

- The constraints $\theta_i \geq 0$ mean that the allocated communication resource must be nonnegative. For example, the transmission power and bandwidth must be nonnegative.
- The constraints $b_{il} \leq b_i \leq b_{iu}$ put limitations on the communication resource that can be allocated to one agent.

Then the resource allocation/scheduling is formulated as an optimization problem, which is given by

$$\begin{array}{lll} & \min_{\{b_i\}} \sum_{i=1}^{M} a_i 2^{-2b_i} & \\ \text{s.t.} & f_i(\mathbf{b},\theta) \leq 0, & i=1,\ldots,m_f \\ & \mathbf{h}_i^T \theta \leq d_i, & i=1,\ldots,m_h \\ & \theta_i \geq 0, & i=1,\ldots,m_\theta \\ & b_{il} \leq b_i \leq b_{iu}, & i=1,\ldots,M. \end{array} \tag{7.47}$$

7.4.2 EFFECTIVE INFORMATION-BASED SCHEDULING

Now we propose the concept of virtual queues, which was proposed in Ref. [143] and serves as an interface between the communications and control subsystems. The key idea is as follows: traditional scheduling algorithms, such as the MaxWeight [40], consider the number of queuing packets in different nodes, since the packets have the same importance in pure data communication networks; however, in the context of a CPS, different packets containing observations on the physical dynamics have different levels of importance for the purpose of control; or equivalently, they bring different amounts of information to the controllers; hence each relay node can compute the equivalent number of bits in the queuing packets and then obtain the length of the virtual queue.

The key challenge in computing the virtual queues is to evaluate the importance of each packet. We will first introduce delay-tolerant Kalman filtering and then provide a detailed algorithm for the virtual queue computation.

Delay-tolerant Kalman filtering

Here, we provide a brief introduction to delay-tolerant Kalman filtering, which was originally proposed in Ref. [122]. It is well known that, in traditional Kalman filtering, the expectation and covariance of **x** are given by

$$\bar{\mathbf{x}}(t|t-1) = \mathbf{A}\bar{\mathbf{x}}(t|t) + \mathbf{B}\mathbf{u}(t), \tag{7.48}$$

$$\Sigma(t|t-1) = \mathbf{A}\Sigma(t-1|t-1)\mathbf{A}^T + \mathbf{W}, \tag{7.49}$$

$$\Sigma(t|t) = [\Sigma^{-1}(t|t-1) + \mathbf{C}^T\mathbf{N}^{-1}\mathbf{C}]^{-1}, \tag{7.50}$$

and

$$\begin{aligned} \bar{\mathbf{x}}(t|t) &= \bar{\mathbf{x}}(t|t-1) \\ &\quad + \Sigma(t|t)\mathbf{C}^T\mathbf{N}^{-1}(\mathbf{y}(t) - \mathbf{C}\bar{\mathbf{x}}(t|t-1)), \end{aligned} \tag{7.51}$$

where the subscripts $m|n$ ($m \geq n$) denote the quantities for time m given observations until time n.

According to Ref. [144], Eqs. (7.50) and (7.51) can be rewritten as

$$\Sigma^{-1}(t|t) = \Sigma^{-1}(t|t-1) + \sum_{m=1}^{M} \mathbf{c}_{\mathrm{m}}^T \mathbf{N}_{mm}^{-1} \mathbf{c}_{\mathrm{m}}, \tag{7.52}$$

where $\mathbf{c}_{\mathrm{m}}$ is the mth row of matrix $\mathbf{C}$ and

$$\Sigma^{-1}(t|t)\bar{\mathbf{x}}(t|t) = \Sigma^{-1}(t|t)\bar{\mathbf{x}}(t|t-1) + \sum_{m=1}^{M} \mathbf{c}_{\mathrm{m}}^T \mathbf{N}_{mm}^{-1} y_{\mathrm{m}}, \tag{7.53}$$

which is called the information form of Kalman filtering [144].

Based on the information form of Kalman filtering, the delay-tolerant Kalman filtering maintains a window lasting for L time slots and a set of buffers. Whenever observations, which could have been delayed, are received, these buffers are updated and then are used to update the estimations and error matrices.

Definition of virtual queues

In delay-tolerant Kalman filtering, different messages have different levels of importance and provide different amounts of information to the controller. We first evaluate the amount of information in each message and then define the virtual queue.

Information bits

Given a measurement y that has not been received and a set of measurements received at the controller $\mathcal{Y}$ ($y \notin \mathcal{Y}$), its amount of information is evaluated by computing the mutual information between y and $\hat{\mathbf{x}}$ [122],[2] i.e.,

$$I(y; \hat{\mathbf{x}}|\mathcal{Y}) = E\left[\log\left(\frac{p(\hat{\mathbf{x}}|\mathcal{Y}, y)}{p(\hat{\mathbf{x}}|y)}\right)\right], \tag{7.54}$$

where $\hat{\mathbf{x}}$ is the system state estimation and the expectation is with respect to the randomness in the observation. Obviously, if the mutual information $I(y; \hat{\mathbf{x}})$ is large, the common information between the observation y and the estimation based on other observations is large, which implies that y can provide much information for system state estimation, once being delivered at the controller, and thus is very important. On the other hand, if the mutual information is small, the measurement y cannot provide significant new information for the purpose of system state estimation. Hence it is reasonable to use $I(y; \hat{\mathbf{x}})$ to measure the information bits (although not rigorously) that the measurement y carries for the purpose of system state estimation.

[2]Note that here $\mathcal{Y}$ is known while y is assumed to be random.

In Ref. [122], it was shown that the mutual information in Eq. (7.54) is given by

$$I(y;\hat{\mathbf{x}}) = \frac{1}{2}E\left[\log\left|\Sigma\Sigma_{\mathcal{Y},y}^{-1}\right|\right], \tag{7.55}$$

where Σ and $\Sigma_{\mathcal{Y},y}$ are the error covariance matrices without and with the observation y, and the expectation is with respect to the randomness of y.

When the value of y is known, e.g., y_0, we claim that the following quantity specifies the number of bits the observation y can bring to the estimation:

$$i(y_0;\hat{\mathbf{x}}) = \frac{1}{2}\log\left|\Sigma\Sigma_{\mathcal{Y},y_0}^{-1}\right|. \tag{7.56}$$

The expression in Eq. (7.56) is intuitive. When $\mathbf{x}$ is scalar, Eq. (7.56) becomes

$$i(y_0;\hat{\mathbf{x}}) = \frac{1}{2}\log\frac{\sigma}{\sigma_{\mathcal{Y},y_0}}, \tag{7.57}$$

where σ is the mean square error (MSE) of $\mathbf{x}$. When i is large, the MSE given $\mathcal{Y}$ and y_0 is much smaller than that given only $\mathcal{Y}$, which implies that the reception of y_0 can significantly reduce the MSE of $\mathbf{x}$.

Based on the discussion of the information bits brought by each measurement, we can define the virtual queue. We assume that a transmitter has R measurements denoted by $y^1, y^2, \ldots, y^R$ in its buffer, all waiting to be transmitted. The set of measurements that the destination controller has received in the history (including the R measurements) is denoted by $\mathcal{Y}$. Then, we assume that the number of bits in its virtual queue is given by

$$Q = \sum_{k=1}^{R} i(y^k;\hat{\mathbf{x}}). \tag{7.58}$$

Distributed scheduling

Based on the concept of virtual queues, we propose a distributed scheduling algorithm for the purpose of stabilizing the physical dynamics in a CPS. We first describe the procedure of evaluating the queue length at each transmitter. Then we modify and apply the back-pressure scheduling algorithm.

In the distributed scheduling, each transmitter (either a source node or an intermediate node) evaluates the length of its virtual queue. To that purpose, it needs to estimate Σ and $\Sigma_{\mathcal{Y},y}$ for a measurement y. Here $\mathcal{Y}$ is the set of measurements the controllers have received. The following two actions will be taken by the node:

- Estimation of $\mathcal{Y}$: The node may not be able to know whether a measurement that it has transmitted has been received by the destination node since the acknowledgment (ACK) could be lost or could be on the way. It can either use the ACKs it has received to estimate $\mathcal{Y}$, which is more conservative, or simple assume that all the measurements it has transmitted have been received, which is more risky. We will use the latter approach in this chapter.

- Computation of Σ and $\Sigma_{\mathcal{Y},y}$: The node carries out delay-tolerant Kalman filtering in order to compute the above two matrices.

Once the virtual queue length is evaluated at each node, they can use the queue length for the scheduling. In this chapter, we use a heuristic algorithm. As we have assumed, before the transmission of a packet, there is a scheduling period. We assume that each node can broadcast its current virtual and real queue lengths to its neighbors during the scheduling period. The detailed design of the queue length broadcast is omitted here. Once receiving the queue lengths, each node can compute the back pressures of itself and its neighbors. Then a node transmits only when there is no neighbor having a higher back pressure. However, it is not straightforward to apply the traditional concept of back pressure.

In traditional queuing networks with multiple flows, the back pressure of node i is given by

$$b_i = \max_{f \in F_i, j=n(f,i)} \mu_{ij}(q_i(f) - q_j(f)), \tag{7.59}$$

where μ_{ij} is the service rate of the link from node i to node j, F_i is the set of flows flowing through node i, $q_i(f)$ is the queue length of flow f at node i, and $n(f,i)$ is the next hop neighbor of node i for flow f. In the context of a CPS with physical dynamics, we have the following concerns:

- Virtual queue and actual queue: In Eq. (7.59), $q_i(f)$ represents the motivation to transmit the packet while $q_j(f)$ is the impedance to prevent packet transmission. In this chapter, we propose to use the virtual queue length for q_i, which measures the desire to transmit, and use the actual queue length for q_j, which means the congestion of the downstream node.
- Multicasting: In the scenario of this chapter, one transmission can possibly send the packet to multiple next-hop neighbors. Then, it is not intuitive what should be used for q_j in Eq. (7.59). We will test three possible options, namely the average, the maximum, or the minimum of the actual queue lengths of the next-hop neighbors.
- Estimation of $\mathcal{Y}$: When evaluating the virtual queue length, the node needs to know the set of received packets at the controllers, namely $\mathcal{Y}$, such that it can evaluate the importance of each packet in the queue. In this chapter, we assume that $\mathcal{Y}$ is the set of packets the node has sent out, with the implicit assumption that all packets have been received by the controllers.

Summarizing the above discussions, we define the back pressure at node i as

$$b_i = \max_f h\left(\{\mu_{ij}\}_{j \in n(i,f)}\right)\left(Q_i(f) - h\left(\{q_{ij}(f)\}_{j \in n(i,f)}\right)\right), \tag{7.60}$$

where the function h is average, max or min, Q is the virtual queue length, and q is the real queue length. Note that it is difficult to derive this definition of back pressure from first principles. However, numerical simulations will demonstrate the validity of this framework. The scheduling algorithm is summarized in Procedure 5.

PROCEDURE 5 DISTRIBUTED SCHEDULING

1: **for** Each time slot **do**
2: **for** Each transmitter having packets to transmit **do**
3: Estimate the set of received measurements $\mathcal{Y}$.
4: **for** Each measurement y in the queue **do**
5: Use delay-tolerant Kalman filtering to estimate the error covariance matrices Σ and $\Sigma_{\mathcal{Y},y}$.
6: Use Eqs. (7.56) and (7.58) to compute the number of equivalent bits.
7: **end for**
8: Compute the total length of the virtual queue and broadcast it.
9: **end for**
10: Broadcast the lengths of the virtual and actual queue lengths.
11: Compute the back pressure using Eq. (7.60).
12: Transmit the measurement having the maximum information, or do not transmit.
13: **end for**

The proposed scheduling algorithm has been tested in the context of smart grids and has achieved good performance [143].

7.5 ROUTING

Routing is an important issue in communication networks; it means finding a path from the source to the destination(s). The following three situations are possible for routing:

- Unicasting: There is only one destination for each packet. The routing procedure needs to determine only one path or, in some cases that require high reliability, one working path and one or more backup paths.
- Multicasting: There are multiple destinations for each packet. Hence the routing procedure needs to determine multiple paths simultaneously, in which a destination node may relay data to other destination nodes.
- Broadcasting: The data needs to be sent to all the nodes in the network, which is a special case of multicasting.

There are many existing practical protocols, and there have been numerous academic studies on routing in networks. However, most of them are focused on the routing in pure data networks, such as the Internet or wireless ad hoc networks. However, there have been very few studies on routing issues in CPSs; due to the following significant differences between pure data communication networks and CPSs, there is a pressing need to study routing in CPSs:

- In pure data communication networks, the goal of routing is usually to minimize the communication delay. However, it has been found that in CPSs a delay distribution with less expected delay or maximum delay may not result in optimal performance.

- In pure data communication networks, the destinations of a source node are predetermined. However, in CPSs, it is not clear which subset of controllers a sensor should send its data to. In the ideal case, a sensor should broadcast its data to all controllers. Due to the limited communication resources, a sensor needs to select a subset of controllers and then carry out the routing procedure. A joint procedure of destination selection and routing is needed in the context of CPSs.

Essentially, these differences originate from different goals of a CPS and a pure data communication network. Although it is reasonable to apply existing routing algorithms to CPSs, it is of much research interest to exploit the awareness of physical dynamics in the routing procedure.

7.5.1 ESTIMATION ORIENTED ROUTING

We first consider the case of unicasting and follow Ref. [24] to discuss the routing procedure for system state estimation. Although we do not consider the control procedure in a CPS, system state estimation is a very important stage in the control of stochastic dynamical systems.

System model

We assume that the system state evolution law is given by the following linear and discrete-time equation:

$$\mathbf{x}(t+1) = \mathbf{A}\mathbf{x}(t) + \mathbf{w}(t), \tag{7.61}$$

where $\mathbf{w}(t)$ is white Gaussian noise with covariance matrix Σ_w. The observation measured by a sensor is given by

$$\mathbf{y}(t) = \mathbf{C}\mathbf{x}(t) + \mathbf{v}(t), \tag{7.62}$$

where $\mathbf{v}$ is also white Gaussian noise with covariance matrix Σ_v. The sensor sends the observations to an estimator, which estimates the system state based on the received observation, via a communication network. It is assumed that the sensor encodes the observation into an n_d-dimensional real vector $\mathbf{s}$ at each time slot. The message sent out by the sensor, $\mathbf{s}(t)$, is a function of all the previous observations at the sensor.

Note that, in digital networks, it is impossible to transmit a real number without any distortion. However, the analysis on quantized real numbers is challenging. Hence we assume that the real vectors can be sent directly, which is reasonable if the number of bits used to encode the vector is large and the quantization error is negligible.

The communication network incurs communication delay (denoted by $d(t)$) of the message $\mathbf{s}(t)$. We assume that the communication delay is dependent on the previous τ delays (hence the memory has a length τ); i.e.,

$$P(d(t)|d(t-1),\ldots,d(0)) = P(d(t)|d(t-1),\ldots,d(t-\tau)). \quad (7.63)$$

It is not assumed that the encoder has knowledge about the previous delays. The estimator uses all the previously received signals to estimate the system state, and the corresponding error covariance is given by

$$\mathbf{P}(t) = E[(\mathbf{x}(t) - \hat{\mathbf{x}}_{\text{dec}}(t))(\mathbf{x}(t) - \hat{\mathbf{x}}_{\text{dec}}(t))^T], \quad (7.64)$$

where $\hat{\mathbf{x}}_{\text{dec}}(t)$ is the estimation of $\mathbf{x}(t)$. Because of the time-varying channel qualities, the error covariance $\mathbf{P}(t)$ can be considered as a random variable, whose moments should be characterized. In Ref. [24], the expectation of $\mathbf{P}$ is studied.

If the packets go through a network to reach the destination, it is assumed that different communication links incur independent delays to the packet. Then the goal of routing is to find a path from the sensor to the controller for minimizing the error covariance in the steady state.

Encoder and decoder

The following encoder and decoder are assumed in Ref. [24]:

- Encoder: At time t the sensor estimates the system state from the observations $\{\mathbf{y}(s)\}_{s\leq t}$ by using Kalman filtering, and sends out the estimate $\hat{\mathbf{x}}(t)$.
- Decoder: At time t, the estimator obtains the estimation of the system state by using different approaches in the following different situations:
  - If no packet arrives, the estimator updates the estimate using

$$\hat{\mathbf{x}}_{\text{dec}}(t) = \mathbf{A}\hat{\mathbf{x}}_{\text{dec}}(t-1). \quad (7.65)$$

  - If a packet with the newest time stamp (newer than all the other previously received packets) is sent out at time m, the estimator updates the system state estimate by using

$$\hat{\mathbf{x}}_{\text{dec}}(t) = \mathbf{A}^{t-m}\hat{\mathbf{x}}(m). \quad (7.66)$$

    Note that there may be other packets with older time stamps arriving. They will be discarded.
  - If one or more packets arrives with time stamps older than a previously received one, the estimator discards these packets and uses Eq. (7.65) to update the system state estimation.

The following theorem guarantees the optimality of the above encoder and decoder scheme:

Theorem 45. *At each time slot, the above encoder-decoder scheme results in the minimum error covariance, compared with all other causal designs.*

Evolution of covariance

The error covariance of system state estimation is denoted by $\Sigma(t)$, which is given by

$$\Sigma(t) = E[(\hat{\mathbf{x}}(t|\{\mathbf{y}(s)\}_{s\leq t}) - \mathbf{x}(t))(\hat{\mathbf{x}}(t|\{\mathbf{y}(s)\}_{s\leq t}) - \mathbf{x}(t))^T]. \quad (7.67)$$

It is easy to verify that the evolution of the error covariance is described by

$$\Sigma(t+1) = f(\Sigma(t)), \tag{7.68}$$

where f is the Lyapunov recursion, which is given by

$$f(\mathbf{X}) = \mathbf{AXA}^T + \mathbf{R}_w. \tag{7.69}$$

We denote by $t_s(k)$ the latest time stamp among all the packets received by time k. Then, it was shown in Ref. [24] that the expected error covariance matrix for $\mathbf{x}(t+1)$ at the estimator is given by

$$E[\mathbf{P}(t+1)] = \sum_{m=-1}^{k} P(t_s(k) = m) f_{k-m}(M(m+1)). \tag{7.70}$$

The following theorem shows a necessary condition for the stability of the encoder-decoder design:

Theorem 46. *A necessary condition for stability is*

$$\limsup_{t\to\infty} (d(0) \le k | d(1) > k-1, \ldots, d(s) > t-s)\rho^2(\mathbf{A}) < 1. \tag{7.71}$$

A straightforward conclusion from the theorem is given as follows:

Corollary 11. *If the delays of different packets are mutually independent, the condition Eq. (7.71) is simplified to*

$$\limsup_{k\to\infty} (1 - F(k))\rho^2(\mathbf{A}) < 1, \tag{7.72}$$

where $F(k)$ is the cumulative distribution function (CDF) of the delay.

An interesting observation is that the system stability is only affected by the packet loss probability and is independent of the delay profile, which is counter-intuitive.

An example in Ref. [24] is provided to show that it is not necessarily optimal to minimize the expected delay during the routing procedure. Consider the following system having scalar dynamics:

$$x(t+1) = 1.2x(t) + w(t), \tag{7.73}$$

where $w(t)$ has variance 1. The initial state has variance $P(0) = 1$. The observation of a sensor is given by

$$y(t) = x(t) + v(t), \tag{7.74}$$

where the noise $v(t)$ has power 1. Consider the following two delay profiles:

$$P_1 : P(d = 5) = 1 \tag{7.75}$$

and

$$P_2: \; P(d=m) = \begin{cases} 1/6, & m=1, \\ 5/6, & m=6. \end{cases} \tag{7.76}$$

Obviously, both delay distributions have the same expectation. However, a simple calculation shows that the error covariances of P_1 and P_2 are 23.88 and 17.34, respectively. If the expectation of P_2 is slightly increased by increasing $P(d = 6)$ (simultaneously decreasing $P(d = 0)$), the error covariance of P_2 is still smaller than that of P_1, although P_2 actually results in a larger expected delay.

To study the relationship between expected delay and error covariance more systematically, we define the modified delay r as a random variable having the following distribution:

$$P(r=-1) = \prod_{j=0}^{k} P(d(t) > j) \tag{7.77}$$

and

$$P(r=m) = P(d(t) \leq t-m) \prod_{j=0}^{t-m-1} P(d(t) > j). \tag{7.78}$$

Then it is easy to verify that

$$E(\mathbf{P}(t+1)) = \sum_{m=-1}^{t} P(r=m) f_{t-m}(\mathbf{M}(m+1)). \tag{7.79}$$

Before proceeding to the next conclusion, we need to define the stochastic dominance:

Definition 21. Consider two scalar real random variables A and B. We say that A stochastically dominates B if

$$P(A \geq x) \geq P(B \geq x), \tag{7.80}$$

for any $x \in (-\infty, \infty)$.

Since the delay distribution is the key consideration in the selection of paths in the routing procedure, the following theorem provides a principle for designing routing algorithms for the purpose of system state estimation:

Theorem 47. *Consider a network in which the sensor is located at node s. If the estimator is located at node d, the optimal path is denoted by P_1. Suppose that node n is a node on P_1 and the portion of path P_1 between s and n is denoted by P_1'. Now suppose that the estimator is moved to node n and the optimal path is P_2. Then the expected costs of system state estimation by the estimator at node n are the same for either P_1' or P_2.*

The following theorem was presented in Ref. [24], and is illustrated in Fig. 7.8:

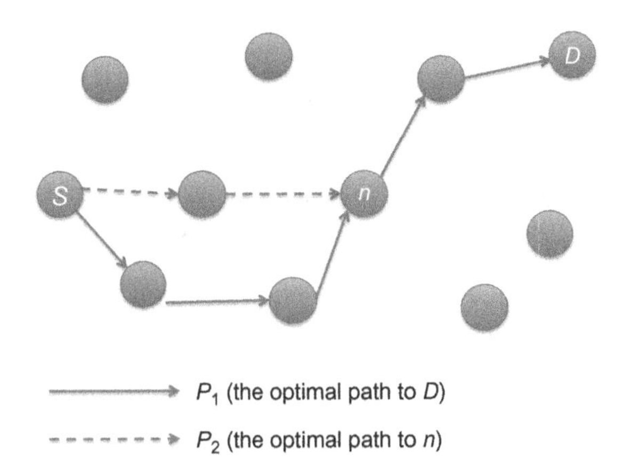

FIG. 7.8

Illustration of the paths in Theorem 47.

Theorem 48. *Two modified delay distributions* $P(r_1 = m)$ *and* $P(r_2 = m)$ *correspond to two delay distributions* $p_1(m)$ *and* $p_2(m)$*, respectively. If* $P(r_1 = m)$ *stochastically dominates* $P(r_2 = m)$*, the expected error covariance attained under* $p_1(m)$ *is no more than that of* $p_2(m)$.

Based on this theorem, the following routing algorithm was proposed in Ref. [24]. Essentially the algorithm is very similar to the shortest path ones such as the celebrated Dijkstra algorithm. The key difference is that the cost in this procedure is related to the cost of system state estimation.

7.5.2 SYSTEM DYNAMICS-AWARE MULTICAST ROUTING

Now we study multicast routing in CPSs, considering the determinations of destinations and routing modes. We can formulate routing as an optimization problem, by employing the theories of hybrid systems and LMIs, and then solve it using heuristic approaches. The introduction follows the study in Ref. [131].

System model

We assume that there are N_c controllers and N_s sensors in the CPS, all being decentralized. For simplicity, we assume that the control action at each controller is scalar, as is the observation at each sensor. This simplifies the mathematical notation and can be straightforwardly extended to the generic case of vector control actions or vector observations.

For simplicity, we assume that the dynamics of the physical subsystem is linear and free of perturbations, and given by

$$\begin{cases} \dot{\mathbf{x}}(t) = \mathbf{A}\mathbf{x}(t) + \mathbf{B}\mathbf{u}(t), \\ \mathbf{y}(t) = \mathbf{C}\mathbf{x}(t), \end{cases} \tag{7.81}$$

PROCEDURE 6 ROUTING FOR SYSTEM STATE ESTIMATION

1: Given the source node s and destination node d.
2: Assign a cost to each node in the network. The initial cost of the source node s is M^* while those of all other nodes are set to ∞.
3: Label every node as unvisited. Set the source node s as the current node.
4: **while** True **do**
5: **for** The current node **do**
6: Check the set of unvisited neighbors of the current node.
7: Initialize the smallest cost.
8: **for** Each unvisited neighbor **do**
9: Check its cost by using the following three steps.
10: Step 1. Denote by p_c the delay distribution from s to the current node and by p_n the delay distribution from the current node to the neighbor. Then the delay distribution from node s to this neighbor is given by

$$p'_c = p_c \otimes p_n,$$

where $\otimes$ is convolution.
11: Step 2. Compute the modified delay distribution using the new delay distribution p'_c.
12: Calculate the cost for this neighbor. Overwrite the smallest cost if the current cost is lower than the current smallest cost.
13: **end for**
14: **if** All neighbors have been considered **then**
15: Label the current node as visited.
16: Fix the current smallest cost.
17: **end if**
18: **end for**
19: **if** All the neighbors of the destination node d have been labeled as visited **then**
20: Break the loop and output the optimal path consisting of all the current nodes.
21: **else**
22: Set the current node to the neighbor having the lowest cost.
23: **end if**
24: **end while**

where $\mathbf{x}$ is an M-vector representing the system state; $\mathbf{u}$ is the N_c-dimensional vector of control actions where u_{n_c} stands for the control action of controller n_c; $\mathbf{y}$ is the N_S-dimensional vector of observations at the sensors where y_{n_s} is the observation at sensor n_S. The dimensions of matrices $\mathbf{A}$, $\mathbf{B}$, and $\mathbf{C}$ are $M \times M$, $M \times N_c$, and $N_S \times M$, respectively. Note that the assumption of linear dynamics is valid for linear systems and is also effective for nonlinear systems when the system state deviates only slightly from an equilibrium point. It is much more challenging to study the general nonlinear case, which is a topic for future work. We also note that we do not consider random perturbations, such as noise in the observations, in the system. The only uncertainty is the initial condition in the system dynamics (otherwise, if the initial state is also known, there is no need for communications). The study of this deterministic system will be extended to stochastic systems in future.

We assume that the N_c controllers and N_s sensors are all equipped with communication interfaces, either wired or wireless. There are also N_r relay nodes in the communication network, which can help with the delivery of observations from the sensors to the controllers. We denote the three types of nodes by $\{n_c\}_{n_c=1,\ldots,N_c}$, $\{n_s\}_{n_s=1,\ldots,N_s}$, and $\{n_r\}_{n_r=1,\ldots,N_r}$, where the subscripts represent the types of nodes. We call the data flow from a sensor to a controller a *connection*. The topology of the communication network composed of the three types of nodes is known in advance. For a wired network, the topology is determined by the existence of a wired link; for a wireless network, the topology is determined by the distances among the nodes. We use the notation $a \sim b$ to signify that nodes a and b are directly connected in the communication network.

We make the following assumptions on the communication network throughout in order to simplify the analysis:

- Fluid traffic: We consider the data flows from sensors to controllers to be continuous; i.e., we ignore the details of sampling, quantization, and possible packet dropout. Although the effects of sampling interval, quantization error, and packet drop probability on the system dynamics have been intensively studied in the area of networked control [19], it renders the analysis prohibitively complicated to consider all these details. This fluid traffic assumption can simplify the analysis and is valid when the sampling rate is high, the quantization error is very small, and the communication channels are of very good quality.
- Bandwidth constraint: We assume that transmitting the data flow of one sensor requires one unit of bandwidth. The bandwidth of the communication link between nodes a and b is assumed to be an integer and is denoted by w_{ab} (in units of bandwidth). Hence the link can support data flows from at most w_{ab} sensors:

$$\sum_{n_s=1}^{N_s} I(n_s, a, b) \leq w_{ab}, \tag{7.82}$$

  where $I(n_s, a, b)$ equals 1 if the data flow of sensor n_s passes through link ab; otherwise, $I(n_s, a, b) = 0$.
- Routing mode switching: We assume that there are a total of Q different routing modes, and that the routing mode can be switched every τ seconds. For simplicity, we consider a simple round-robin switching policy; i.e., the routing mode is selected in the order $1, 2, \ldots, Q$. The performance can be improved by adaptively selecting the routing mode. However, the mode decision needs to be carried out in a decentralized manner.

Single mode routing

We first consider the case in which there is only one routing mode (or path design). We assume that linear feedback control is employed for the CPS, which is given by

$$\mathbf{u}(t) = \mathbf{K}\mathbf{y}(t), \tag{7.83}$$

where $\mathbf{K}$ is a constant feedback gain matrix. Note that $\mathbf{K}$ is dependent on the routing mode. Since we consider decentralized control, the matrix $\mathbf{K}$ has a special structure, i.e.,

$$\mathbf{K}_{ij} = 0, \tag{7.84}$$

if there is no connection between sensor j and controller i, such that the control action u_i is independent of the observation y_j. Equivalently, the nonzero elements of the ith row of $\mathbf{K}$ correspond to the sensors having connections to controller i. For instance, in the example in Fig. 7.9, the matrix $\mathbf{K}$ is given by

$$\mathbf{K} = \begin{pmatrix} K_{11} & K_{12} & K_{13} \\ K21 & K_{22} & 0 \\ 0 & 0 & K_{33} \end{pmatrix}. \tag{7.85}$$

Substituting Eq. (7.83) into Eq. (7.81), we obtain the system dynamics, which is given by

$$\begin{aligned}\dot{\mathbf{x}}(t) &= \mathbf{A}\mathbf{x}(t) + \mathbf{BKC}\mathbf{x}(t) \\ &= \tilde{\mathbf{A}}\mathbf{x}(t),\end{aligned} \tag{7.86}$$

where $\tilde{\mathbf{A}} \triangleq \mathbf{A} + \mathbf{BKC}$.

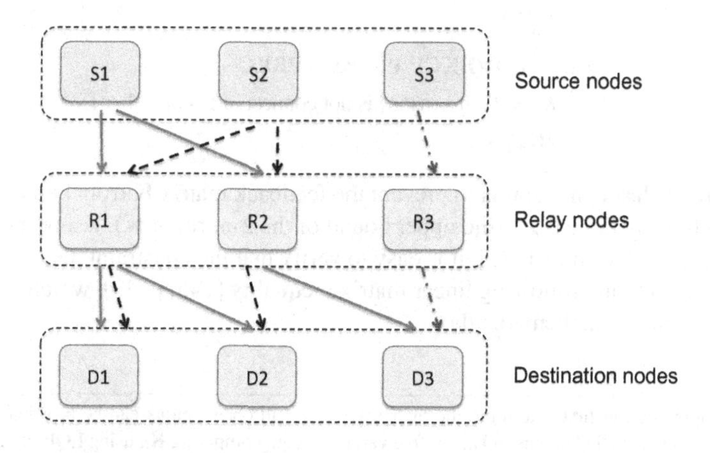

FIG. 7.9

An example of a communication network and connections.

Then we consider how to optimize the single routing mode. The key challenge is how to determine the set of destinations. Given an arbitrary routing mode, if the feedback gain matrix $\mathbf{K}$ is determined in advance,[3] we can simply check the eigenvalues of $\tilde{\mathbf{A}}$: if the real parts of all eigenvalues are in the left half of the complex plane, the system is stable (here the stability means that all trajectories of the dynamics converge to a unique equilibrium point [133]); otherwise, it is unstable. In this chapter, we consider the nontrivial design of $\mathbf{K}$ given the routing mode. First we obtain the following proposition stating a sufficient condition for system stability. The proof is very simple and a very similar one can be found on page 30 of Ref. [145].

Proposition 4. *If there exists a matrix* $\mathbf{K}$ *with the nonzero element pattern corresponding to a given routing mode, and a positive definite matrix* $\mathbf{P}$*, such that the following LMI holds*[4]*:*

$$\mathbf{A}^T\mathbf{P} + (\mathbf{BKC})^T\mathbf{P} + \mathbf{PA} + \mathbf{PBKC} < 0, \tag{7.87}$$

then the system is stable for the given routing mode.

Proposition 4 provides a sufficient condition for judging whether a given routing mode can stabilize the system. The difficulty is how to find suitable matrices $\mathbf{P}$ and $\mathbf{K}$. We take an approach similar to that in Ref. [145] via the following two steps:

1. Computation of $\mathbf{P}$: Assume that $\mathbf{A}$ is unstable.[5] Suppose that there exists a $\beta > 0$ such that $\mathbf{A} - \beta\mathbf{I}$ is unstable and there exists a positive definite matrix $\mathbf{P}$ such that

$$(\mathbf{A} - \beta\mathbf{I})\mathbf{P} + \mathbf{P}(\mathbf{A} - \beta\mathbf{I})^T = -\mathbf{I}. \tag{7.88}$$

   Note that the rationale of Eq. (7.88) is to make the feasible set of the optimization problem Eq. (7.89), which will be discussed forthwith, nonempty.
2. Computation of $\mathbf{K}$: Given $\mathbf{P}$, we obtain $\mathbf{K}$ by considering the following optimization problem:

$$\begin{aligned} &\max_{\mathbf{K}} \gamma \\ &\text{s.t. } \mathbf{A}^T\mathbf{P} + (\mathbf{BKC})^T\mathbf{P} + \mathbf{PA} + \mathbf{PBKC} + \gamma\mathbf{I} < 0 \\ &K_{ij} = 0, \text{ if sensor } j \text{ is not connected to controller } i \\ &\|\mathbf{K}\|_2 \le c_K, \end{aligned} \tag{7.89}$$

   where the last constraint is to prevent the feedback matrix $\mathbf{K}$ from being prohibitively large (c_K is the upper bound of the 2-norm of $\mathbf{K}$). Using the Schur complement formula [133], it is easy to verify that the constraint $\|\mathbf{K}\|_2 \le c_K$ is equivalent to the following linear matrix inequality [145, p. 33], which is easier to manipulate mathematically:

[3] For example, we can first determine the matrix $\mathbf{K}_0$ when all observations can be broadcasted to all controllers such that all elements in $\mathbf{K}_0$ are free variables, e.g., computing $\mathbf{K}_0$ using LQR control. Then we obtain $\mathbf{K}$ for the given routing mode by eliminating the elements from unobserved sensors.

[4] For a symmetric matrix $\mathbf{X}$, $\mathbf{X} < 0$ means that $\mathbf{X}$ is negative definite.

[5] Otherwise, there is no need for control; the system can converge to zero by itself.

$$\begin{pmatrix} -c_K\mathbf{I} & \mathbf{K}^T \\ \mathbf{K} & -\mathbf{I} \end{pmatrix} < 0. \tag{7.90}$$

It is easy to verify that, if the optimal value of Eq. (7.89) is positive, then the condition Eq. (7.87) holds. Note that the optimization problem in Eq. (7.89) can be readily solved using the theory of LMI.

Based on the above discussion, we are able to check the stability of the system given the routing mode; meanwhile, we can also construct the stabilizing feedback gain matrix $\mathbf{K}$ (if possible), as a byproduct. Now the challenge is how to find a routing mode such that the optimal γ is positive (thus stabilizing the system). Formulated as an optimization problem, the search for a routing mode stabilizing the system dynamics is to maximize γ; i.e.,

$$\max_{\mathcal{R}} \gamma(\mathcal{R})$$
$$\text{s.t. } \mathcal{R} \text{ satisfies the bandwidth constraint,} \tag{7.91}$$

where $\mathcal{R}$ stands for a routing mode and γ is a function of $\mathcal{R}$ which can be obtained in Eq. (7.89). The difficulty is that it is too complicated to analytically describe the relationship between the routing mode and the objective function γ. Currently, we still do not have an analytic approach to find the stabilizing routing mode. Moreover, it is prohibitively complicated to carry out an exhaustive search. In the next subsection, we will propose a heuristic algorithm to search for such a routing mode in a greedy manner.

Multiple routing modes

Now we consider the case in which the system can switch among multiple routing modes. Recall that there is a total of Q routing modes. We denote by $\mathbf{K}_q$ the feedback gain matrix corresponding to the qth routing mode. Then the system dynamics can be written as

$$\dot{\mathbf{x}}(t) = \tilde{\mathbf{A}}_{q(t)}\mathbf{x}(t), \tag{7.92}$$

where $\tilde{\mathbf{A}}_{q(t)} \triangleq \mathbf{A} + \mathbf{B}\mathbf{K}_{q(t)}\mathbf{C}$ and $q(t)$ is the routing mode at time t. Obviously, (8.19) represents the dynamics of a hybrid system with Q modes in which $\mathbf{x}(t)$ is the continuous system state and $q(t)$ is the discrete system state.

It is possible that a single routing mode may not stabilize the CPS or may not be rigorously shown to stabilize the system dynamics, which will be illustrated in the numerical results. In this case, we need to consider multiple routing modes and let the communication network switch among these modes. For simplicity, we assume that the feedback gain matrix $\mathbf{K}_q$ corresponding to routing mode q is obtained from Eq. (7.89). It may result in a better performance if the matrices $\{\mathbf{K}_q\}_{q=1,\ldots,Q}$ are optimized jointly. However, the optimization will be much more complicated and thus will not be studied in this book. The separate design of $\mathbf{K}_q$ is also reasonable since a larger γ means greater stability; even if the system is unstable, a larger $\gamma < 0$

implies slower divergence of the system state. Hence we adopt the matrices obtained in the single routing mode case.

The following proposition provides a sufficient condition for system stability when there are multiple routing modes. Note that the proof is similar to that of Theorem 1 in Ref. [132]. Hence the detailed proof is omitted due to limitations of space.

Proposition 5. *Suppose that the feedback gain matrix* $\mathbf{K}_q$ *and the mode sequence* $1, 2, \ldots, Q, 1, 2, \ldots$ *have been fixed. If there exist* Q *positive definite matrices* $\mathbf{P}_1, \ldots, \mathbf{P}_Q$ *such that (recall that* τ *is the time interval between two switches of routing modes)*

$$(e^{\tau \tilde{A}_q})^T \mathbf{P}_q e^{\tau \tilde{A}_q} - \mathbf{P}_{q-1} < 0, \quad \forall q = 2, \ldots, Q, \tag{7.93}$$

and

$$(e^{\tau \tilde{A}_1})^T \mathbf{P}_1 e^{\tau \tilde{A}_1} - \mathbf{P}_Q < 0, \tag{7.94}$$

the system is stable.

Since $\{\tilde{\mathbf{A}}_q\}_{q=1,\ldots,Q}$ have been fixed, the LMIs in Eqs. (7.93) and (7.94) can be easily verified and can be rewritten as the following optimization problem:

$$\begin{aligned} &\max_{\mathbf{P}_q, q=1,\ldots,Q} \sum_{q=1}^{Q} \gamma_q \\ &\text{s.t. } (e^{\tau \tilde{A}_q})^T \mathbf{P}_q e^{\tau \tilde{A}_q} - \mathbf{P}_{q-1} + \gamma_{q-1} < 0, \quad \forall q = 2, \ldots, Q, \\ &\quad (e^{\tau \tilde{A}_1})^T \mathbf{P}_1 e^{\tau \tilde{A}_1} - \mathbf{P}_Q + \gamma_Q < 0. \end{aligned} \tag{7.95}$$

The rationale of the formulation of the optimization problem is as follows. When we maximize the sum of γ_q, we are driving the variables γ_q to be positive. Once all γ_q become positive, the sufficient conditions in Proposition 5 are satisfied.

Then the problem of selecting the routing modes is formulated as the following optimization problem:

$$\begin{aligned} &\max_{\{\mathcal{R}_q\}_{q=1,\ldots,Q}} \sum_{q=1}^{Q} \gamma_q \\ &\text{s.t. } \mathcal{R}_q \text{ is feasible}, q = 1, \ldots, Q, \end{aligned} \tag{7.96}$$

where $\mathcal{R}_q$ is the qth routing mode. The rationale is to choose the routing modes to maximize the objective functions in Eq. (7.96), thus trying to make the conditions in Proposition 5 valid.

The proposed framework has been applied in the context of voltage control in microgrids in Ref. [131]. The performance is demonstrated to be better than traditional approaches.

7.6 CONCLUSIONS

In this chapter, we have studied the scheduling and routing problems in communication networks of CPSs. The following conclusions can be drawn from the studies:

- Scheduling: Essentially the scheduling of the communication resource is a joint optimization together with the continuous state of physical dynamics. Hence the theory of hybrid systems is a natural approach to handle the coexistence of discrete and continuous system states. The main challenge is the complexity of the optimization and analysis, which hinders applications in practical systems. The sensitivity to parameter and model errors is also a serious concern.
- Routing: It has been shown that the expected delay of data packets does not uniquely determine the performance of a CPS. Hence the distribution of the delay should be taken into consideration. It has also been shown that the routing procedure can be made aware of the system dynamics, through the framework of LMI. Again, the complexity and robustness are major concerns.

CHAPTER

Physical layer design

8

8.1 INTRODUCTION

We now consider the lowest layer of the communication network in cyber physical system (CPSs), namely the physical layer. Usually the main tasks in the physical layer consist of modulation and coding, which were introduced in Chapter 2. In modern communication systems, modulation and coding are usually independent of upper layer activities. However, as we will see, if the modulation and coding take into consideration the current physical dynamic state, this can improve the performance of controlling the physical dynamics, at the expense of a more complex system design. In this chapter, we ignore the complexity of system design, which is an important concern in practical system design, and study how the physical layer can be made adaptive to the physical dynamics and how much performance gain we can achieve compared with existing communication systems.

8.1.1 MODULATION

As mentioned previously, the task of modulation is to convert information bits into physical symbols. Modern communication systems, particularly wireless systems, can make the modulation scheme adaptive to the communication channel condition, thus making a tradeoff between the communication rate (a denser constellation results in a higher rate) and reliability (a denser constellation results in less reliable transmission). In the context of a CPS, the different communication rates and reliabilities can be interpreted as different packet delays and different packet loss rates. In Section 8.2 we will study the effects of delay and packet loss rate on control performance and then propose a scheme to select the appropriate modulation scheme in an adaptive manner.

8.1.2 CODING

In communications, coding consists of source and channel coding. Both will be discussed in the context of CPSs in this chapter.

Communications for Control in Cyber Physical Systems. http://dx.doi.org/10.1016/B978-0-12-801950-4.00008-1

Source coding

We know that the purpose of source coding is to convert the original information source into information bits. The main challenge of source coding in a CPS is how to adapt the source coding scheme to the physical dynamics. In Section 8.3 we consider the point-to-point case, in which there is a single sensor and a single controller. Then the source coding is extended to the distributed case in Section 8.4, where we consider multiple sensors.

Channel coding

Channel coding is used to protect the communication from noise and interference. The challenge of channel coding in a CPS is how to embed the channel coding procedure into the control procedure. In Section 8.5 we consider existing channel codes but integrate the decoding procedure into the system state estimation. In Section 8.6 the channel codes are specially designed for the purpose of control in a CPS. Since control can be considered as a special case of computing, the channel codes are designed in the context of distributed computing in Section 8.7.

8.2 ADAPTIVE MODULATION

There have been many studies on adaptive modulation in wireless communication systems. They mostly consider time-varying wireless channels. A transmitter can change its modulation scheme according to the current wireless channel conditions. For example, when the channel gain is good, the transmitter can use a denser constellation for the modulation (e.g., changing from BPSK to 64QAM), in order to better utilize the good channel quality to transmit more data; on the other hand, when the channel becomes worse, the transmitter can adjust to a more sparse constellation such that the communication is still reliable despite the bad channel condition.

Adaptive modulation in a CPS serves a different purpose. Essentially, a tradeoff is needed between the delay and reliability of communications in a CPS. In the context of a CPS, the requirements of the communication delay and reliability (e.g., the packet drop rate) may change with the system state of physical dynamics. It is desirable to make the modulation take into consideration the system state.

An example in the physical layer is given in Fig. 8.1, in which using QPSK or 8PSK for modulation generates different physical dynamics with different feedback delays and packet drop rates. Here it is assumed that the same number of bits is used to describe an observation. Hence when QPSK is used, the sensor needs more delays to transmit one packet while incurring less demodulation errors; when 8PSK is used, the delay is decreased while the reliability is worse. It is therefore desirable to study how the modulation should be adaptive to the system state of physical dynamics. Note that the proposed analysis and design are also valid for adaptive channel coding in a CPS.

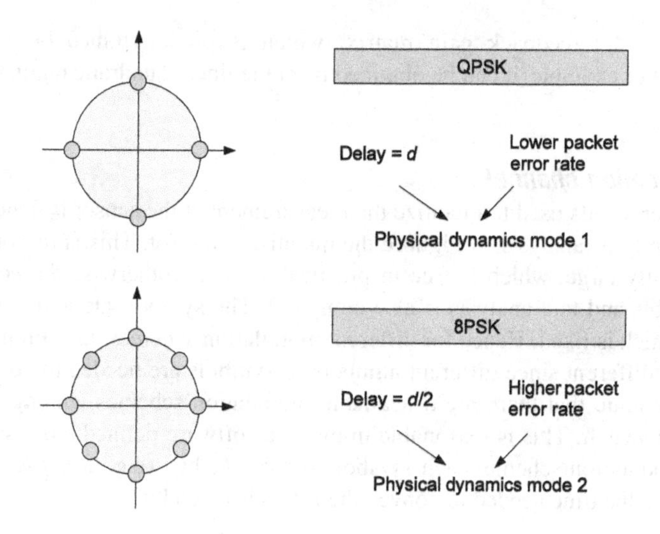

FIG. 8.1

Illustration of adaptive modulation.

In this section, we follow Ref. [146] to introduce algorithms for adaptive modulation in CPSs, which take into consideration the physical dynamics system state.

8.2.1 SYSTEM MODEL

Physical dynamics

Continuous-time physical dynamics are considered, whose evolution is described by the following differential equation:

$$\dot{\mathbf{x}}(t) = \mathbf{A}\mathbf{x}(t) + \mathbf{B}\mathbf{u}(t), \tag{8.1}$$

where $\mathbf{x}(t)$ is the N-dimensional system state, $\mathbf{u}(t)$ is the M-dimensional control action, and $\mathbf{A}$ and $\mathbf{B}$ are both system parameters which are assumed to be perfectly known. Here the physical dynamics are deterministic. The analysis of stochastic systems is much more complicated, and is beyond the scope of our discussion. A sensor is used to measure the system state $\mathbf{x}$. It is assumed that there is no measurement noise. Then the sensor sends the measurements in a digital communication channel to the controller. We assume that linear feedback control is used by the controller in the ideal case; i.e.,

$$\mathbf{u}(t) = -\mathbf{K}\mathbf{x}(t), \tag{8.2}$$

where **K** is the feedback gain matrix, which is predetermined before system operation. For example, it can be obtained from the linear quadratic regulator (LQR) controller.

Communication channel

The number of bits used to quantize the measurement at the sensor is denoted by B. For simplicity of analysis, we ignore the quantization error. This is reasonable if B is sufficiently large, which is true in practical systems (otherwise the control will be unreliable and that analysis is also complex). The symbol rate is R symbols per second, which is fixed. Hence for different modulation schemes, the communication delays are different since different numbers of symbols are needed to convey the B bits. We assume that there are S available modulation schemes, among which the sensor can switch. This is reasonable in modern software-defined radio systems. In the sth modulation scheme, each symbol contains b_s bits (e.g., a 8PSK contains 3 bits). Hence the time needed to convey the B bits is given by

$$T_s = \frac{B}{b_s R}, \tag{8.3}$$

for the sth modulation scheme.

It is possible that the sensor cannot switch the modulation scheme freely, since usually the switching time is nonnegligible. Therefore we assume that the modulation scheme can be changed every T_m seconds. The time period of sampling for measurement is denoted by $\mathcal{T}$. The following assumptions are used for simplicity of analysis:

- We assume $\mathcal{T} > T_s$, for all possible s; i.e., any measurement can always be sent out (but not necessarily reach the controller) before the next sample arrives.
- We assume that $t_m = T_m/\mathcal{T}$ is an integer; i.e., each modulation scheme can be used for t_m samples after being activated.

Since noise exists in the communication channel, it is possible that packet loss occurs. The packet loss rate of the sth modulation scheme is denoted by $P_{e,s}$. For simplicity, we assume that all the transmission errors can be detected (but not corrected). Since the communication is highly real-time, we assume that there is no retransmission in the communications even if a packet is dropped.

Suppose that the sth scheme starts from time 0. The control action $\mathbf{u}$ at time t is determined by linear feedback control:

$$\mathbf{u}(t) = -\mathbf{K}\mathbf{x}(\tau_s(t)), \tag{8.4}$$

where $\tau_s(t)$ is the time of the nearest sample of the system state:

$$\tau_s(t) = \max\{(n-1)T_s | nT_s \leq t\}. \tag{8.5}$$

Compared with Eq. (8.2), we replace the instantaneous feedback $\mathbf{x}(t)$ with the closest observation $\mathbf{x}(\tau_s(t))$, by assuming that the error between $\mathbf{x}(t)$ and $\mathbf{x}(\tau_s(t))$ is small, which is reasonable if the system dynamics do not change very rapidly.

When transmission error occurs, the following two possible approaches are considered to synthesize $\mathbf{u}$:

- Zero observation upon transmission failure (ZOTF): When $\mathbf{x}(t)$ is not successfully received, we set $\mathbf{x}(t) = 0$ and then obtain $\mathbf{u}(t) = 0$; i.e., the controller takes zero control action.
- Most recent observation upon transmission failure (MROTF): When $\mathbf{x}(t)$ is not successfully received, the controller uses the most recent sample of $\mathbf{x}$ as a reasonable estimation of $\mathbf{x}(t)$.

8.2.2 IMPACT OF MODULATION ON SYSTEM DYNAMICS

According to the system model, when the sth modulation scheme is selected, the corresponding delay is T_s and the packet loss rate is $P_{e,s}$. When the sensor uses denser modulation, it results in a smaller delay but a higher packet loss rate; and vice versa. Hence it is necessary to analyze the impacts of delay and packet loss on the system dynamics.

Impact of communication delay

In Ref. [16], the impact of communication delays on the system dynamics has been analyzed by assuming that the delay is shorter than the sampling period. Essentially, the analysis converts the controlled system with communication delay into an equivalent discrete-time system, whose state is defined as

$$\mathbf{z}(k) = (\mathbf{x}(k\mathcal{T}), \mathbf{u}((k-1)\mathcal{T})), \tag{8.6}$$

and the evolution law of dynamics is given by

$$\mathbf{z}(k+1) = \Phi_s \mathbf{z}(k), \tag{8.7}$$

where, when the sth modulation scheme is used, the matrix Φ_s is given by

$$\Phi_s = \begin{pmatrix} \Psi - \Gamma_0(T_s)\mathbf{K} & \Gamma_1(T_s) \\ -\mathbf{K} & 0 \end{pmatrix}, \tag{8.8}$$

where

$$\begin{cases} \Psi = e^{\mathbf{A}\mathcal{T}}, \\ \Gamma_0(T_s) = \int_0^{\mathcal{T}-T_s} e^{\mathbf{A}s} ds \mathbf{B}, \\ \Gamma_1(T_s) = \int_{\mathcal{T}-T_s}^{\mathcal{T}} e^{\mathbf{A}s} ds \mathbf{B}. \end{cases} \tag{8.9}$$

Impact of delay and packet loss

We add the impact of packet losses to that of the communication delay. Since we assume two possible control actions upon a packet loss, we discuss the impact of packet loss in the following two situations:

- ZOTF criterion: In this case, the controller carries out zero control action if a packet of measurement is lost. Then there are two possible modes for the control feedback gain matrix, namely

$$\mathbf{K}_r = k\mathbf{K}, \quad r = 0, 1, \tag{8.10}$$

  where $k = 0$ (or 1) means transmission failure (or transmission success). Thus the matrix Φ_s in Eq. (8.7) should be replaced with ${\Phi_s}^{r(k)}$ where $\mathbf{K}$ is replaced with $\mathbf{K}_{r(k)}$. Therefore when the sth modulation scheme is used, the system dynamics is governed by

$$\mathbf{z}(k+1) = \Phi_s^{r(k)}\mathbf{z}(k). \tag{8.11}$$

- MROTF criterion: In this case, if a packet is lost, the controller uses the most recently received packet for system state estimation. For the analysis a new variable $\mathbf{x}_c(t)$ is defined, which is given by

$$\mathbf{x}_c(k) = r(k)\mathbf{x}(k) + (1 - r(k))\mathbf{x}_c(k). \tag{8.12}$$

  We can verify that $\mathbf{x}_c(k)$ is the most recent received observation no later than time k.
  Then the equivalent system state $\mathbf{z}$ is defined as

$$\mathbf{z}(k) = (\mathbf{x}^T(k), \mathbf{x}_c(k), \mathbf{u}(k-1))^T, \tag{8.13}$$

  in which the timing of the variables is illustrated in Fig. 8.2.

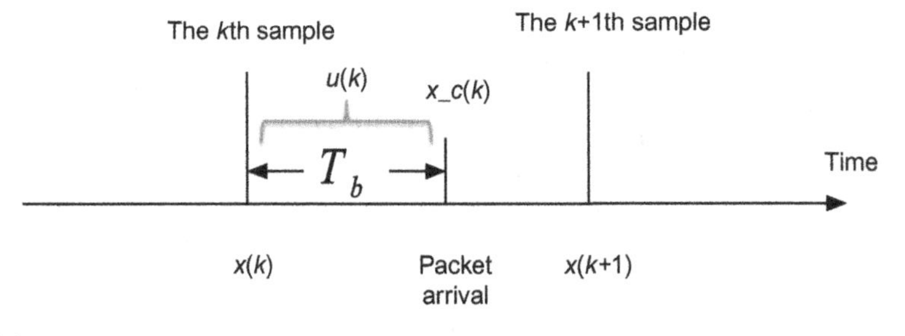

FIG. 8.2

Illustration of the timing of the variables.

The equivalent system state satisfies the same expression as in Eq. (8.11), in which the expression of $\Phi_s^{r(k)}$ is given by

$$\Phi_s^{r(k)} = \begin{pmatrix} \Psi & -\Gamma_0(T_s)\mathbf{K} & \Gamma_1(T_s) \\ r(k)\mathbf{I} & (1-r(k))\mathbf{I} & 0 \\ 0 & -\mathbf{K} & 0 \end{pmatrix}. \quad (8.14)$$

Markovian jump process

The system described in Eq. (8.11) is essentially a Markovian jump process [147], in which the state $r(k)$ forms a Markov chain. On the other hand, the state s (namely the selected modulation scheme) is a discrete state while it can be controlled by the sensor. Given a fixed s, the system stability is judged by the following theorem, as a straightforward conclusion from the theorem in Ref. [148]:

Theorem 49. *The system described in Eq. (8.11) is mean-square stable if and only if there exists a matrix* $\mathbf{G} > 0$ *such that*

$$\mathbf{G} - P_{e,s}(\Phi_s^0)^T \mathbf{G} \Phi_s^0 - (1 - P_{e,s})(\Phi_s^1)^T \mathbf{G} \Phi_s^1 > 0. \quad (8.15)$$

Beyond the system stability, we can also consider the cost of the dynamics. We define the cost as

$$\begin{aligned} J_N &= \sum_{k=0}^{N-1} \mathbf{x}^T(k)\mathbf{Q}\mathbf{x}(k) \\ &= \sum_{k=0}^{N-1} \mathbf{z}^T(k)\tilde{\mathbf{Q}}\mathbf{z}(k), \end{aligned} \quad (8.16)$$

where $\tilde{\mathbf{Q}} = \text{diag}(\mathbf{Q}, 0, 0)$. As a straightforward conclusion of Theorem 1 in Ref. [149], the expected cost is given by

$$J_N(\mathbf{z}(0), r(0), s) = \mathbf{x}^T(0)\mathbf{P}_{0,r(0),s}\mathbf{x}(0), \quad (8.17)$$

where

$$\begin{aligned} \mathbf{P}_{t,j,s} &= P_{e,s}(\Phi_s^0)^T \mathbf{P}_{t,0,s}\Phi_s^0 \\ &+ (1 - P_{e,s})(\Phi_s^1)^T \mathbf{P}_{t,1,s}\Phi_s^1 + \tilde{\mathbf{Q}}, \end{aligned} \quad (8.18)$$

with $\mathbf{P}_{N,j,s} = \mathbf{Q}$. Note that $\mathbf{z}(0)$ and $r(0)$ are known initial conditions.

8.2.3 HYBRID SYSTEM MODELING

As explained in Chapter 7, a hybrid system is a powerful tool to model systems with both discrete and continuous states. Again, we list the linear switching dynamics, for instance:

$$\begin{cases} \mathbf{x}(t+1) = \mathbf{A}_{k(t)}\mathbf{x}(t) + \mathbf{B}_{k(t)}\mathbf{u}(t), \\ \mathbf{y}(t) = \mathbf{C}_{k(t)}\mathbf{x}(t). \end{cases}$$

where $\mathbf{x}$ is the continuous system state, k is the discrete system state, $\mathbf{u}$ is the control action, $\mathbf{y}$ is the observation vector, and $\mathbf{A}$, $\mathbf{B}$ and $\mathbf{C}$ are system parameters, which are dependent on the discrete system state. Then in our context, we define the discrete system state as $k(t) = (r(t), s(t))$; i.e., the discrete state represents the status of transmission and the current modulation scheme. Note that $r(t)$ is random in nature and uncontrollable, while the sensor can control $s(t)$.

We then consider two types of adaptive modulation schemes, namely System Parameter Aware Adaptive Modulation (SPAAM) and System State Aware Adaptive Modulation (SSAAM). In SPAAM the modulation scheme is adaptive to only the system parameters, not the state of the dynamics. Hence it is fixed throughout the operation, unless the system parameters are changed. In SSAAM the modulation is adaptive to the instantaneous system state of physical dynamics, and thus is time varying. Below we discuss both types of adaptive modulation.

SPAAM

As indicated by the name, in SPAAM the modulation scheme is adaptive only to the system parameters and is fixed regardless of the current system state of physical dynamics. We assume that the objective of SPAAM is the stability of system dynamics. Then by applying the conclusion in Theorem 49, we obtain the following optimization problem:

$$\begin{aligned} &\max_{s,\gamma,\mathbf{G}} \gamma \\ &\text{s.t. } \mathbf{G} - P_{e,s}(\Phi_s^0)^T \mathbf{G}\Phi_s^0 \\ &\quad - (1 - P_{e,s})(\Phi_s^1)^T \mathbf{G}\Phi_s^1 - \gamma \mathbf{I} > 0 \\ &\mathbf{G} > 0, \end{aligned} \tag{8.19}$$

where γ represents the eigenvalue of the corresponding matrix.

Below is an intuitive explanation for the above optimization: the goal is to maximize the minimal eigenvalue on the left-hand side of Eq. (8.15). If the maximum γ is positive, then the corresponding modulation can stabilize the system dynamics. A simple algorithm for the optimization is to carry out an exhaustive search for all modulation schemes. For each s, we apply the theory of linear matrix inequalities (LMI) [133] to find the optimal γ and $\mathbf{G}$. Finally, we choose the optimal s that maximizes the objective function γ. Since the number of possible modulation schemes is usually quite limited, an exhaustive search is computationally efficient.

SSAAM

As indicated by the name SSAAM, the modulation scheme is selected adaptively to the current system state of physical dynamics. We have assumed that the modulation scheme can be changed every T_{m} seconds (or equivalently t_{m} samples). The set of modulation schemes that the sensor can choose from is denoted by Ω. The strategy of adaptive modulation is denoted by π; essentially it is a mapping from the system state $\mathbf{x}$ to the modulation scheme within Ω. In terms of hybrid systems, the modulation scheme selection is equivalent to the control of the discrete system state in the corresponding hybrid system.

The basic elements of SSAAM, as an optimization problem, are as follows:

- Cost function: A quadratic cost is considered over T_f scheduling periods (equivalently $T_f t_{\mathrm{m}}$ samples), which is given by

$$J = \sum_{t=1}^{T_f t_{\mathrm{m}}} \mathbf{x}^T(t)\mathbf{Q}\mathbf{x}(t). \tag{8.20}$$

  Thus the optimal adaptive modulation strategy is to minimize the cost given the current system state; i.e.,

$$\pi^* = \arg\min_{\pi} J(\pi). \tag{8.21}$$

- Dynamic programming (DP): This is a powerful approach to optimize the above cost function. As a standard procedure of DP, we define the following cost-to-go function at the tth decision as the minimal cost in the future:

$$J_t(\mathbf{x}) = \sum_{r=(t-1)t_{\mathrm{m}}+1}^{T_f t_{\mathrm{m}}} \mathbf{x}^T(r)\mathbf{Q}\mathbf{x}(r)\Big|_{\mathbf{x}((t-1)t_{\mathrm{m}})=\mathbf{x}}. \tag{8.22}$$

  By solving the following Bell equation, we can obtain the cost-to-go function in a recursive manner:

$$\begin{aligned} J_t(\mathbf{x}) = \min_{s} & \sum_{r=(t-1)t_{\mathrm{m}}+1}^{t t_{\mathrm{m}}} \mathbf{x}^T(r)\mathbf{Q}\mathbf{x}(r)\Big|_{\mathbf{x}((t-1)t_{\mathrm{m}})=\mathbf{x}} \\ & + J_{t+1}(\mathbf{x}')\big|_{\mathbf{x}((t-1)t_{\mathrm{m}})=\mathbf{x},s}, \end{aligned} \tag{8.23}$$

  where $\mathbf{x}'$ is the initial system state in the next decision period, which is determined by the current system state $\mathbf{x}$ and the action s (the selection of modulation). The cost-to-go function can be calculated from the last time slot $T_f t_{\mathrm{m}}$, which is trivial.
- Approximate dynamic programming (ADP): The main challenge of DP is that it is difficult to represent the continuous function $J_t(\mathbf{x})$, as many DP problems do. An effective approach to handle this challenge is to apply ADP. In Ref. [146], we approximate the cost-to-go function by using quadratic functions given by

$$J_t(\mathbf{x}) = \mathbf{x}^T \Sigma_t \mathbf{x}, \tag{8.24}$$

  where Σ_t is a positive-definite function. Then the computation of the cost-to-go function is equivalent to estimating Σ_t. Due to limitations of space, we are unable to discuss how to estimate Σ_t. The detailed algorithm can be found in Ref. [146]. Numerical results have shown that the proposed ADP approach can effectively handle the challenge and results in good performance for controlling the physical dynamics.

Both SPAAM and SSAAM have been demonstrated to achieve good performances in the context of smart grids [146].

8.3 SOURCE CODING IN A CPS: POINT-TO-POINT CASE

Source coding encodes the original information (either continuous or discrete valued) into discrete information bits. The main purpose is to reduce redundancy and thus transmit fewer bits (thus using fewer communication resources). In many existing CPSs, the measurements (e.g., the frequency and phase measurements at phasor measurement units (PMUs) in smart grids) are quantized with certain precisions and then sent to the controller. However, such a straightforward scheme may waste communication resources for the following reasons:

- Ignoring temporal redundancy: The measurements at different times are usually correlated. For example, in power grids, the frequency and phase measurements are continuous in time and do not change very rapidly. Hence successive measurements are similar to each other, and it is not necessary to fully quantize each measurement in an independent manner. A possibly more efficient approach is to quantize increments in the frequency and phase.
- Ignoring spatial redundancy: When there are multiple sensors, the measurements at different sensors are usually correlated, unless the sensors happen to measure orthogonal subspaces. The redundancies among different sensors can be fully utilized to decrease the source coding rate, even though there are no communications among the sensors (and thus no explicit coordination among them).

In this section we assume a single sensor and consider only temporal redundancy. Spatial redundancy will be considered in the next section. In traditional communication systems, the temporal redundancy can be compressed by carrying out joint source coding for all the source information symbols. However, such a source coding scheme requires a long latency since the coding needs to be carried out until all the source information symbols have been gathered. A long communication latency is intolerable for real-time feedback control in a CPS. Hence it is of key importance to study real-time source coding, in which the source information symbols are encoded in a sequential manner.

8.3.1 SYSTEM MODEL

We assume that the information source generates a discrete-time random process $\{X_t\}_{t=-\infty}^{\infty}$. The encoder generates bit sequence $\{Z_t\}_{t=-\infty}^{\infty}$ via the mapping

$$Z_k = h_k(\{X_t\}_{t=-\infty}^{\infty}). \tag{8.25}$$

The decoder tries to regenerate $\{X_t\}_{t=-\infty}^{\infty}$ from the received bit sequence $\{Z_t\}_{t=-\infty}^{\infty}$, namely

$$\hat{X}_k = g_k(\{Z_t\}_{t=-\infty}^{\infty}). \tag{8.26}$$

For notational simplicity, we rewrite the decoder output as

$$\hat{X}_t = f_k(\{X_t\}_{t=-\infty}^{\infty}). \tag{8.27}$$

Definition 22. The encoder $\{f_k\}_{k=-\infty}^{\infty}$ is called causal if

$$f_k(\{X_t\}_{t=-\infty}^{\infty}) = f_k(\{Y_t\}_{t=-\infty}^{\infty}) \tag{8.28}$$

whenever $\{X_t\}_{t=-\infty}^{k} = \{Y_t\}_{t=-\infty}^{k}$.

Obviously, the output of the decoder is dependent only on the received bits and is thus nonanticipating.

8.3.2 STRUCTURE OF CAUSAL CODERS

Given the requirement of causality, we follow Ref. [150] to introduce the typical categories of causal source encoders, which are illustrated in Fig. 8.3.

Sliding block causal coders

In this case the output of a decoder at time t is given by

$$\hat{X}_t = f_t(\{X_k\}_{k=-\infty}^{t}). \tag{8.29}$$

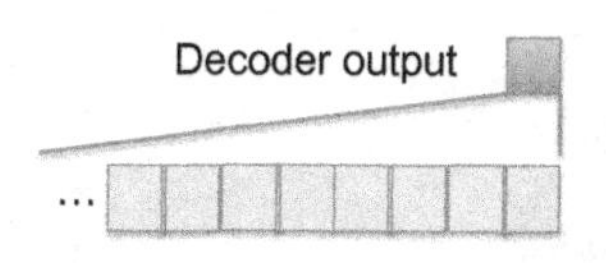

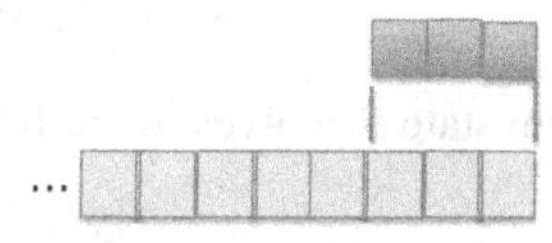

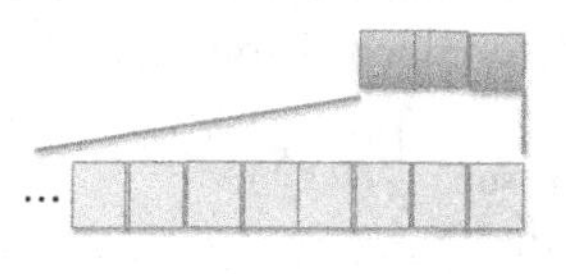

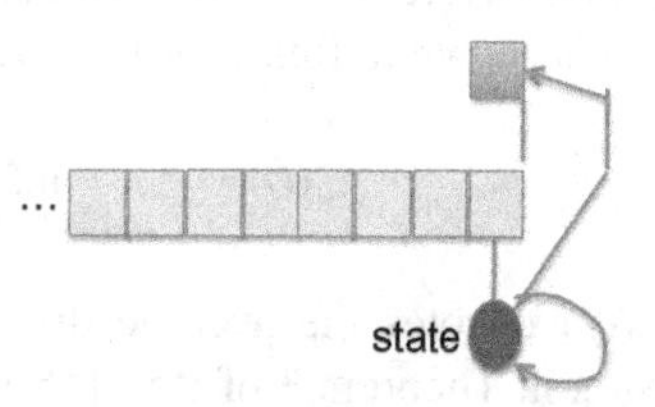

FIG. 8.3

Illustration of coder structures.

If the coder has finite memory, then we have

$$\hat{X}_t = f_t(\{X_k\}_{k=t-M}^{t}), \tag{8.30}$$

where M is the memory length.

Block causal coders

In such coders, the information is encoded and decoded in a block manner, namely

$$\hat{X}_{kN+1}^{kN+N} = f(X_{kN+1}^{kN+N}), \tag{8.31}$$

where k is the index of the block (or codeword). Obviously, such a coding scheme will cause an average delay $N/2$.

Block stationary coders

Block stationary coders can be considered as a combination of sliding block causal coders and block causal coders. The output is given by

$$\hat{X}_{kN+1}^{kN+N} = f(X_{-\infty}^{kN+N}), \tag{8.32}$$

where the decoder output is in blocks and the decoding procedure uses all the previous received bits.

Finite-state causal coders

A challenge of sliding block causal coders and block stationary coders is that infinite memory is needed to store all the previously received bits. To alleviate this problem while not discarding the history, a state S can be defined for the coders, such that the decoder output is a function of the state and the newest information symbol, namely

$$\hat{X}_k = f_x(S_k, X_k), \tag{8.33}$$

and the system state S evolves in the following law:

$$S_k = f_s(S_{k-1}, X_k). \tag{8.34}$$

Rate distortion

In the context of causal coding, the rate-distortion function $r_c(D)$ is defined to be the minimum average transmission rate such that the average distortion is no larger than D. It is shown in Ref. [150] that $r_c(D)$ is determined as follows:

$$r_c(D) = \inf_{f \text{ causal}, \, d(f) \leq D} \limsup_{K \to \infty} \frac{1}{K} H(\hat{X}_1^K), \tag{8.35}$$

where $d(f)$ denotes the average distortion caused by the coder f. An interesting conclusion in Theorem 3 of Ref. [150] is that the causal rate-distortion functions in the above coder schemes are all the same when the information source is memoryless. However, it is still an open problem to compare the rate-distortion functions when the information source has memory.

8.3.3 FINITE-STATE TRANSCEIVER

As mentioned above, to avoid the requirement of infinite memory and the performance loss caused by discarding the history, an effective approach is to make both the transmitter and receiver have finitely many states, such that the history can be incorporated into the evolution of the system states. This type of transceiver has been discussed in Ref. [151]. Here we provide a brief introduction to the main framework and conclusions.

Transceiver structure

We assume that the encoder and decoder have states T and R, respectively. Upon receiving a source information symbol X_t, the transmitter sends out a coded symbol D_t (with possibly N_D values) and updates its state:

$$\begin{cases} T_{t+1} = \tau(X_t, T_t), \\ D_t = \delta(X_t, T_t). \end{cases} \tag{8.36}$$

Upon receiving the coded symbol D_t, the receiver updates its own state and outputs a symbol Y_t:

$$\begin{cases} R_{t+1} = \rho(D_t, R_t), \\ Y_t = \eta(D_t, R_t). \end{cases} \tag{8.37}$$

It is assumed that the information source $\{X_t\}_t$ is a stationary ergodic stochastic process, and the encoding/decoding mappings are deterministic. The distortion function is denoted by $\psi(x, y)$. We assume that the following limit exists:

$$\epsilon_k^2 = \lim_{n \to \infty} E\psi(X_{n-k}, Y_n). \tag{8.38}$$

Major problem and conclusions

Ref. [151] mainly focused on how to find the encoding and decoding mechanisms τ, δ, ρ, and η in order to minimize the expected distortion ϵ_k^2. Although there is no determining conclusion for generic cases, many conclusions on the transceiver structure have been obtained.

First it has been shown that, given sufficiently long delay k, the finite-state transceiver can achieve the traditional rate-distortion function with unlimited codeword length.

Theorem 50. *Denote by $R(d)$ the traditional rate-distortion function with unlimited codeword length. Assume that $\{X_t\}$ has positive entropy. Then for any $\epsilon > 0$, if $\log N_D > R(d)$, we can find a finite-state transceiver such that*

$$E\Psi(X_{n-k}, Y_n) \le d + \epsilon, \quad \forall n > k; \tag{8.39}$$

if $\log N_D < R(d)$, any finite-state transceiver satisfies $\epsilon_k^2 > d$ for all possible k values.

Then it is of major interest to study how to choose the encoding and decoding mechanisms τ, δ, ρ, and η such that the average distortion ϵ_k^2 is minimized, given the delay k. For notational simplicity, we define

$$p_{X_{n-k},T_n,D_n,R_n}(x,t,d,r) = Pr(X_{n-k} = x, T_n = t, D_n = d, R_n = r). \tag{8.40}$$

We assume that the initial states T_1 and R_1 are known, or the distributions are given. If τ, δ, and ρ are properly chosen, the distribution p_{X_{n-k},T_n,D_n,R_n} will converge to a stationary one:

$$\lim_{n\to\infty} p_{X_{n-k},T_n,D_n,R_n}(x,t,d,r) = p^k_{X,T,D,R}(x,t,d,r). \tag{8.41}$$

Or the distribution will converge to a periodic structure with period L:

$$\lim_{n\to\infty} p_{X_{m-k},T_m,D_m,R_m}(x,t,d,r) = p^k_{j,X,T,D,R}(x,t,d,r), \tag{8.42}$$

where $m = nL + j$. In this case, we can define the average distribution over one period:

$$p^k_{X,T,D,R}(x,t,d,r) = \frac{1}{L}\sum_{j=1}^{L} p^k_{j,X,T,D,R}(x,t,d,r). \tag{8.43}$$

The definitions of $p^k_{X,T,D,R}(x,t,d,r)$ in both Eqs. (8.41) and (8.43) are called the steady-state distribution. Hence the average distortion is given by

$$\epsilon_k^2 = \sum_{x,d,r} p^k_{X,D,R}(x,d,r)\psi(x,\eta(d,r)), \tag{8.44}$$

where

$$p^k_{X,D,R}(x,d,r) = \sum_t p^k_{X,T,D,R}(x,t,d,r). \tag{8.45}$$

Using the Bayesian rule, we have

$$\begin{aligned}\epsilon_k^2 &= \sum_{d,r} p^k_{D,R}(d,r)\sum_x p^k_{X|D,R}(x|d,r)\psi(x,\eta(d,r))\\ &= \sum_{d,r} p^k_{D,R}(d,r)F_{d,r}(\eta(d,r)),\end{aligned} \tag{8.46}$$

where

$$F_{d,r}(y) = \sum_x p^k_{X|D,R}(x|d,r)\psi(x,y). \tag{8.47}$$

Let $y_0 = \arg\max_y F_{d,r}(y)$. Then the following theorem shows that the optimal decoder should adopt y_0.

Theorem 51. *Suppose that the decoding delay is k. Then the optimal decoding function is given by*

$$\eta(d, r) = y_0(d, r), \tag{8.48}$$

where $y_0(d, r)$ minimizes the function $F_{d,r}(y)$ in Eq. (8.47).

Given the optimal decoder η, the next task is to design the optimal functions of τ, δ, and ρ. However, the search space of these functions is enormously large. Hence the optimal design is still an open problem, even for special cases such as Markov information sources. However, for some special cases, such as the one discussed subsequently, we can obtain more information.

8.3.4 CHANNEL FEEDBACK

In Ref. [152], the special case of Markov information source and perfect channel feedback has been discussed. The following assumptions are adopted for the system under study:

- The information source $\{X_t\}$ is Markovian; i.e.,

$$p(X_{t+1}|X_t, X_{t-1}, \ldots) = p(X_{t+1}|X_t). \tag{8.49}$$

  Many practical information sources can be modeled as Markovian.
- All the communication channel outputs are fed back to the transmitter. This is reasonable in a CPS with continuous-valued physical dynamics, since the physical dynamics can encode the communication channel output and convey it to the sensor.

Denote by x_n, y_n, z_n the output of decoder, and the input and output of the communication channel. Then the outputs of the encoder and decoder are given by

$$\begin{cases} u_n = c_n(x^n, y^{n-1}), \\ z_n = h_n(z_{n-1}, y_n). \end{cases} \tag{8.50}$$

The following theorem shows the structure of the optimal codes in the above framework:

Theorem 52. *There exists an optimal code c^* satisfying*

$$c^*(x^n, y^{n-1}) = \gamma_n(x_n, z_{n-1}). \tag{8.51}$$

When the receiver has perfect memory (i.e., it can remember all the previously received bits), the conclusion is sharpened in the following theorem:

Theorem 53. *Suppose that $z_n = y^n$. Then there exists an optimal code c^* satisfying*

$$c_n^*(x^n, y^{n-1}) = \psi_n^*(x_n, \xi_{n-1}^*(y^{n-1})), \tag{8.52}$$

for certain functions ψ^*, where $\xi_n^* = \xi^{c*}$.

When the communication channel is symmetric, the conclusion can be further simplified. First we define the concept of a symmetric channel. Consider a transition matrix Q between two finite sets U and Y. We say that Q is of type S if, for every $f : U \to U$, there exists a transition matrix $A : Y \to Y$ such that

$$Q(y|f(u)) = \sum_{y'} A(y|y')Q(y'|u), \tag{8.53}$$

for all $u \in U$ and $y \in Y$. We say that a memoryless channel is symmetric if its transition matrices $\{Q_n\}_n$ are all of type S. Then we have the following conclusion:

Theorem 54. *If the channel is symmetric* ($X = U$), *and* $z_n = y^n$, *then*

$$c_n^*(x^n, y^{n-1}) = x_n, \quad \forall n \geq 1, \tag{8.54}$$

is optimal.

8.3.5 SEQUENTIAL QUANTIZATION

A detailed quantization scheme, which meets the requirement of causality, is proposed in Ref. [153]. Here we provide a brief introduction to the corresponding algorithm.

System model

We consider a discrete-time information source $\{X_n\}$, which is ergodic and Markovian. The information source $\{X_n\}$ cannot be observed directly. An associated observation process $\{Y_n\}$ can be measured. The conditional distribution of Y given X is denoted by $\psi(Y|X)$. We assume that X_n and Y_n are n-dimensional and d-dimensional real vectors, respectively. The evolution law of X_n and Y_n is given by

$$\begin{aligned} P(X_{n+1} \in A, Y_{n+1} \in B|X^n, Y^n) &= \int_{A\times B} p(X_n, dx, dy) \\ &= \int_A \int_B \phi(y, z|X_n)dydz, \end{aligned} \tag{8.55}$$

for any $A \subset R^n$ and $B \subset R^d$. Since given X_n the distribution of X_{n+1} and Y_{n+1} is independent of X^{n-1} and Y^n, such an information source is called a Markov source. An explicit description of the source and observation processes is given by

$$\begin{cases} X_{n+1} = g(X_n, \xi_n), \\ Y_{n+1} = h(X_n, \xi_n'), \end{cases} \tag{8.56}$$

where g and h are the mechanisms of both processes, and ξ_n and ξ_n' are random perturbations that are mutually independent and are independent of X_n. We further assume that the evolution of the system state X_n is continuous with respect to the initial value. In more detail, if two realizations X_n and X'_n, with initial values $X_0 = x$ and $X'_0 = y$, have the same driving random noises $\{\xi_n\}$, we have

$$E[\|X_n - X'_n\|] \leq K\beta^n \|x - y\|, \quad n \geq 0, \tag{8.57}$$

for some constants K and β.

Vector quantizer

A vector quantizer is assumed. The set of alphabets is denoted by $\Sigma = \{\alpha_1, \ldots, \alpha_N\}$. The quantized version of the observation process $\{Y_n\}$ is given by $\{q_n\}$, which takes values in Σ.

We denote the set of all possible finite subsets of R^d by D, which satisfies the following condition: there exists an $M > 0$ and a $\Delta > 0$ such that

- $x \in A \in D$ implies $\|x\| \leq M$;
- if $x = (x_1, \ldots, x_d)$ and $y = (y_1, \ldots, y_d)$ are both within $A \in D$ and $x \neq y$, then we have $|x_i - y_i| > \Delta$ for all i.

Intuitively, A denotes the set of representative vectors for the quantizer. The first requirement means that all the representative vectors have bounded norms. The second requirement means that all the dimensions of any pair of representative vectors should be well separated.

Then the quantizer is given by

$$\begin{cases} q_{n+1} = i_{Q_n}(l_{Q_n}(Y_{n+1})), \\ Q_n = \eta_n(q^n), \end{cases} \tag{8.58}$$

where Q_n is the set of representative vectors which change with time, η_n maps from all the quantization outputs to the set of representative vectors, l_{Q_n} maps from the observation to the nearest representative vector, and i_{Q_n} maps from the representative vector to the index in Σ.

The goal of the quantizer is to optimize the choice of η_n such that the average entropy rate of $\{q_n\}$ and the average distortion are jointly optimized. The cost function can be formulated as

$$\limsup_{n\to\infty} \frac{1}{n} \sum_{m=0}^{n-1} E[H(q_{m+1}|q^{\mathrm{m}}) + \lambda\|Y_{\mathrm{m}} - \bar{q}_{\mathrm{m}}\|^2], \tag{8.59}$$

where λ is the weighting factor and $\bar{q}_{\mathrm{m}} = i^{-1}_{Q_{m-1}}(q_{\mathrm{m}})$.

Equivalent control problem

We now rewrite the cost function in Eq. (8.59) into an alternative form, which is reduced to a control problem. First we define

$$\pi_n(A) = P(X_n \in A|q^n), \tag{8.60}$$

which is the probability distribution of X_n given the previous quantizer outputs, and can be computed recursively. We further define

$$h_a(\pi, A) = \int \pi(dx) f_n(x, A), \tag{8.61}$$

where

$$f_n(x, A) = \int \psi(y|x) I[i_A(l_A(y)) = a] dy. \tag{8.62}$$

The intuitive meaning of $f_n(x, A)$ is the probability of generating the quantization output a when the true system state is x and the set of representative vectors is A. $h_a(\pi, A)$ is the probability of outputting a when the distribution of x is π and the representative vector set is A.

We also define

$$\hat{f}(x, A) = \int \psi(y|x) \|y - l_A(y)\|^2 dy, \tag{8.63}$$

which is the expected cost when the true system state is x and the set of representative vectors is A, and

$$k(\pi, A) = -\sum_a h_a(\pi, A) \log h_a(\pi, A), \tag{8.64}$$

which means entropy of the quantizer output when the distribution of x is π and the set of representative vectors is A, and

$$r(\pi, A) = \int \pi(dx) \hat{f}(x, A), \tag{8.65}$$

which is the expected cost when the distribution of x is π and the set of representative vectors is A.

Then the cost in Eq. (8.59) can be written as

$$\limsup_{n \to \infty} \frac{1}{n} \sum_{m=0}^{n-1} E[k(\pi_{\mathrm{m}}, Q_{\mathrm{m}}) + \lambda r(\pi_{\mathrm{m}}, Q_{\mathrm{m}})]. \tag{8.66}$$

In Ref. [153], a technical mechanism is proposed to make the corresponding probabilities bounded from below by a small number. Here we skip this step and simply assume that all the probabilities have been bounded by a small constant.

Now the problem becomes to find the control policy mapping from the a posteriori probability π_n to the quantization policy (i.e., the representative vectors) Q_n, which is denoted by $Q_n = v(\pi_n)$. Thus the source coding problem is transformed to a Markov decision problem.

Dynamic programming

The cost function in Eq. (8.66) is the long-term average of costs, whose solution is involved. Hence we first convert the cost function to a discounted one, which is given by

$$J_\alpha(\pi_0, \{Q_n\}) = \sum_{m=0}^{\infty} \alpha^m E[k(\pi_\mathrm{m}, Q_\mathrm{m}) + \lambda r(\pi_\mathrm{m}, Q_\mathrm{m})]. \tag{8.67}$$

The standard procedure of DP can then be applied. First we define the value function as

$$V_\alpha(\pi_0) = \inf_{v} J_\alpha(\pi_0, \{Q_n\}). \tag{8.68}$$

Using Bellman's equation, the value function satisfies

$$V_\alpha(\pi) = \min_{A}[k(\pi_\mathrm{m}, Q_\mathrm{m}) + \lambda r(\pi_\mathrm{m}, Q_\mathrm{m}) + \alpha \int V_\alpha(\pi')P(d\pi'|A, \pi)], \tag{8.69}$$

where $P(d\pi'|A, \pi)$ is the conditional probability of the a posteriori probability given the quantization scheme represented by A and the current a posteriori probability π. The corresponding optimal strategy can be obtained via

$$v(\pi) = \arg\min_{A}[k(\pi_\mathrm{m}, Q_\mathrm{m}) + \lambda r(\pi_\mathrm{m}, Q_\mathrm{m}) + \alpha \int V_\alpha(\pi')P(d\pi'|A, \pi)]. \tag{8.70}$$

To remove the discounting in Eq. (8.67), it was shown in Ref. [153] that we can let $\alpha \to 1$ and obtain the optimal policy for Eq. (8.66). The details are omitted here. One challenge for the DP here is that both π_n and Q_n are continuous and thus the strategy v is infinitely dimensional. It still remains an open problem to efficiently solve the Bellman equation in Eq. (8.69). One possible approach is to use ADP and obtain suboptimal solutions.

Once the optimal strategy of sequential quantization is found, the estimations of X and Y are given by the maximum a posteriori (MAP) ones, i.e.,

$$\hat{X}_n = \arg\max \pi_n(\cdot) \tag{8.71}$$

and

$$\hat{Y}_n = \arg\max \int I\{i_{Q_n l} = q_n P(\cdot, z|x)dz\pi_{n-1}(dx)\}. \tag{8.72}$$

8.4 SOURCE CODING IN A CPS: DISTRIBUTED CASE

In this section, we follow Ref. [154] to study the distributed source coding in a CPS with multiple sensors, in which we exploit both temporal and spatial redundancies in the measurements at different sensors.

8.4.1 DISTRIBUTED CODING IN TRADITIONAL COMMUNICATIONS

We first introduce distributed source coding in traditional communications. Consider two information sources whose information symbols are represented by $X_1(t)$ and $X_2(t)$. It is assumed that there are no communications between the two sources. An astonishing conclusion in information theory is that the two information sources can achieve lossless source coding when the sum of the two source coding rates equals the entropy of the information sources [155]; i.e.,

$$R_1 + R_2 = H(X_1, X_2), \tag{8.73}$$

as if the two information sources were a single information source (which can be achieved if there is a communication link between the two sources with sufficiently large capacity), although they are actually distributed. This is the well-known Splepian-Wolf coding scheme. Similar conclusions are obtained for the case of side information at the decoder (i.e., Wyner-Ziv coding) and distributed lossy source coding (Berger-Tung bounds).

The information-theoretic argument is then extended to practical distributed source coding schemes by employing parity check-based channel coding [156–160] or nested lattice coding [161,162]. A survey can be found in Ref. [163]. However, in the context of a CPS, we face the following new challenges, which require a reexamination of the distributed source coding:

- Complicated relationships: The relationships among the distributed observations can be more complicated than in many existing studies in which $X = Y + N$, where X is the binary information source, Y is side information, and N is binary random noise.
- Control-aware distortion: The distortion function can be directly related to the purpose of controlling the CPS, which is more complex than simple distortion criteria such as minimum mean square error (MMSE) used in many lossy source coding scenarios.
- Time correlations: Many existing studies assume independent observations in time. However, the observations are usually correlated in the time domain in a CPS due to the system state evolution.

Due to the above new challenges in CPSs, there is a pressing necessity to study the distributed source coding in CPSs.

8.4.2 SYSTEM MODEL

In this book, we consider discrete-time physical dynamics. The N-dimensional system state of physical dynamics, denoted by $\mathbf{x}$, has the evolution law:

$$\mathbf{x}(t+1) = f(\mathbf{x}(t), \mathbf{u}(t), \mathbf{n}(t)), \tag{8.74}$$

where $\mathbf{u}$ is the control action and $\mathbf{n}$ is random noise.

For simplicity, we consider two sensors, 1 and 2. The corresponding observations are denoted by $\mathbf{y}_1$ and $\mathbf{y}_2$, whose dimensions are denoted by N_1 and N_2, respectively. They are functions of the system state and random observation noise, i.e.,

$$\mathbf{y}_i(t) = g_i(\mathbf{x}(t), \mathbf{w}_i(t)), \quad i = 1, 2. \tag{8.75}$$

It is assumed that the function g_i is invertible and forms a one-to-one mapping. For the generic case, it is much more difficult to describe the relationship between $\mathbf{y}_1$ and $\mathbf{y}_2$, which is beyond the scope of this book. If we fix the noise term $\mathbf{w}_i(t)$; i.e.,

$$\mathbf{x}(t) = g_i^{-1}(\mathbf{y}_i(t), \mathbf{w}_i(t)), \tag{8.76}$$

then the observations at the two sensors are statistically correlated, which is characterized by the following equation:

$$\mathbf{y}_2 = g_2(g_1^{-1}(\mathbf{y}_1(t), \mathbf{w}_1(t)), \mathbf{w}_2(t)). \tag{8.77}$$

We say that $\mathbf{y}_1$ and $\mathbf{y}_2$ are separable if there exist a function h_{ij} and noise $\mathbf{w}_{ij}$ (as a function of $\mathbf{w}_1$ and $\mathbf{w}_2$) such that

$$\mathbf{y}_j = h_{ij}(\mathbf{y}_i) + \mathbf{w}_{ij}, \tag{8.78}$$

which means that there is no nonlinear coupling between $\mathbf{y}_i$ and the noise.

A special (but very typical) case is linear system dynamics, which is described by linear equations:

$$\begin{cases} \mathbf{x}(t+1) = \mathbf{A}\mathbf{x}(t) + \mathbf{B}\mathbf{u}(t) + \mathbf{n}(t), \\ \mathbf{y}(t) = \mathbf{C}\mathbf{x}(t) + \mathbf{w}(t), \end{cases} \tag{8.79}$$

where $\mathbf{y}(t) = (\mathbf{y}_1^T(t), \mathbf{y}_2^T(t))^T$ and $\mathbf{w}(t) = (\mathbf{w}_1^T(t), \mathbf{w}_2^T(t))^T$ and $\mathbf{C} = (\mathbf{C}_1^T, \mathbf{C}_2^T)^T$. For simplicity, we assume that both $\mathbf{C}_1$ and $\mathbf{C}_2$ have full column ranks. Then it is easy to verify that

$$\mathbf{y}_1 = \mathbf{C}_1(\mathbf{C}_2^T\mathbf{C}_2)^{-1}\mathbf{C}_2^T(\mathbf{y}_2 - \mathbf{w}_2) + \mathbf{w}_1, \tag{8.80}$$

which is obviously separable.

The following assumptions are used for the communication system:

- There is no communication between the two sensors.
- There is no communication error during transmission.

The transmission rates of information bits of the two sensors are denoted by R_1 and R_2, respectively.

8.4.3 DISTORTION AND QUANTIZATION

Before we discuss the details of distributed source coding for CPSs, we consider the basics for the single sensor case. Since we have special requirements for the communications in a CPS, we need to reexamine the distortion and quantization of source coding in a CPS.

Distortion

In traditional source coding, a typical distortion criterion is the mean square error (MSE), which can also be applied in a CPS. However, the eventual goal of a CPS is not to convey the measurements; instead, it is to stabilize the system and minimize the cost of physical dynamics operation. Therefore we propose a new criterion for the distortion in a CPS.

First we notice that the system state estimation error may have different effects on system stability. In stable subspace, the error of system state estimation will be automatically damped and then vanish; in unstable subspace, the error may be amplified and make the whole system unstable. Therefore we first decompose the system state space by decomposing the matrix **A** to

$$\mathbf{A} = \mathbf{U}^T \Lambda \mathbf{U}, \tag{8.81}$$

where **U** is unitary and $\Lambda = \mathrm{diag}(\lambda_1, \ldots, \lambda_N)$, where λ_n is the nth eigenvalue. For simplicity, we assume that $\lambda_i \neq \lambda_j$, if $i \neq j$. Also for simplicity, we assume that the matrix **A** is of full rank. We consider the following dichotomy for the eigenvalues:

- $|\lambda_n| > 1$: in this subspace, the physical dynamics is not stable; the error of system state estimation may be amplified;
- $|\lambda_n| < 1$: in this subspace, the system state of physical dynamics inherently converges to zero (if no random perturbation is considered); any system state estimation error will be naturally suppressed; hence even if there is no control in this subspace, the system state is still stable.

The decomposition is illustrated in Fig. 8.4. We denote by Ω the set of unstable subspaces. For simplicity, we assume that the first N_0 subspaces are unstable.

Using the above decomposition, the system can be decomposed into N uncorrelated subsystems:

$$z_n(t+1) = \lambda_n z_n(t) + v_n(t) + \tilde{n}(t), \tag{8.82}$$

where $\mathbf{z} = (z_1, \ldots, z_N) = \mathbf{Ux}$, $\tilde{\mathbf{n}} = (\tilde{n}_1, \ldots, \tilde{n}_N) = \mathbf{Un}$, and $\mathbf{v} = (v_1, \ldots, v_N) = \mathbf{UBu}$.

We assume that a "blow-up" control policy is used by the controller, which intuitively means that the controller simply removes the deviation from the desired operational point and does not consider the cost of the control action (not like the strategy in linear quadratic Gaussian (LQG) control, where the cost of control action is also considered in the cost function). The corresponding control action in the nth subspace is given by

$$v_n(t) = -\lambda_n \hat{z}_n(t), \tag{8.83}$$

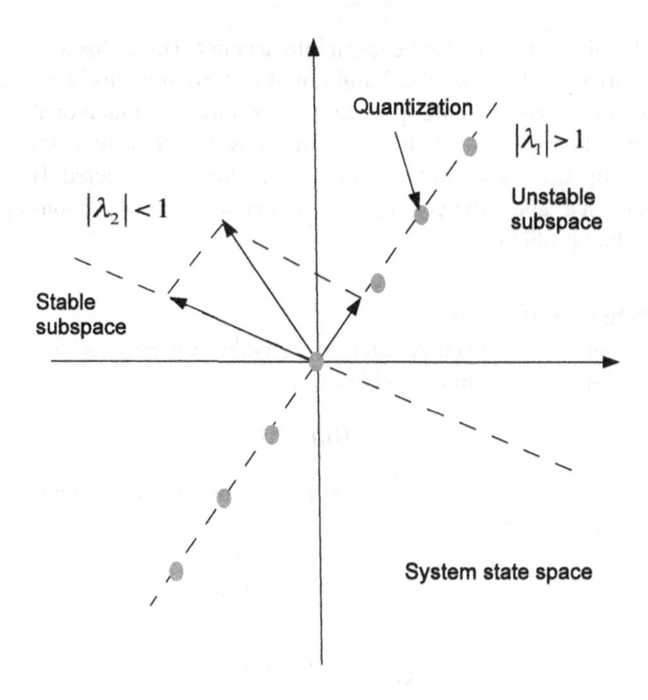

FIG. 8.4

Illustration of unstable and stable subspaces.

where $\hat{z}_n$ is the estimation of z_n. We assume that the matrix **UB** has a full row rank, which assures the existence of **v**. When there is no random noise (namely $\tilde{n}(t) = 0$) and the state estimation $\hat{z}_n$ is perfect, the blow-up control action drives the system state back to zero, which is assumed to be desired. Then the variance of system state in the nth subspace is given by

$$V[z_n] = E[\tilde{n}_n^2] + |\lambda_n|^2 E[\epsilon_n^2], \tag{8.84}$$

where ϵ is the estimation error of z_n.

Based on the above discussion, we define the distortion as

$$\mathcal{D} = \sum_{n \in \Omega} V[z_n], \tag{8.85}$$

namely the sum variance of the system state in the unstable subspaces. The fluctuations in the stable subspaces are not taken into consideration. Note that Eq. (8.85) is a heuristic definition for the distortion. It emphasizes the unstable subspaces and totally omits the stable ones, since the latter can be automatically stabilized and do not affect the system stability. Below are some other possibilities as regards distortion:

- The stable subspaces can also be taken into account. This helps to better suppress the total variance. However, the definition of distortion in this book increases the dimension of the system state space and incurs more computational cost.
- Typical criteria in controls such as quadratic cost functions (e.g., the cost function in the framework of an LQG system) can be considered. However, we are still not clear about the connection between these cost functions and the source coding problem.

Lattice-based quantization

An effective approach for quantization is to apply the concept of a lattice. A lattice is defined as the following infinite set [162,164]:

$$\Sigma = \{\mathbf{Gi}, \mathbf{i} \in \mathbf{Z}^n\}, \tag{8.86}$$

where $\mathbf{Z} = (\ldots, -2, -1, 0, 1, 2, \ldots)$, n is the dimension, and $\mathbf{G}$ is an $n \times n$ matrix, which is called the generation matrix.

We then study how to form lattices for the observations $\mathbf{y}_1$ and $\mathbf{y}_2$. We consider a higher dimension lattice generated by $\mathbf{G}_1$ and $\mathbf{G}_2$ (i.e., the generating matrices for $\mathbf{y}_1$ and $\mathbf{y}_2$); i.e.,

$$\mathbf{G}_{12} = (\mathbf{G}_1^T, \mathbf{G}_2^T)^T \tag{8.87}$$

and

$$\begin{aligned}\Sigma_{12} &= \{\mathbf{G}_{12}\mathbf{i}, \mathbf{i} \in \mathbf{Z}^{N_1+N_2}\} \\ &= \{((\mathbf{G}_1\mathbf{i}_1)^T, (\mathbf{G}_2\mathbf{i}_2)^T)^T, \mathbf{i}_1 \in \mathbf{Z}^{N_1}, \mathbf{i}_2 \in \mathbf{Z}^{N_2}\}.\end{aligned} \tag{8.88}$$

The next task is to design the generating matrices $\mathbf{G}_1$ and $\mathbf{G}_2$. We apply the distortion function defined in Eq. (8.85). It is assumed that the system state estimation based on $\mathbf{y}_1$ and $\mathbf{y}_2$ is obtained from

$$\hat{\mathbf{x}} = (\mathbf{C}_1^T\mathbf{C}_1 + \mathbf{C}_2^T\mathbf{C}_2)^{-1}(\mathbf{C}_1^T, \mathbf{C}_2{}^T)(\mathbf{y}_1^T, \mathbf{y}_2^T)^T. \tag{8.89}$$

We map the space to the unstable subspaces of the system state. Then the lattice generated by matrix $\mathbf{G}_x$ is given by

$$\mathbf{G}_x = \tilde{\mathbf{U}}\tilde{\mathbf{C}}\mathbf{G}_{12}, \tag{8.90}$$

where $\tilde{\mathbf{U}}$ consists of the eigenvectors in $\mathbf{U}$ associated with the unstable subspaces, and $\tilde{\mathbf{C}} = (\mathbf{C}_1^T\mathbf{C}_1 + \mathbf{C}_2^T\mathbf{C}_2)^{-1}(\mathbf{C}_1^T, \mathbf{C}_2{}^T)$.

We want to equalize the mean square quantization errors in each unstable subspace. To that end, the lattice is designed to be parallel to the axes, and the distances between two neighboring lattice points are designed to be proportional to $\left\{\frac{1}{\lambda n}\right\}_{n\in\Omega}$. This means that, in a particular unstable subspace, the quantization error variance is inversely proportional to the corresponding eigenvalue, since the eigenvalue scales the impact of the quantization error on the system state. Hence the generating matrix $\mathbf{G}_x$ is designed to be

$$\mathbf{G}_x = \Delta \begin{pmatrix} \frac{1}{\lambda_1} & 0 & \cdots & 0 \\ 0 & \frac{1}{\lambda_2} & \cdots & 0 \\ \vdots & \vdots & \ddots & \vdots \\ 0 & 0 & \cdots & \frac{1}{\lambda_{N_0}} \end{pmatrix}, \tag{8.91}$$

where Δ is a constant to control the tradeoff between coding rate and quantization errors. This procedure is illustrated in Fig. 8.5 when $\mathbf{y}_2$ (i.e., $\mathbf{C}_2 = 0$) is ignored.

For lattices Σ_1 and Σ_2, we require

$$E[\mathbf{y}_1|\mathbf{y}_2] \in \Sigma_1, \quad \forall \mathbf{y}_2 \in \Sigma_2. \tag{8.92}$$

If $\mathbf{y}_1$ and $\mathbf{y}_2$ are separable, we have

$$h_{21}(\mathbf{y}_2) + E[\mathbf{w}_{12}] \in \Sigma_1, \quad \forall \mathbf{y}_2 \in \Sigma_2. \tag{8.93}$$

8.4.4 CODING PROCEDURE

Given the above schemes of distortion and lattice quantization, we have converted the continuously valued information sources to discrete ones. Then we propose an approach to implement the source coding in a distributed manner without explicit collaborations between the sensors.

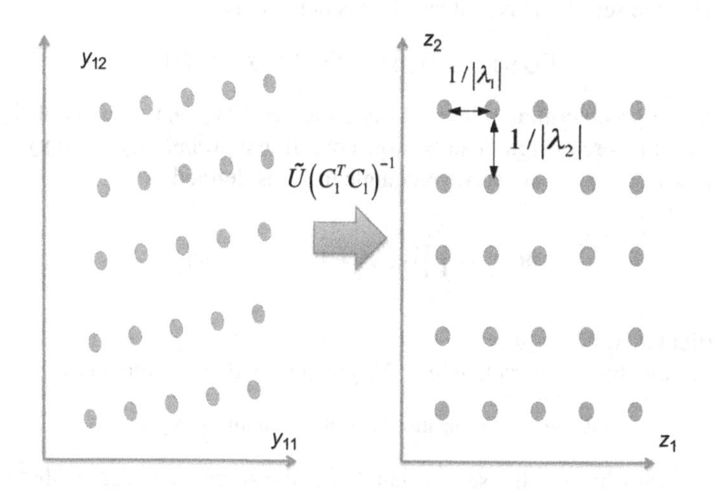

FIG. 8.5

Illustration of transformation to the desired lattice.

Slepian-Wolf coding

We first revisit the well-known scheme of Slepian-Wolf coding proposed in 1973. Consider two information sources whose symbols are denoted by $X_1(t)$ and $X_2(t)$. There is no communication between the two information sources. The surprising conclusion of Slepian-Wolf coding is that it can achieve asymptotically lossless coding when the sum of the transmission rates of the two sources equals the entropy of the information sources [155]; i.e.,

$$R_1 + R_2 = H(X_1, X_2), \tag{8.94}$$

as if the two information sources can perfectly collaborate as a single information source, although there is no explicit communication and coordination between the two information sources.

Coloring-based coding

We now apply the principle of Slepian-Wolf coding in the context of distributed source coding for a CPS. The main challenge is that the coding delay in a CPS cannot be tolerated while the Slepian-Wolf coding needs an infinite delay. Moreover, the random codebook in Slepian-Wolf coding is infeasible in practice, while the encoding/decoding complexities must be addressed in our proposed coding scheme.

In this book, we adopt the approach of coloring-based coding. It is well known that the source coding for computing functions with side information is equivalent to the problem of coloring a graph. We apply the same idea to distributed source coding in a CPS.

For simplicity of analysis, it is assumed that sensor 2 sends its full quantized version of $\mathbf{y}_2$. This provides side information for the transmission and decoding of $\mathbf{y}_1$ from sensor 1 and thus can decrease the communication requirement of sensor 1. For each $\mathbf{y}_2$, the set of related values of $\mathbf{y}_1$ is defined as

$$\Gamma(\mathbf{y}_2, \gamma_1) = \{\mathbf{y}_1 | \mathbf{y}_1 \in \Sigma_1, P(\mathbf{y}_1 | \mathbf{y}_2) > \gamma_1\}, \tag{8.95}$$

where γ_1 is a predetermined threshold and the set $\Gamma(\mathbf{y}_1, \gamma_1)$ consists of the lattice points in Σ_1 that are of significantly large conditional probability given $\mathbf{y}_2$.

An $n_1 \times n_2 \times \cdots \times n_{N_1}$ super-rectangle in Σ_1 is defined as

$$\mathbf{R} = \mathbf{G}_1 \prod_{p=1}^{N_1} [y_p, y_p + 1, \ldots, y_p + n_p], \tag{8.96}$$

for a particular $\{y_p\}_{p=1,\ldots,N_1}$.

The compatible super-rectangle in Σ_1 given $\mathbf{y}_2$ and γ_1 is defined as

$$\mathbf{R}_1(\mathbf{y}_2, \gamma_1) = \text{the minimum } \mathbf{R} \text{ containing } \Gamma(\mathbf{y}_1, \gamma_1). \tag{8.97}$$

The reason why we choose a rectangle for the shape of the set is that it results in an efficient coloring scheme and the corresponding source coding, which will be detailed later. Once γ_1 is properly selected, we assume that the probability of $\mathbf{y}_1$

falling out of $\mathbf{R}_1(\mathbf{y}_2, \gamma_1)$ is negligible. For the generic case, the compatible super-rectangles of different $\mathbf{y}_2$ values may have different sizes. However, for the case of separable $\mathbf{y}_1$ and $\mathbf{y}_2$, the super-rectangles of different $\mathbf{y}_2$ values have the same sizes (with different centers); i.e.,

$$\mathbf{R}_1(\mathbf{y}_2, \gamma) = h_{21}(\mathbf{y}_2) + \bar{\mathbf{R}}(\gamma_1), \tag{8.98}$$

where $\bar{\mathbf{R}}(\gamma_1)$ is the basic rectangle defined as

$$\bar{\mathbf{R}}_1(\gamma) = \text{the minimum } \mathbf{R} \text{ containing } \Gamma(h_{12}^{-1}(\mathbf{0}), \gamma_1). \tag{8.99}$$

The definition of super-rectangles is illustrated in Fig. 8.6, in which $\mathbf{y}_2^0$ in Σ_2 corresponds to $\mathbf{y}_1^0$ in Σ_1 (i.e., $\mathbf{y}_1^0$ is the value of $\mathbf{y}_1$ when there is no noise in the observation). A super-rectangle is given by the dotted lines. Hence with a large probability, $\mathbf{y}_1$ lies in the area specified by the dotted lines.

We have assumed that the precise information of quantized $\mathbf{y}_2$ has been transmitted to the fusion center, as the side information for sensor 1. Since with a large

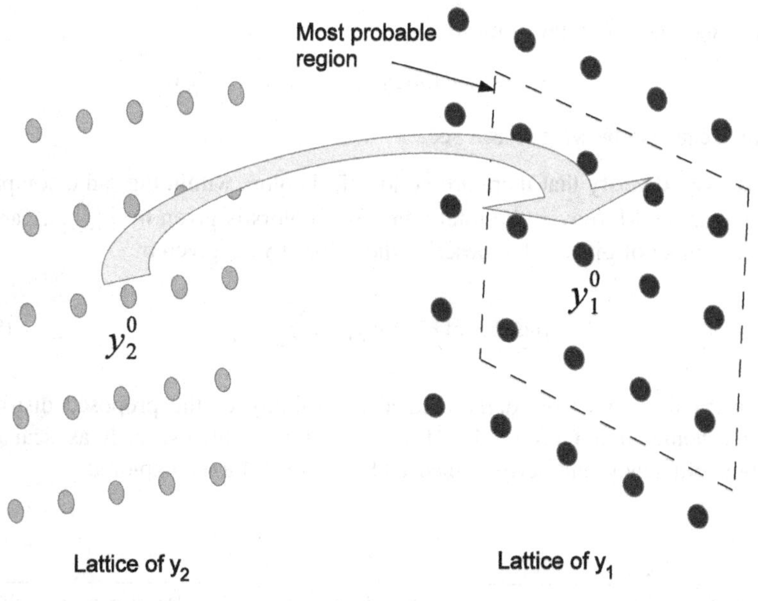

FIG. 8.6

Illustration of super-rectangles.

probability $\mathbf{y}_1$ falls in the compatible super-rectangle $\mathbf{R}_1(\mathbf{y}_2, \gamma_1)$,[1] sensor 1 only needs to specify the point within $\mathbf{R}_1(\mathbf{y}_2, \gamma_1)$. If the labeling of each point in Σ_1 is considered as coloring a set of elements, two different points in $\mathbf{R}_1(\mathbf{y}_2, \gamma_1)$ should have different colors; otherwise, the same color causes decoding confusion.

Hence the following requirements are needed for the coloring problem:

- There are no different points having the same color in the same compatible super-rectangle.
- The mappings to and from the color should be computationally efficient.

Although algorithms in graph theory can be applied to minimize the number of colors used in the coding of $\mathbf{y}_1$ with the guarantee of different colors, the mapping between lattice points and colors can be irregular and needs look-up tables, which substantially increases the computational complexity. Hence we propose a simple but efficient approach of coloring:

- First we determine the maximal sizes in different dimensions of a compatible super-rectangle; i.e.,

$$n_k^* = \max\{n_k | n_k \text{ is the length of the } k\text{th dimension of a compatible super-rectangle}\}. \quad (8.100)$$

- Then for a lattice point in Σ_1 which can be written as

$$\mathbf{y}_1 = \mathbf{G}_1(n_1, \ldots, n_{N_1})^T, \quad n_k \in Z, \quad (8.101)$$

  we assign the following color to it:

$$c(\mathbf{y}_1) = (\text{mod}(n_1, n_1^*), \ldots, \text{mod}(n_{N_1}, n_{N_1}^*)), \quad (8.102)$$

  which can also be written as a vector.

  It is easy to verify that there are no identical colors within the same compatible super-rectangle. Moreover, the total number of colors is given by $\prod_{k=1}^{N_1} n_k^*$ and the average number of bits used to describe the color of $\mathbf{y}_1$ is given by

$$E[\text{number of bits for } \mathbf{y}_1] = \sum_{k=1}^{N_1} \log_2 n_k^*. \quad (8.103)$$

Numerical results have demonstrated the validity of the proposed distributed coding scheme in a CPS [154]. However, more problems, such as scalability, algorithm efficiency, and performance analysis, are still to be explored.

[1]When $\mathbf{y}_1$ falls out of $\mathbf{R}_1(\mathbf{y}_2, \gamma_1)$, the controller claims that there is a decoding error and discards the received message. Such a decoding error does not cause much impact on the control since the probability is very small and the controller can estimate the system state using previous history.

8.4.5 ADAPTATION TO PHYSICAL DYNAMICS

One of the challenges of communications for a CPS is that the distribution of sensor observations is time variant. Hence we need to incorporate the change of the distribution into the distributed source coding scheme.

Prediction

Due to the change of distribution in the physical dynamics, it is necessary for the controller to predict the most probable range of the observation in the next time slot. Suppose that all the received messages from the two sensors before time slot t are given by

$$\mathbf{c}_i^{t-1} = (\ldots, c_i(t-2), c_i(t-1)), \quad i = 1, 2. \tag{8.104}$$

The distribution of observations in the next time slot (before receiving the observation in the next time slot) is given by the conditional probability $P(\mathbf{y}|\mathbf{c}_1^{t-1}, \mathbf{c}_2^{t-1})$. We assume that the fusion center broadcasts the received messages back to the sensors such that the sensors share with the controller the same information on $P(\mathbf{y}|\mathbf{c}_1^{t-1}, \mathbf{c}_2^{t-1})$. For the case of no feedback, the sensors and controller do not have the same set of information, which may make the design much more complicated.

Take the linear system in Eq. (8.79) with Gaussian noise, for instance. We can use Kalman filtering to obtain an approximation of $P(\mathbf{y}|\mathbf{c}_1^{t-1}, \mathbf{c}_2^{t-1})$, namely

$$\bar{\mathbf{x}}(t|t-1) = \mathbf{A}\bar{\mathbf{x}}(t-1|t-1) + \mathbf{B}\mathbf{u}(t) \tag{8.105}$$

and

$$\Sigma(t|t-1) = \mathbf{A}\Sigma(t-1|t-1)\mathbf{A}^T + \mathbf{W}, \tag{8.106}$$

where $\bar{\mathbf{x}}(t|t-1)$ is the expected system state at time t, $\Sigma(t|t-1)$ is the expected covariance matrix, $\bar{\mathbf{x}}(t-1|t-1)$ is the system estimation at time $t-1$, and $\Sigma(t-1|t-1)$ is the covariance matrix at time $t-1$. Then the distribution of $\mathbf{y}$ is given by

$$P(\mathbf{y}|\mathbf{c}_1^{t-1}, \mathbf{c}_2^{t-1}) \sim \mathcal{N}(\mathbf{C}\bar{\mathbf{x}}(t|t-1), \mathbf{C}\Sigma(t|t-1)\mathbf{C}^T + E[\mathbf{w}\mathbf{w}]^T). \tag{8.107}$$

Note that the above prediction consists of quantization errors due to the digital communications, which is negligible if the lattice is sufficiently dense in space.

Range adaptation

We make the assumption that both the sensors and the fusion center share the same estimation of the distribution $P(\mathbf{y}|\mathbf{c}_1^{t-1}, \mathbf{c}_2^{t-1})$, due to the feedback from the controller. We define the most probable region of $\mathbf{y}_2(t)$ based on the prediction to be

$$\Gamma_2(t, \gamma_2) = \{\mathbf{y}_2 \in \Sigma_2 | P(\mathbf{y}|\mathbf{c}_1^{t-1}, \mathbf{c}_2^{t-1}) > \gamma_2\}, \tag{8.108}$$

and the most probable super-rectangle, denoted by $\mathbf{R}_2(t, \gamma_2)$, to be the minimum super-rectangle in Σ_2 that includes $\Gamma_2(t, \gamma_2)$. Then similarly to the coloring scheme of $\mathbf{y}_1$ in distributed source coding, we define m_k^* as the maximum length of the kth dimension of all most probable super-rectangles of $\mathbf{y}_2$. Then the coding of $\mathbf{y}_2$ is given by

$$c(\mathbf{y}_2) = (\text{mod}(m_1, m_1^*), \ldots, \text{mod}(m_{N_2}, m_{N_2}^*)), \tag{8.109}$$

where $(m_1, \ldots, m_{N_2})$ is the coordinate of $\hat{\mathbf{y}}_2$ in Σ_2 that is closest to $\mathbf{y}_2$; i.e.,

$$\mathbf{y}_1 = \mathbf{G}_1(m_1, \ldots, m_{N_1})^T, \quad m_k \in Z. \tag{8.110}$$

Based on the above discussion, the procedures for design and online operation of distributed source coding in a CPS are summarized in Procedures 7 and 8.

PROCEDURE 7 DISTRIBUTED SOURCE CODING IN A CPS: DESIGN

1: Decompose the vector space of the system state into unstable and stable subspaces using eigenstructure decomposition, as in Eq. (8.81).
2: Determine the lattice of $\mathbf{z}$ using Eq. (8.91).
3: Determine the lattice-generating matrices $\mathbf{G}_1$ and $\mathbf{G}_2$ using Eqs. (8.90) and (8.92).
4: Determine the thresholds γ_1 and γ_2.
5: Determine the maximal sizes of compatible super-rectangle and most probable super-rectangle.
6: Color the lattice points of $\mathbf{y}_1$ and $\mathbf{y}_2$ using Eqs. (8.102) and (8.109).

PROCEDURE 8 DISTRIBUTED SOURCE CODING IN A CPS: CODING AND DECODING

1: **for** Each time slot **do**
2: Both the controller and sensor predict the distribution of $\mathbf{y}$ using Eq. (8.107).
3: Both the controller and sensor 2 determine the most probable region of $\mathbf{y}_2$ using Eq. (8.108).
4: Sensor 1 sends the report using the color of the lattice point closest to $\mathbf{y}_1$.
5: Sensor 2 sends the report using the color of the lattice point closest to $\mathbf{y}_2$.
6: The controller uses the most probable region to determine $\mathbf{y}_2$ according to the report from sensor 2.
7: The controller uses the compatible super-rectangle to determine $\mathbf{y}_1$ according to the report from sensor 1 and the recovered $\mathbf{y}_2$.
8: The controller feeds back the received messages to both sensors.
9: **end for**

8.5 PHYSICAL DYNAMICS-AWARE CHANNEL DECODING

In this section, we use existing coding schemes and outline a decoding procedure that considers the physical dynamics in a CPS, following the study in Ref. [165].

8.5.1 MOTIVATION

Since the reliability of communication is usually very high in CPSs (e.g., in the wide area monitoring system in smart grids, the packet error rate should be less than 10^{-5} [166]) while the communication channel may experience various degradations (e.g., the harsh environments for communications in smart grids), it is important to use channel coding to protect the information transmission.

A straightforward approach for coding in a CPS is to follow the traditional procedure of separate quantization, coding, transmission, decoding, and further processing, which is adopted in conventional communication systems (Fig. 8.7). However, separate channel decoding and system state estimation may not be optimal when there exist redundancies in the transmitted messages. For example, the transmitted codeword at time t, $\mathbf{b}(t)$, is generated by the observation on the physical dynamics, $\mathbf{y}(t)$. Due to the time correlation of system states, $\mathbf{y}(t)$ is correlated with $\mathbf{y}(t+1)$, thus being able to provide information for decoding the codeword in the next time slot $t+1$. Hence if the decoding procedure is independent of the system state estimation, the permanence will be degraded due to the loss of redundancy. One may argue that the redundancy can be removed in the source coding procedure (e.g., encoding only the innovation vector in Kalman filtering if the physical dynamics is linear with Gaussian noise [167]). However, this requires substantial computation, which a sensor may not be able to carry out. Moreover, the sensor may not have the overall system parameters for the extraction of innovation information. Hence it is more desirable to shift the burden to the controller, which has more computational power and more system information; then the inexpensive sensor focuses on transmitting the raw data.

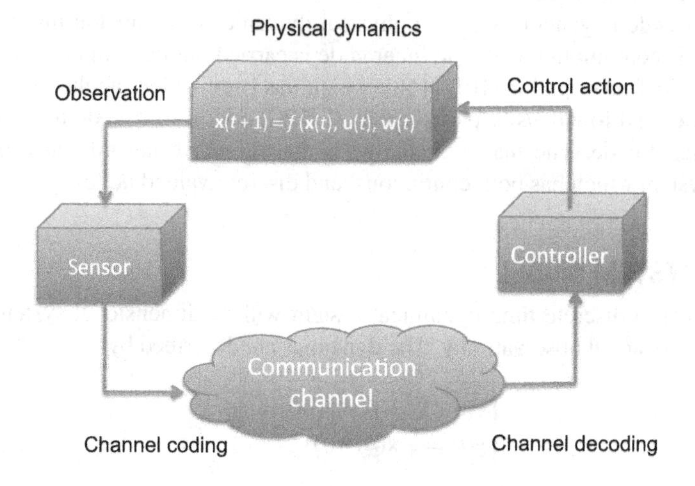

FIG. 8.7

Illustration of the components in a CPS.

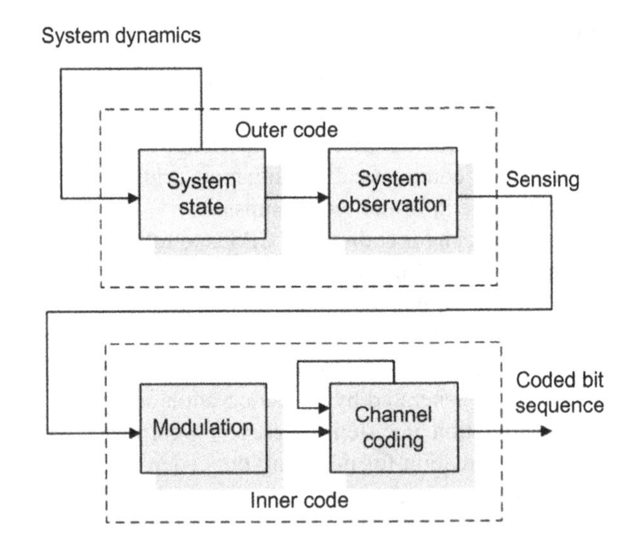

FIG. 8.8

Illustration of the equivalent concatenated code for system state evolution and channel coding.

In this section, we propose an algorithm for joint channel decoding and state estimation in a CPS. We consider traditional channel codes with memory (e.g., convolutional codes). Then the physical dynamics system state evolution and the channel codes can be considered as an equivalent concatenated code, as illustrated in Fig. 8.8, in which the outer code is the generation of system observation (we call it a "nature encoded" codeword) while the inner code is the traditional channel coding. The outer code is generated by the physical dynamics in an analog manner (since the value is continuous) while the inner code is carried out by a digital encoder. We will use belief propagation (BP) [168], which has been widely applied in decoding turbo codes and low-density parity-check (LDPC) codes, to decode this equivalent concatenated code. The major challenge is that the Bayesian inference is over a hybrid system which has both continuous- and discrete-valued nodes.

8.5.2 SYSTEM MODEL

We consider a discrete-time dynamical system with N-dimensional system state $\mathbf{x}$ and K-dimensional observation $\mathbf{y}$. The dynamics are described by

$$\begin{cases} \mathbf{x}(t+1) = f(\mathbf{x}(t), \mathbf{u}(t), \mathbf{n}(t)), \\ \mathbf{y}(t) = g(\mathbf{x}(t), \mathbf{w}(t)), \end{cases} \tag{8.111}$$

where $\mathbf{u}$ is the control action taken by the controller, and $\mathbf{n}$ and $\mathbf{w}$ are both random perturbations, whose probability distributions are assumed to be known.

A special case is linear system dynamics, which is described by

$$\begin{cases} \mathbf{x}(t+1) = \mathbf{A}\mathbf{x}(t) + \mathbf{B}\mathbf{u}(t) + \mathbf{n}(t), \\ \mathbf{y}(t) = \mathbf{C}\mathbf{x}(t) + \mathbf{w}(t), \end{cases} \tag{8.112}$$

where $\mathbf{n}$ and $\mathbf{w}$ are both Gaussian noises with zero means and covariance matrices Γ_n and Γ_w, respectively.

We assume that $\mathbf{y}(t)$ is observed by a sensor.[2] It quantizes each dimension of the observation using B bits, thus forming a KB-dimensional binary vector which is given by

$$\mathbf{b}(t) = (b_1(t), b_2(t), \ldots, b_{KB}(t)). \tag{8.113}$$

We assume that the sensor simply carries out a scalar quantization for each dimension and does not use compression to remove the redundancy between $\mathbf{y}(t)$ and $\mathbf{y}(t-1)$. It is straightforward to extend this to the case of vector quantization. However, it is nontrivial to extend it to the case of partial compression of the redundancy, which is beyond the scope of this study.

The information bits $\mathbf{b}(t)$ are put into an encoder to generate a codeword $\mathbf{c}(t)$. We assume binary phase shift keying (BPSK) modulation for the communication from the sensor to the controller. which is given by

$$\mathbf{r}(t) = 2\mathbf{c}(t) - 1 + \mathbf{e}(t), \tag{8.114}$$

where $2\mathbf{c}(t) - 1$ converts the alphabet $\{0, 1\}$ to the antipodal signal $\{-1, +1\}$, and $\mathbf{e}(t)$ is the additive white Gaussian noise with zero expectation and normalized variance σ_e^2. In this chapter, we do not consider fading, since within one codeword the channel is usually stationary and the channel gain can be incorporated (together with the transmit power) into the noise variance σ_e^2.

8.5.3 JOINT DECODING

As we have explained, there exists information redundancy between the system state in different time slots, as well as the received bits in the communications. Hence we can apply the framework of BP for the procedure of joint decoding and system state estimation.

A brief introduction to Pearl's BP

We first provide a brief introduction to the BP algorithm, in particular a version of Pearl's BP. We use an example to illustrate the procedure. The details and formal description can be found in Ref. [168].

[2] In this study, we consider the single sensor case. The same analysis and algorithm can be extended to the case of multiple sensors.

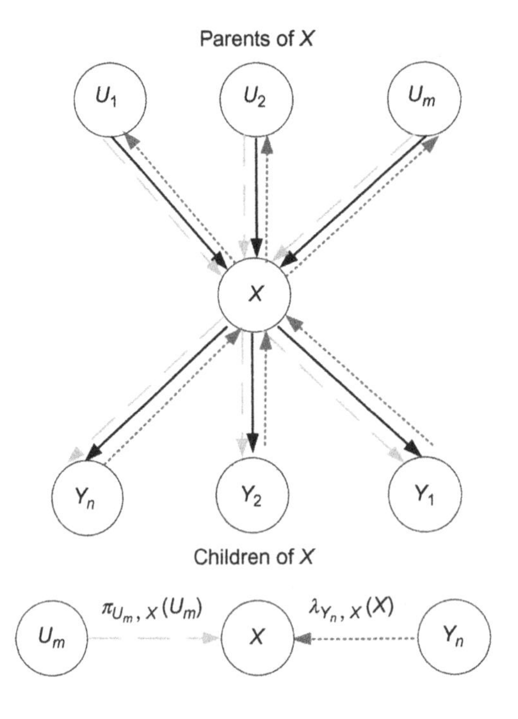

FIG. 8.9

Illustration of message passing in BP.

Take Fig. 8.9 for instance, where a random variable X has parents $U_1, U_2, \ldots, U_M$ and children $Y_1, Y_2, \ldots, Y_N$. Here the parent nodes are characterized by the following equality of conditional probability:

$$P(X|\text{all other random variables}) = P(X|U_1, U_2, \ldots, U_M), \tag{8.115}$$

that is, given the random variables in the parent nodes, X is independent of all nonparent nodes. As indicated by the name, we say that a node is a child of X if X is a parent of it. The message passing of Pearl's BP is illustrated by dotted arrows and dashed arrows in the figure. The dashed arrows transmit π-messages from a parent to its children. For instance, the message passing from U_m to X is $\pi_{U_\text{m},X}(U_\text{m})$, which is the prior information of U_m given that all the information U_m has been received. The dotted arrows transmit λ-messages from child to parent. For instance, the message passed from Y_n to X is $\lambda_{Y_n,X}(X)$, which is the likelihood of X given that the information Y_n has been received. After X receives all π-messages (i.e., $\pi_{U_\text{m},X}(U_\text{m})$ from its parents $U_1, U_2, \ldots, U_M$) and all λ-messages (i.e., $\lambda_{Y_n,X}(X)$, X from its children $Y_1, Y_2, \ldots, Y_N$), X updates its local belief information $BEL_X(\mathbf{x})$, and transmits λ-messages $\lambda_{X,U_\text{m}}(U_\text{m})$ to its parents and π-messages $\pi_{X,Y_n}(\mathbf{x})$ to its children. The expressions for these messages are given by

$$\pi_X(\mathbf{x}) = \sum_{\mathbf{U}} p(\mathbf{x}|\mathbf{U}) \prod_{m=1}^{M} \pi_{U_m,X}(U_m), \tag{8.116}$$

$$\gamma_X(\mathbf{U}) = \sum_{\mathbf{x}} \prod_{n=1}^{N} \lambda_{Y_n,X}(\mathbf{x}) p(\mathbf{x}|\mathbf{U}), \tag{8.117}$$

$$BEL_X(\mathbf{x}) = \alpha \times \prod_{n=1}^{N} \lambda_{Y_n,X}(\mathbf{x}) \times \pi_X(\mathbf{x}), \tag{8.118}$$

$$\lambda_{X,U_m}(U_m) = \sum_{\mathbf{U}, \neq U_m} \gamma_X(\mathbf{U}) \times \prod_{j \neq m} \pi_{U_j,X}(U_j), \tag{8.119}$$

$$\pi_{X,Y_n}(\mathbf{x}) = \pi_X(\mathbf{x}) \times \prod_{i \neq n} \lambda_{Y_i,X}(\mathbf{x}), \tag{8.120}$$

where $\mathbf{U} = (U_1, U_2, \ldots, U_M)$ and $\mathbf{Y} = (Y_1, Y_2, \ldots, U_N)$.

In the initialization stage of Pearl's BP, the initial values are defined as

$$\lambda_{X,U}(\mathbf{u}) = \begin{cases} p(\mathbf{x}_0|\mathbf{u}), & X \text{ is evidence}, \quad \mathbf{x} = \mathbf{x}_0, \\ 1, & X \text{ is not evidence} \end{cases} \tag{8.121}$$

and

$$\pi_{X,Y}(\mathbf{x}) = \begin{cases} \delta(\mathbf{x}, \mathbf{x}_0), & X \text{ is evidence}, \quad \mathbf{x} = \mathbf{x}_0, \\ p(\mathbf{x}), & X \text{ is not evidence}. \end{cases} \tag{8.122}$$

Iterative decoding

Based on the tools of a Bayesian network, we can derive the iterative decoding procedure. Fig. 8.10 shows the procedure of message passing in the decoding of a CPS: $\mathbf{x}_{t-2}$ summarizes all the information obtained from previous time slots and transmits π-message $\pi_{\mathbf{x}_{t-2},\mathbf{x}_{t-1}}(\mathbf{x}_{t-2})$ to $\mathbf{x}_{t-1}$. The BP procedure can be implemented in either synchronous or asynchronous manner. To accelerate the procedure, we implement the asynchronous Pearl BP. The updating order and message passing in one iteration are given in Procedure 9.

The algorithm has been tested in the context of voltage control in smart grids [165]. Numerical results have shown that the performance of both decoding and system state estimation is substantially improved, when compared with traditional separate decoding and estimation. One potential challenge in joint decoding and estimation is the possibility of error propagation; i.e., the decoding error or estimation error will be propagated to the next time slot, since the decoding and estimation procedures in different time slots are correlated. An effective approach to handle error propagation is to monitor the performance and restart the decoding/estimation procedure once performance degradation is detected.

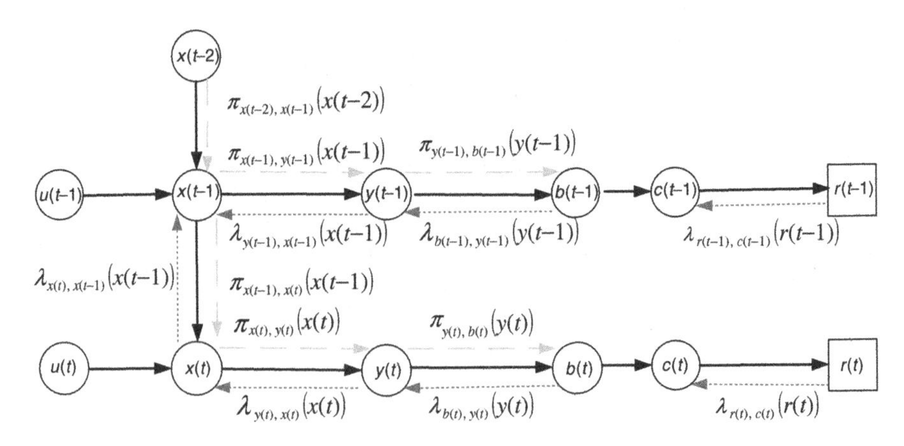

FIG. 8.10

Bayesian network structure and message passing for a CPS.

PROCEDURE 9 BP-BASED CHANNEL DECODING

1: Set the maximal iteration times N
2: **for** Each time slot t **do**
3: Receive $\mathbf{r}(t)$.
4: **for** Iteration time $i \leq N$. **do**
5: Pass message $\mathbf{x}_{t-1} \rightarrow 213\ \mathbf{y}_{t-1}$; i.e., the sensor observation procedure.
6: Pass message $\mathbf{y}_{t-1} \rightarrow 213\ \mathbf{b}_{t-1}$; i.e., the quantization procedure.
7: Pass message $\mathbf{b}_{t-1} \rightarrow 213\ \mathbf{y}_{t-1}$; i.e., the observation reconstruction procedure.
8: Pass message $\mathbf{y}_{t-1} \rightarrow 213\ \mathbf{x}_{t-1}$; i.e., the system estimation procedure.
9: Pass message $\mathbf{x}_{t-1} \rightarrow 213\ \mathbf{x}_{t}$; i.e., the system state evolution.
10: Pass message $\mathbf{x}_{t} \rightarrow 213\ \mathbf{y}_{t}$; i.e., the observation procedure.
11: Pass message $\mathbf{y}_{t} \rightarrow 213\ \mathbf{b}_{t}$; i.e., the quantization procedure.
12: Pass message $\mathbf{b}_{t} \rightarrow 213\ \mathbf{y}_{t}$; i.e., the observation reconstruction procedure.
13: Pass message $\mathbf{y}_{t} \rightarrow 213\ \mathbf{x}_{t}$; i.e., the system state estimation procedure.
14: Pass message $\mathbf{x}_{t} \rightarrow 213\ \mathbf{x}_{t-1}$; i.e., the trace back procedure.
15: $\mathbf{x}_{t-1}$ updates information.
16: **end for**
17: Use the belief of $\mathbf{b}(t)$ as the prior information for the decoding with $\mathbf{r}(t)$.
18: **end for**

8.6 CONTROL-ORIENTED CHANNEL CODING

Most channel coding schemes are designed for common purpose data communications. Although they can also be used in the context of controlling a CPS, it is not clear whether these coding schemes designed for pure data communications are optimal for the purpose of control. There have been some studies on designing specific channel codes for control systems [169,170]. In this section we give a brief introduction to these efforts, which may provide insights for further communication system designs in CPSs, although they have not been implemented in real systems.

8.6.1 TRAJECTORY CODES

Trajectory codes for the purpose of automatic control were proposed in Ref. [169]. The main feature is the online error correction capability for real-time estimation and control.

System model

We assume that the communication channel has binary input and output. The transmitter has a time-varying state x_t, which is represented by the vertex of a graph G. The edges in the graph G show the possibility of state transitions. It is assumed that G is known to both the transmitter and receiver. This setup is similar to the random walk on graphs.

The time is divided into rounds and channel use times. In each round, the state of the transmitter makes one move in G, and then the transmitter can use the communication channel M times. The encoded bits are based on all the history of the transmitter state. The goal of the design is to find a coding scheme that enables the receiver to obtain accurate estimation of the transmitter's state with a large probability. Hence the transmitter can be considered as the combination of physical dynamics and communication transmitter. The coding rate is given by

$$\rho = \frac{\log \Delta}{M}, \tag{8.123}$$

where Δ is the maximum out-degree or in-degree of nodes in the graph G.

The receiver outputs an estimation on x_t, which is denoted by $\hat{x}_t$, based on all the received coded bits. The error of decoding is represented by the shortest length between x_t and $\hat{x}_t$ in the graph G, which is denoted by $d_G(x, \hat{x})$. The communication scheme has an error exponent κ if

$$P(d_G(x, \hat{x}) \geq l) \leq \exp(-\kappa l). \tag{8.124}$$

It was shown in Ref. [169] that the communication is online efficient if the time and space complexities of coding and decoding are of the order of $(\log t)^{O(1)}$. If the coding scheme has a positive rate, positive error exponent and is online efficient, then we say that the code is asymptotically good.

Note that the problem considered above is more like a system estimation problem. The control procedure is not considered; i.e., the system state x_t evolves independently of the communication performance. However, the system state is closely related to the control problems, since the error of system state estimation substantially determines the performance of control; meanwhile the system state x_t can be considered as the system state in which the impact of control has been removed.

Concept of trajectory codes

The trajectory of the transmitter's state can be represented by a path in the corresponding graph G, whose initial time is denoted by t_0. If two trajectories, γ and γ', have the same length, the same starting time and the same initial vertex, we denote this by $\gamma \sim \gamma'$. The distance between two trajectories indicates the number of different states in the two trajectories and is denoted by $\tau(\gamma, \gamma')$.

A trajectory code, denoted by χ, maps from a trajectory to a certain alphabet in a concatenation manner:

$$\chi(\gamma) = \left(\chi(\gamma(t_0+1), t_0+1), \ldots, \chi(\gamma(t_0+t), t_0+t)\right). \tag{8.125}$$

The Hamming distance between two equal-length codewords is denoted by h. The relative distance of a trajectory code is denoted by

$$\delta = \inf_{\gamma \sim \gamma'} \frac{h(\chi(\gamma), \chi(\gamma'))}{\tau(\gamma, \gamma')}, \tag{8.126}$$

which implies the capability of distinguishing two different trajectories of the transmitter's state. A trajectory code is said to be asymptotically good if it has a positive rate and provides a positive relative distance.

Obviously, the larger the relative distance δ, the more capable the receiver is of estimating the transmitter's state. When δ is large, even if there are some transmission errors, the remaining discrepancy between two codewords can still be used to distinguish the two different state trajectories. Then the major challenge is the existence and construction of an asymptotically good trajectory code. The following theorem guarantees the existence of correspondence:

Theorem 55. *Every graph has an asymptotically optimal trajectory code. A trajectory code with $\delta < 1$ is always feasible.*

The details of the proof are given in Ref. [169]. Here we provide a sketch of the proof, which is based on the probabilistic approach and random coding, similarly to Shannon's random coding argument [33]. First we consider a random coding scheme; i.e., the transmitter randomly selects an output from the output alphabet for each possible input (i.e., a state, or equivalently a vertex in G). Consider two trajectories γ_1 and γ_2, which satisfy $\gamma_1 \sim \gamma_2$ but do not have common vertices except for the initial one. We call these two trajectories twins. It is easy to verify that

$$\inf_{\gamma_1 \sim \gamma_2} \frac{h(\chi(\gamma), \chi(\gamma'))}{\tau(\gamma, \gamma')} = \inf_{\gamma_1, \gamma_2 \text{ are twins}} \frac{h(\chi(\gamma), \chi(\gamma'))}{\tau(\gamma, \gamma')}. \tag{8.127}$$

Then we define the event

$$A_\gamma = \left\{\gamma = (\gamma_1, \gamma_2) | \gamma_1 \sim \gamma_2, \frac{h(\chi(\gamma), \chi(\gamma'))}{\tau(\gamma, \gamma')}) \le \delta\right\}. \tag{8.128}$$

If we can prove

$$\cap_\gamma A_\gamma^c \neq \phi, \tag{8.129}$$

then there exists at least one asymptotically good trajectory code. In Ref. [169], Eq. (8.129) is proved by citing the Lovasz Local Lemma.

Construction of trajectory codes

Although the existence of trajectory codes has been proved, it is still not clear how to construct these codes. In Ref. [169], two approaches are proposed to construct trajectory codes for the special case of d-dimensional grids. In this book, we provide a brief introduction to the first method.

Note that the grid graph is determined by the set of vertices $V_{n,d}$ to be

$$V_{n,d} = \{-n/2+1, n/2\}^d. \tag{8.130}$$

Two nodes with distance 1 have a connecting edge. The goal of the construction procedures is to design a trajectory code

$$\chi : V_{n,d} \times \{1, \ldots, n/2\} \rightarrow S, \tag{8.131}$$

which is asymptotically good.

The first approach of construction is based on two codes, where $n_1 \in \Theta(\log n)$ and k is the least even integer no less than $\frac{12}{1-\delta} + 4$.

- Block code: Consider a block code $\eta : V_{n,d} \rightarrow R_1^{n_1}$, where R_1 is a particular finite alphabet. Intuitively, η encodes each vertex in $V_{n,d}$ to a block code with codeword length n_1. We assume that the block code η has a positive rate and a relative distance $(1+\delta)/2$. The coding and decoding procedures can be computed in a time of order $n_1^{O(1)}$. η can be rewritten as

$$\eta(x) = (\eta_1(x,1), \ldots, \eta_1(x,n_1)), \tag{8.132}$$

  where $\eta_1 : V_{n,d} \times \{1, \ldots, n_1\} \rightarrow R_1$.
- Recursive code: We assume that $\chi_1 : V_{kn_1,\ldots,d} \times \{1, \ldots, kn_1/2\} \rightarrow S_1$ is a trajectory code with relative distance $(1+\delta)/2$, where S_1 is a finite alphabet. The construction of this code will be provided later.

Now we cover the space $V_{n,d}$ by overlapping tiles. The tile placed at vertex $x = (x_1, \ldots, x_d, x_{d+1}) \in V_{n,d}$ is the mapping defined as

$$\sigma_x(y) = (\chi_1(y-x), \eta_1(x_1, \ldots, x_d, (y_{d+1} - x_{d+1} \bmod n_1))), \tag{8.133}$$

which maps from the domain

$$\left(\prod_{i=1}^{d}(x_1 - kn_1/2+1, \ldots, x_i + kn_1/2)\right) \times (x_{d+1}+1, \ldots, x_{d+1}+kn_1/2), \tag{8.134}$$

to $S_1 \times R_1$.

We assume that n can be divided by kn_1. Then we pick out the tiles placed at the locations x having the following form:

$$x = (n_1 k z_1 + n_1 a_1, \ldots, n_1 k z_{d+1} + n_1 a_{d+1}), \tag{8.135}$$

where $\{z_i\}_i$ satisfy

$$(z_1, \ldots, z_{d+1}) \in \left\{-\frac{n}{2kn_1} + 1, \ldots, \frac{n}{2kn_1}\right\}^d \times \left\{1, \ldots, \frac{n}{2kn_1}\right\}, \tag{8.136}$$

and

$$(a_1, \ldots, a_{d+1}) \in \left\{-\frac{k}{2} + 1, \ldots, \frac{k}{2}\right\}^d \times \{0, \ldots, k-1\}. \tag{8.137}$$

The set of these tiles, which actually form a covering of $V_{n,d}$, are labeled using $(a_1, \ldots, a_{d+1})$. Hence there are a total of k^{d+1} such sets. This procedure is illustrated in Fig. 8.11, in which we set $n_1 = 4$ and $k = 2$.

The trajectory code is then given by

$$\chi(y) = \{\chi_{(a_1,\ldots,a_{d+1})}(y)\}_{(a_1,\ldots,a_{d+1})}, \tag{8.138}$$

that is, for each state y, the coder output is the concatenation of all the coder outputs at the tiles satisfying Eq. (8.135).

The intuition behind the complicated procedure of construction is given below:

- The coder for long trajectories (of the order of n) is based on the coders for short trajectories (of the order of $\log n$).

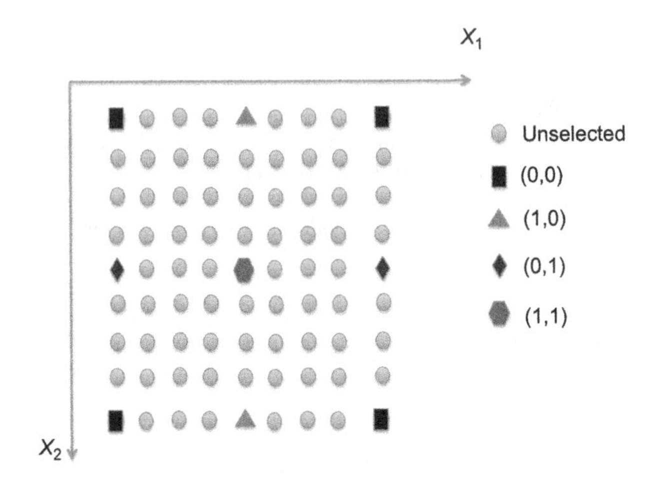

FIG. 8.11

Illustration of the tiles for designing trajectory codes.

- When the transmitter state is at a particular point of R^d, the coder output is determined by the output of the tiles close to it.
- The coder output of each tile in Eq. (8.133) consists of the recursive coder output, which encodes the transmitter state with respect to the center point of the tile, and the output of the block coder, which encodes the location of the center point of the tile.

It was shown in Ref. [169] that the above construction of trajectory codes can achieve the relative distance $\delta < 1$, whose proof is involved. Then the only unsolved problem is the construction of the coder χ_1 for each tile. Since the problem size is much smaller, it is possible to obtain the coder by using exhaustive search, thus solving the whole problem.

8.6.2 ANYTIME CHANNEL CODES

In this section, we follow the discussion in Ref. [170] to show that it is possible to design the channel codes directly for control.

System model

We consider a linear system with system state $\mathbf{x}$ and observations $\mathbf{y}$, which are given by

$$\mathbf{x}(t+1) = \mathbf{F}\mathbf{x}(t) + \mathbf{G}\mathbf{u}(t) + \mathbf{w}(t) \tag{8.139}$$

and

$$\mathbf{y}(t) = \mathbf{H}\mathbf{x}(t) + \mathbf{v}(t), \tag{8.140}$$

where $\mathbf{u}(t)$ is the control action, $\mathbf{w}(t)$ and $\mathbf{v}(t)$ are bounded noise processes. It is assumed that $(\mathbf{F}, \mathbf{G})$ are controllable and $(\mathbf{F}, \mathbf{H})$ are observable.

At time t, the sensor observes $\mathbf{y}(t)$ and generates a k-bit message, which is a function of all existing observations. Suppose that n bits can be transmitted for the k information bits. Hence the data transmission rate is k/n.

Anytime reliability

According to the discussion in Chapter 5, the communication should satisfy the anytime reliability in order to stabilize the linear system. We say that an encoder-decoder pair can achieve (R, β, d_0)-reliability over a particular communication channel if

$$P(\min\{\tau : \hat{b}(\tau|t) \neq b_\tau\} = t - d + 1) \leq \eta 2^{-n\beta d}, \quad \forall d \geq d_0, \quad \forall t, \tag{8.141}$$

where $\hat{b}(\tau|t)$ denotes the estimation of b_τ (the message transmitted at time τ) given the received messages until time t, and η is a constant. Intuitively Eq. (8.141) indicates that the probability of decoding error decreases exponentially with elapsed time.

Note that this definition of anytime reliability is not literally the same as the original one in Ref. [13]. In Ref. [13], the encoder-decoder pair can achieve α-anytime reliability if

$$P(\hat{b}(\tau|t) \neq b(\tau)) \leq K2^{-\alpha(t-\tau)}, \tag{8.142}$$

for any t and τ. However, it is easy to see that these two definitions are equivalent. First, if Eq. (8.141) holds, then we have

$$\begin{aligned} P(\hat{b}(\tau|t) \neq b(\tau)) &\leq \sum_{d=t-\tau}^{\infty} P(\min\{\tau : \hat{b}(\tau|t) \neq b_\tau\} = t-d+1) \\ &\leq \sum_{d=t-\tau}^{\infty} \eta 2^{-n\beta d} \\ &= \frac{\eta 2^{-n\beta(t-\tau)}}{1-2^{-n\beta}}, \end{aligned} \tag{8.143}$$

which implies β-anytime reliability. The first inequality arises because the event that b_τ is incorrectly decoded belongs to the event that the first incorrectly decoded message is b_s, where $s \leq \tau$.

On the other hand, if the encoder-decoder pair is α-anytime reliable according to Eq. (8.142), we have

$$\begin{aligned} P(\min\{\tau : \hat{b}(\tau|t) \neq b_\tau\} = t-d+1) &\leq P(\hat{b}(t-d+1|t) \neq b(t-d+1)) \\ &\leq K2^{-\alpha d}, \end{aligned} \tag{8.144}$$

where the first inequality arises because the event that the first decoding error happens for b_{t-d} belongs to the event that b_{t-d} is incorrectly decoded. According to the conclusion in Ref. [13, Theorem 5.2], the anytime reliability can guarantee the stability of linear systems with sufficiently small outage probability.

Linear tree codes

Ref. [170] proposed the use of linear tree codes to design the anytime reliable code. Note that the concept of tree codes will be explained in the next section. Due to the requirement of causal coding, the output of linear tree codes c_r at time slot r is given by

$$c_r = f_r(b_{1:r}) = \sum_{k=1}^{r} G_{rk} b_k, \tag{8.145}$$

where b_k is the k-bit information symbol generated at time k, the subscript r is the current time index, and $\{G_{rk}\}$ are the $n \times k$ generating matrices. This coding procedure is illustrated in Fig. 8.12.

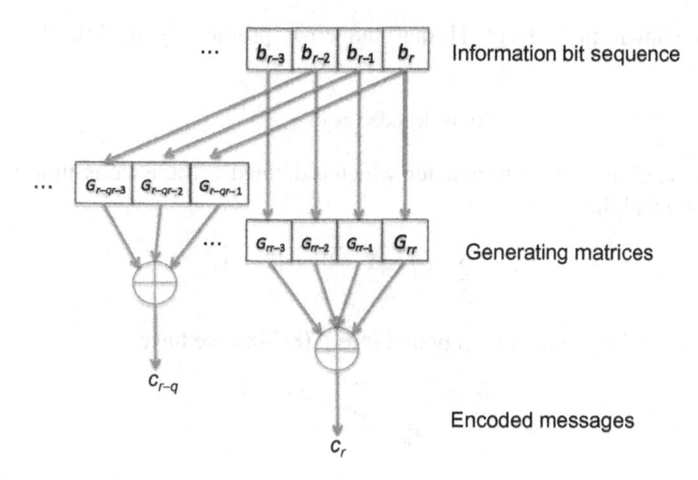

FIG. 8.12

Illustration of the linear tree code.

Hence the overall generating matrix can be written as

$$\mathbf{G}_{n,R} = \begin{pmatrix} G_{11} & 0 & \cdots & \cdots & \cdots \\ G_{21} & G_{22} & 0 & \cdots & \cdots \\ \vdots & \vdots & \ddots & \vdots & \vdots \\ G_{r1} & G_{r2} & \cdots & G_{rr} & \cdots \\ \vdots & \vdots & \vdots & \vdots & \ddots \end{pmatrix}. \tag{8.146}$$

The parity check matrix of $\mathbf{G}_{n,R}$ is denoted by $\mathbf{H}_{n,R}$, which satisfies

$$\mathbf{G}_{n,R}\mathbf{H}_{n,R} = 0. \tag{8.147}$$

It is easy to verify that $\mathbf{H}_{n,R}$ also has the lower triangular matrix form.

Maximum likelihood decoding

We assume that maximum likelihood (ML) decoding is used. Since we want to test whether the communication is anytime reliable, we focus on the probability of the event that the earliest wrongly decoded message is d time slots ago. Denoting this probability by $P^e_{t,d}$, we use the union bound to upper bound $P^e_{t,d}$:

$$P^e_{t,d} \leq \sum_{c \in C_{t,d}} P(0 \text{ is decoded as } c), \tag{8.148}$$

where $C_{t,d}$ is the set of codewords whose earliest nonzero symbol occurs at time $t-d$, and we assume that an all-zero sequence is transmitted.

It was shown in Ref. [171] that the error probability of ML decoding is bounded by

$$P(0 \text{ is decoded as } c) \leq \zeta^{\|c\|}, \tag{8.149}$$

where ζ is the Bhattacharya parameter, which is defined as (here we assume a discrete channel output alphabet $\mathcal{Z}$)

$$\xi = \sum_{z \in \mathcal{Z}} \sqrt{p(z|X=0)p(z|X=1)}. \tag{8.150}$$

Hence according to the union bound in Eq. (8.148), we have

$$P^e_{t,d} \leq \sum_{w^t_{\min,d} \leq w \leq nd} N^t_{w,d} \zeta^2, \tag{8.151}$$

where w is the weight of decoder output c, $N^t_{w,d}$ is the number of codes in $C_{t,d}$ having weight w, and $w_{\min,d}$ is the minimum weight of the codes in $C_{t,d}$. The reason why $w \leq nd$ is that there is no error before the $t-d$th message. This motivates the following definition:

Definition 23. A linear tree code in Eq. (8.145) is said to have (α, θ, d_0)-anytime distance if:

- The parity check matrix $\mathbf{H}_{n,R}$ has a full rank for all $t > 0$.
- $w^t_{\min,d} \geq \alpha$ and $N^t_{w,d} \leq 2^{\theta w}$, for all $t > 0$ and $d \geq d_0$.

The first requirement on the full rank of $\mathbf{H}_{n,R}$ is to guarantee that the encoding procedure is invertible. The second condition implies

$$P^e_{t,d} \leq \eta 2^{-\alpha\left(\log_2\left(\frac{1}{\zeta}\right)-\theta\right)}, \tag{8.152}$$

where

$$\eta = \left(1 - 2^{\log_2\left(\frac{1}{\zeta}\right)-\theta}\right)^{-1}. \tag{8.153}$$

Based on the above discussion, we have the following proposition:

Proposition 6. *If the linear tree code uses ML decoding and has (α, θ, d_0)-anytime distance, then it is (R, β, d_0)-anytime reliable, for channels having Bhattacharya parameter ζ where*

$$\beta = \alpha\left(\log\left(\frac{1}{\zeta}\right) - \theta\right). \tag{8.154}$$

Code construction

Now the challenge is how to construct a linear code satisfying the requirement of (α, θ, d_0)-anytime distance. In Ref. [170], the following Toeplitz structure was considered:

$$\mathbf{H}_{n,R}^{TZ} = \begin{pmatrix} H_1 & 0 & \cdots & \cdots & \cdots \\ H_2 & H_1 & 0 & \cdots & \cdots \\ \vdots & \vdots & \ddots & \vdots & \vdots \\ H_r & H_{r-1} & \cdots & H_1 & \cdots \\ \vdots & \vdots & \vdots & \vdots & \ddots \end{pmatrix}, \tag{8.155}$$

which assures the causality of the coding procedure.

Then a random construction approach for $\mathbf{H}_{n,R}^{TZ}$ is proposed in Procedure 10.

PROCEDURE 10 CONSTRUCTING THE LINEAR TREE CODE

1: Select a fixed full rank binary matrix for H_1.
2: **for** $k = 2, 3, \ldots$ **do**
3:     **for** Each element in H_k **do**
4:         Select 1 with probability p and 0 with probability $1 - p$.
5:     **end for**
6: **end for**

The main theorem in Ref. [170] shows that the above construction results in an anytime reliable code with a large probability.

Theorem 56. *The probability that the construction procedure in Procedure 10 results in a code with (α, θ, d_0)-anytime distance is bounded by*

$$P(\textit{the resulting code has } (\alpha, \theta, d_0)\textit{-anytime distance}) \geq 1 - 2^{-\Omega(nd_0)}, \tag{8.156}$$

where

$$\alpha < H^{-1}(1 - R\log(1/(1 - \bar{p}))) \tag{8.157}$$

and

$$\theta > -\log[(1 - \bar{p})^{-(1-R)} - 1], \tag{8.158}$$

with $\bar{p} = \min(p, 1 - p)$.

The detailed proof of this theorem in Ref. [170] is quite involved. In this book, we provide an intuitive sketch of the proof. Consider time slot t. Suppose that an all-zero codeword is sent and consider a codeword $\mathbf{c} \neq 0$. If $H_{r,R}^{t}\mathbf{c} = 0$, then $\mathbf{c}$ may be confused with the correct one, which results in a decoding error. Due to the requirement of anytime reliability, we are interested in the locations of errors (or equivalently the nonzero elements of $\mathbf{c}$). Since we require that the messages before time $t - d$ be correctly decoded, we have

$$\begin{cases} \mathbf{c}(\tau) = 0, & \forall \tau \leq t - d, \\ \mathbf{c}(t - d + 1) \neq 0 \\ \mathbf{H}^t_{n,R}\mathbf{c} = 0, \end{cases} \tag{8.159}$$

where $\mathbf{c} = (\mathbf{c}_0, \ldots, \mathbf{c}_t)$. This is equivalent to

$$\begin{pmatrix} H_1 & 0 & \cdots & \cdots & \cdots \\ H_2 & H_1 & 0 & \cdots & \cdots \\ \vdots & \vdots & \ddots & \vdots & \vdots \\ H_r & H_{r-1} & \cdots & H_1 & \cdots \end{pmatrix} \begin{pmatrix} c_{t-d+1} \\ c_{t-d+2} \\ \vdots \\ c_t \end{pmatrix} = \begin{pmatrix} 0 \\ 0 \\ \vdots \\ 0 \end{pmatrix}, \tag{8.160}$$

which can be rewritten as

$$\begin{pmatrix} C_{1-d+1} & 0 & \cdots & \cdots & \cdots \\ C_{t-d+2} & C_{t-d+1} & 0 & \cdots & \cdots \\ \vdots & \vdots & \ddots & \vdots & \vdots \\ C_t & C_{t-1} & \cdots & C_{t-d+1} & \cdots \end{pmatrix} \begin{pmatrix} h_1 \\ h_2 \\ \vdots \\ h_d \end{pmatrix} = \begin{pmatrix} 0 \\ 0 \\ \vdots \\ 0 \end{pmatrix}, \tag{8.161}$$

where $h_i = \mathrm{vec}(H_i^T)$ and $C_i = \mathrm{diag}(c_i^T, \ldots, c_i^T)$. Since H_1 has been fixed in the construction procedure, we have

$$\begin{pmatrix} C_{1-d+1} & 0 & \cdots & \cdots & \cdots \\ C_{t-d+2} & C_{t-d+1} & 0 & \cdots & \cdots \\ \vdots & \vdots & \ddots & \vdots & \vdots \\ C_{t-1} & C_{t-2} & \cdots & C_{t-d+1} & \cdots \end{pmatrix} \begin{pmatrix} h_2 \\ h_3 \\ \vdots \\ h_d \end{pmatrix} = \begin{pmatrix} C_{t-d+2} \\ C_{t-d+3} \\ \vdots \\ C_t \end{pmatrix} h_1, \tag{8.162}$$

and $C_{t-d+1}h_1 = 0$. We abbreviate Eq. (8.162) to

$$\mathbf{Ch} = \mathbf{C}'\mathbf{h}_1. \tag{8.163}$$

Hence the higher dimensional vector $\mathbf{h}$ cannot be arbitrarily chosen due to the linear constraint in Eq. (8.163). So $\mathbf{h}$ must be selected within a lower dimensional subspace. Since $\{h_i\}$, $i = 2, \ldots, t$, are randomly chosen, the corresponding probability that $\mathbf{h}$ falls in a lower dimensional subspace can be bounded by the following lemma:

Lemma 8. *Consider a randomly selected m-dimensional vector* $\mathbf{v}$ *in* $\{0, 1\}^m$ *with probability*

$$P(V) = p^{\|\mathbf{v}\|}(1 - p)^{m - \|\mathbf{v}\|}, \tag{8.164}$$

that is, with probability p *(or* $1-p$*), 1 (or 0) is selected for each dimension of* $\mathbf{v}$*. Then the probability that* $\mathbf{v}$ *lies in an l-dimensional subspace* $\mathbf{U}$ *(*$l \leq m$*) is bound by*

$$P(\mathbf{U}) \leq \max(p, 1 - p)^{m-l}. \tag{8.165}$$

The detailed evaluation of the upper bound will lead to the desired conclusion.

8.6.3 CONVOLUTIONAL LDPC FOR CONTROL

In the previous subsection, it is rigorously shown that the proposed coding scheme can stabilize the physical dynamics in a CPS. However, the following two issues in the coding scheme hinder its application in real practice:

- The parity check matrices are generated randomly, so it is difficult to control the decoding error probability.
- With a large probability, the matrix H_r is nonzero, which means that even the bits with a long time lapse still have an impact on the current coding procedure; hence all the previous information bits need to be stored, thus requiring infinite memory.

To handle these two challenges, some practical coding schemes have been proposed. In this book we briefly introduce the convolutional LDPC codes [172,173].

It was proposed in Ref. [172] that the LDPC convolutional codes can be used for assuring anytime reliability for the purpose of controlling dynamics in a CPS. Originally LDPC convolutional codes were proposed for pure data communications, rather than for CPSs. Essentially, the parity check matrix in the LDPC convolutional codes has the following form:

$$\mathbf{H}_{\infty} = \begin{pmatrix} \mathbf{H}_0(1) & \cdots & \cdots & \cdots & \cdots \\ \mathbf{H}_1(1) & \mathbf{H}_0(2) & \cdots & \cdots & \cdots \\ \vdots & \mathbf{H}_1(2) & \vdots & \cdots & \cdots \\ \mathbf{H}_{m_s}(1) & \vdots & \ddots & \mathbf{H}_0(t) & \vdots \\ \vdots & \mathbf{H}_{m_s}(2) & \cdots & \mathbf{H}_1(t) & \vdots \\ \vdots & \vdots & \vdots & \vdots & \ddots \\ \cdots & \cdots & \cdots & \mathbf{H}_{m_s}(t) & \vdots \\ \vdots & \vdots & \vdots & \vdots & \ddots \end{pmatrix}, \tag{8.166}$$

where m_s is the memory length. It is easy to verify that the generating matrix $\mathbf{G}$ has a similar structure; hence the output of the tth message is given by

$$\mathbf{c}(t) = \sum_{s=t-m_s+1}^{t} \mathbf{b}(s)\mathbf{H}_{t-s}. \tag{8.167}$$

Such a scheme handles the above two challenges in the following manner:

- There have been plenty of studies on how to optimize the LDPC codes; moreover, efficient decoding algorithms have been found for the LDPC codes, which makes the LDPC convolutional codes fairly practical.
- The convolutional structure makes the encoding procedure causal. Moreover, the finite memory requires only bounded memory.

It was shown in Ref. [172] that the proposed convolutional LDPC code can achieve anytime reliability. The details can be found therein.

8.7 CHANNEL CODING FOR INTERACTIVE COMMUNICATION IN COMPUTING

The channel coding schemes in the previous sections are dedicated to system state estimation and control in a CPS. We can also consider state estimation and control as a problem of computing: the agents (sensors or controllers) in a CPS have local parameters or observations on the physical dynamics and want to compute the control actions as functions of the local numbers. Hence the agents can communicate by using an existing protocol of interactive communications (e.g., a sensor sends an observation to a controller; or two controllers exchange their control actions and observations). Such a protocol can always be designed when there is no communication error. Since a noisy communication channel can incur transmission errors, it is important to study how to carry out the interactive communication protocol designed for noiseless communication channels. Although the study on coding for interactive communication in computing is not confined to the purpose of control, it is of significant importance for us to design channel coding schemes in a CPS. Hence in this section, we follow the seminal work of Schulman [174] to introduce the studies on this topic.

8.7.1 SYSTEM MODEL

For simplicity, we consider two agents A and B, which have local discrete arguments z_A and z_B, respectively. They are required to compute a function $f(A, B)$. The two agents can communicate with each other, one bit per transmission. The simplest approach is for A to transmit z_A to B and then B calculates $f(z_A, z_B)$. However, in many situations, fewer bits can be transmitted to obtain $f(z_A, z_B)$, which is encompassed in research on communication complexity [94,95].

We assume that the protocol for computing $f(z_A, z_B)$ has been designed, given the assumption that there is no communication error. Now our challenge is how to carry out the communication for computing with a given probability of transmission errors, when the communication channel is noisy (thus the probability of transmission error is nonzero).

8.7.2 TREE CODES

The construction of a communication protocol for computing the function $f(z_A, z_B)$ is based on the concept of tree codes. A d-ary tree with depth n is a tree in which each nonleaf node has d child nodes and the number of levels is n. Then a d-ary tree code over alphabet S with distance parameter α and depth n can be represented by a d-ary tree with depth n, in which each edge corresponds to an element in S. A codeword

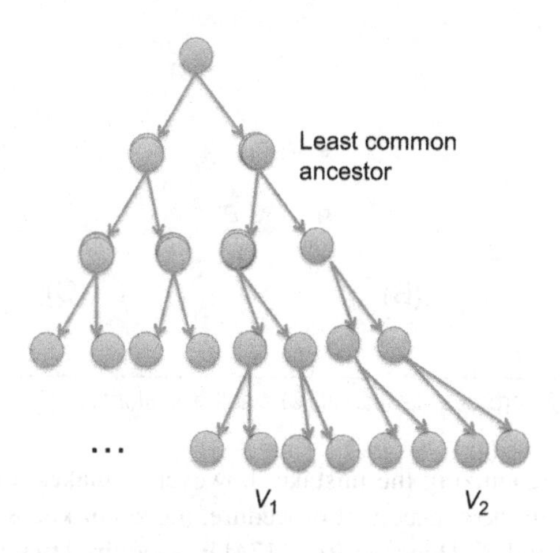

FIG. 8.13

Illustration of tree code.

is generated by tracing from the root to a certain node in the tree and concatenating the outputs of the edges. We denote by $W(v)$ the codeword generated by node v in the tree. Consider two nodes v_1 and v_2 with the same depth h in the tree, whose least common ancestor has depth $h-l$. Then $\Delta(W(v_1), W(v_2)) \geq \alpha l$, where $\Delta(\cdot, \cdot)$ is the Hamming distance. An illustration of the tree code is given in Fig. 8.13, where $h = 5$ and $l = 3$. Hence if $\alpha = 0.5$, we have $\Delta(W(v_1), W(v_2)) \geq 2$.

Construction of potent tree codes

In Ref. [174], the existence of good tree code with large relative Hamming distance is only assumed; the detailed construction of the code is not discussed. It was not until recent years that the explicit construction of the tree code was discussed [175], in which some intuitions were provided at the beginning. Since the random coding scheme has achieved great success in traditional information theory, the first idea for the code construction is to try the random coding scheme; i.e., we randomly assign output alphabets to each edge in the tree. This looks reasonable, since we require two paths in the tree with large discrepancies (as illustrated in Fig. 8.14A); otherwise it is hard to distinguish the two trajectories VP' and VP. Since the outputs are randomly assigned, the output sequences should have a large Hamming distance with a large probability.

However, this is not the only requirement. We should let short divergent paths in the tree have sufficiently large Hamming distances. This nontrivial requirement is illustrated in Fig. 8.14B. Suppose that the transmission errors make the pebbles diverge from the correct path to wrong locations A, A', B, B'. Assume that these short divergent paths have small relative Hamming distances. Then it is possible that the agents keep making mistakes due to the small relative Hamming distances. For example, the agent may be distracted to the wrong location A and return to the

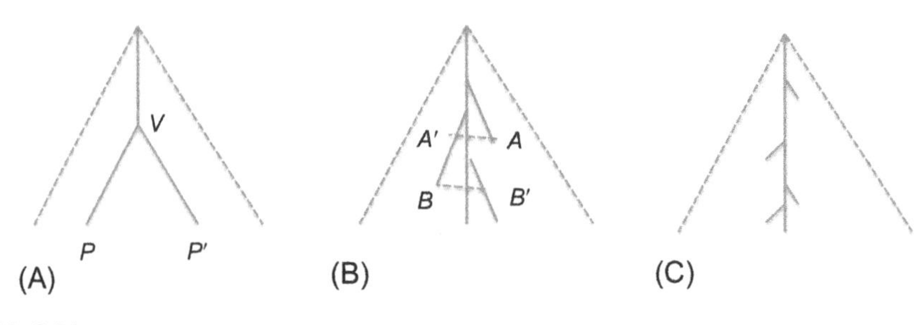

FIG. 8.14

(A) large deviations (B) crossed deviations (C) small deviations.

correct path after recognizing the mistake; however, it makes a mistake soon after and diverges to A'. In the subsequent procedure, the agent keeps falling into wrong locations such as B and B'. Hence in Ref. [174] it is required that all paths have large Hamming distances.

However, it is not necessary to require all paths to have large Hamming distances, which incurs too many constraints. It is argued in Ref. [175] that we can allow some branches in the tree to have small relative Hamming distances. As illustrated in Fig. 8.14C, each path from the root to a leaf is allowed to have some small diverged branches that have small relative Hamming distances. It is shown in Ref. [175] that this requirement can assure the success of protocol simulation.

To quantify the above requirement on the tree code, we need the following definition of a potent tree code:

Definition 24. Consider two nodes u and v with the same depth h in the tree, whose least common ancestor is w with depth $h-l$. We say that u and v are α-bad nodes if

$$\Delta(W(u), W(v)) < \alpha l. \tag{8.168}$$

The paths from w to u and v are called α-bad paths. The interval $[h-l, h]$ is called an α-bad interval.

Consider a tree code with depth N. If there is a path Q whose union of all α-bad intervals has a total length no less than ϵN, then we say that this tree is an (ϵ, α)-bad tree. Otherwise, it is an (ϵ, α)-potent tree code.

For a detailed construction of potent tree code, it is proposed in Ref. [175] to use the ϵ-biased sample space, which is defined as follows:

Definition 25. Consider a sample space S on n bits. For any element $\mathbf{x} = (x_1, \ldots, x_n)$ in X and any nonzero binary sequence $\mathbf{a} = (a_1, \ldots, a_n)$, we define $y(\mathbf{x}, \mathbf{a}) = \sum_{i=1}^{n} x_i a_i$, where the addition is binary. X is said to be ϵ-biased with respect to a linear test if it satisfies

$$|P(y(X, \mathbf{a}) = 0) - P(y(X, \mathbf{a}))| \leq \epsilon, \tag{8.169}$$

where X is randomly selected from S, for any nonzero $\mathbf{a}$.

It is shown in Ref. [175] that an ϵ-biased sample space can be constructed as follows: consider a prime $p > (n/\epsilon)^2$; a point in S is given by a number x in $[0, 1, \ldots, p-1]$ mapped to the n-bit sequence $(r_0(x), \ldots, r_{n-1}(x))$, where

$$r_i(x) = \frac{1 - \chi_p(x+i)}{2}, \tag{8.170}$$

and $\chi_p(x)$ denotes the quadratic character of $x \pmod p$.

Once an ϵ-biased sample space S is constructed, we can use it to build an (ϵ, α)-potent tree code. For simplicity, we assume that the output of each edge in the tree is binary. Since there are a total of d^N edges in a d-ary tree, the outputs of all edges in the tree can be described by a d^N-bit sequence: we sort the edges in the tree in a predetermined order and then assign each element in the sequence to an edge. This forms a d-ary small biased tree code.

Small biased tree codes have a useful property defined as follows:

Definition 26. A sample space for n bits, $x_1, \ldots, x_n$, is called (ϵ, k)-independent if

$$|P((x_{i_1}, \ldots, x_{i_k}) = \xi) - 2^{-k}| \leq \epsilon,$$

for any k indices $i_1, \ldots, i_k$ and any k-bit sequence ξ.

The intuitive explanation of the property of (ϵ, k)-independence is that the corresponding sequence is very close to uniformly distributed for any k-subset. It is shown in Ref. [175] that if a sample space is ϵ-biased then it is also $((1-2^{-k})\epsilon, k)$-independent, for any k. This property guarantees that, with a large probability, the small divergent branches in a small biased tree code have reasonable relative Hamming distances, since their bit assignments are "very random."

In Ref. [175], the following theorem guarantees that the construction of small biased tree codes leads to potent tree codes, with a large probability:

Theorem 57. *Consider ϵ and α in $(0, 1)$. With probability $1 - 2^{-\Omega(N)}$, a d-ary small biased tree code having depth N is (ϵ, α)-potent if the alphabet size is larger than $(2d)^{2+2/\epsilon}/(1-\alpha)$.*

The detailed proof is given in Ref. [175]. Here we provide a brief summary of the basic idea of the proof. Consider a node v. It is possible that there exists another node u having the same depth such that u forms a bad interval of l. To that end, we need to check the last l bits leading to v and u, denoted by $W_l(v)$ and $W_l(u)$, respectively. Due to the property of small biasedness in Definition (26), the Hamming distance between $W_l(v)$ and $W_l(u)$ should be large with a large probability. As will be shown later, potent tree codes can accomplish the task of iterative communications.

Explicit construction of tree codes

In the above discussion, the construction of potent tree codes is based on a random coding scheme. It is of substantial theoretic importance since it can prove the existence of potent tree codes. However, it is of no use in practice due to the complexity caused by the lack of efficient coding and decoding schemes.

A breakthrough in the deterministic and efficient construction of tree codes was made by Braverman [176]. The corresponding cost of computation time is subexponential; i.e., the time costs of code construction, encoding, and decoding are all 2^{n^ϵ} for a tree code with size n, where ϵ is a small number.

In this book, we focus on the deterministic construction of tree code. The corresponding coding and decoding complexities can be found in Ref. [176]. The basic idea of tree code construction is to combine multiple small-sized tree codes and then obtain a large-scale tree code. To that end, we introduce an operation called tree code product, i.e., we construct a tree code with depth $d_1 \times d_2$ from two tree codes having depths d_1 and d_2.

First we assume that a tree code with depth d has been constructed. Here we assume that d is small such that we can use exhaustive search to find the tree code. Then we convert it to a local tree code of depth $D \gg d$ and locality l. Note that a local tree code with depth D and locality l, as well as other parameters, is defined as follows:

Definition 27. A local tree code, with depth d, distance α, locality l, input alphabet size σ_i, and output alphabet size σ_o, is a σ_i regular tree with depth d. In the tree, each edge is labeled with a letter from σ_o. This defines the coding mapping $\mathcal{T}$. For any three codewords w, w_1, and w_2 over the input alphabet, such that

$$\begin{cases} |w_1| = |w_2| \leq l, \\ |w| + |w_1| \leq d, \\ w_1(q) \neq w_2(1), \end{cases} \tag{8.171}$$

we have

$$\Delta(\mathcal{T}(w \circ w_1), \mathcal{T}(w \circ w_2)) \geq \alpha |w_1|, \tag{8.172}$$

where $\circ$ means the operation of concatenation.

Remark 17. Eq. (8.172) means that, for two input sequences, the distance of their coding outputs is proportional to the depth of their diverged paths in the tree. This is the same as the definition of the tree code. The difference from the tree code is that we only require this property for the sequences with small disparities (i.e., the depth of the subtree rooted at the least ancestor is at most l). This is why we call it "local" tree code.

The construction of the local tree code with a much larger depth D is carried out by repeating the mapping $\mathcal{T}$ provided by the original tree code with a much smaller depth d. It is shown in Ref. [176] that this construction satisfies the requirement for local tree code.

Based on the conversion from a tree code to a local tree code with much larger depth, we can combine two tree codes with smaller depths to obtain a new tree code with a larger depth, which is stated in the following theorem:

Theorem 58. *Consider two tree codes $\mathcal{T}_I$ and $\mathcal{T}_O$, whose depths, distances, and alphabet sizes are given by $(d_1, \alpha_1, \sigma_i, \sigma_{o1})$ and $(d_2, \alpha_2, \sigma_i, \sigma_{o2})$, respectively. Then one can construct a product tree code $\mathcal{T}_P$ having parameters $(d, \alpha, \sigma_i, \sigma_o)$ which satisfy*

$$\begin{cases} \alpha = \min(\alpha_1, \alpha_2/10), \\ d = \frac{d_1 d_2}{4}, \\ \sigma_o = (\sigma_{o1}\sigma_{o2})^{O(1)}. \end{cases} \tag{8.173}$$

The procedures of code construction and encoding are illustrated in Fig. 8.15. Essentially, the encoding is a concatenation of an inner code $\mathcal{T}_i'$ and an outer code $\mathbf{T}_o'$. The inner code $\mathcal{T}_i'$ is a local tree code, which is obtained from $\mathcal{T}_i$. It takes care of the divergent paths whose divergence is short. Then the coding output is fed into the outer code $\mathbf{T}_o'$, which takes care of longer paths. $\mathcal{T}_o'$ is obtained from an intermediate code $\mathcal{T}_o''$. Simply speaking, $\mathcal{T}_o''$ is obtained by spreading $\mathcal{T}_o$ over different blocks. Then the output of $\mathcal{T}_o''$ is protected by traditional error correction codes, thus forming $\mathcal{T}_o'$. The details of the construction can be found in Ref. [176].

8.7.3 PROTOCOL SIMULATION

We explain the protocol based on the tree codes, as proposed in Ref. [174].

The protocol for computing the function $f(z_A, z_B)$, denoted by π, can be represented by a 4-tree, in which the trajectory of a communication for computing can be

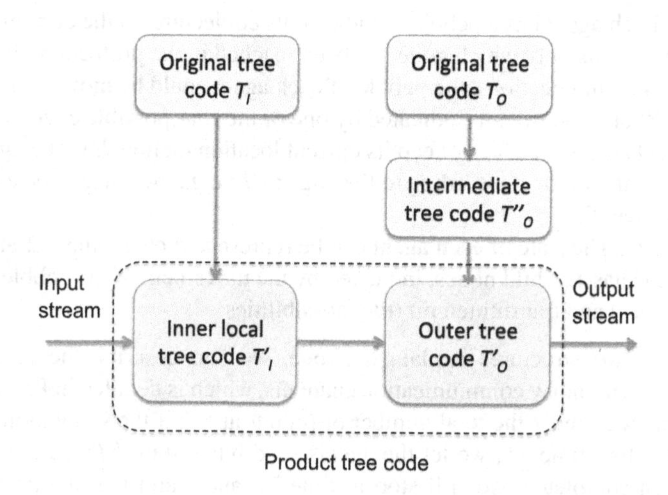

FIG. 8.15

Illustration of the procedure of constructing the product tree code.

represented by a path from the root to a leaf node. Each level of edges means one round of communications. Note that agents A and B transmit simultaneously within one round. For each node in the tree, the four child nodes led by the four edges 00, 01, 10, and 11 represent the state after this round of communications. For example, the node led by edges 00 is the state if the bits transmitted in this round are both zero. This procedure can be illustrated by the first two rounds:

- In the first round, agents A and B transmit bits $\pi_A(z_A, \phi)$ and $\pi_B(z_B, \phi)$, respectively. The two bits are denoted by $m_1 = \pi(z, \phi)$, which generates the four child nodes.
- In the second round, the agents A and B transmit bits $\pi_A(z_A, \pi(z, \phi))$ and $\pi_B(z_B, \pi(z, \phi))$, respectively. The two transmitted bits are denoted by m_2.

In a generic round, say round i, the agents A and B transmit bits $\pi_A(z_A, \{m_i\}_{j=1,\ldots,i-1})$ and $\pi_B(z_B, \{m_i\}_{j=1,\ldots,i-1})$, respectively. This procedure can be represented by a trajectory in a tree denoted by $\mathcal{T}$. At node v of $\mathcal{T}$, agent A (or B) transmits bit $\pi_A(z_A, v)$ (or $\pi_B(z_B, v)$).

Given a tree code and the protocol π for noiseless communication channels, we can design the protocol subject to noisy channels by simulating the protocol π. The basic idea is that both agents estimate each other's current state by using the received bits. If the two agents have different estimations of the current location in $\mathcal{T}$, due to errors in the transmissions, they will begin to find that the received bits are possibly different from what they expect. At this time, they will trace back in the tree $\mathcal{T}$ and try to synchronize their estimated location in $\mathcal{T}$.

To that end, the following structures are needed at both agents A and B:

- Pebble: Each agent has a pebble to indicate its conjecture on the current location in the tree $\mathcal{T}$, namely which node has been reached in the protocol π. In each round of communications, the pebble of each agent could be moved upstream (action $\mathcal{B}$) or downstream (indicated by one of the four possible edges, 00, 01, 10, and 11) of the tree $\mathcal{T}$, or keep its current location (action $\mathcal{H}$). The agent also transmits either 0 or 1 according to the edge in $\mathcal{T}$; e.g., the current location of the protocol tree $\mathcal{T}$.
- Agent state: The state of each agent can be represented by another 12-ary tree $\mathcal{Y}$. Each node has 12 child nodes, indicated by the movement of the pebble (six possibilities) and transmitted bit (two possibilities).

With the two structures explained above, we can describe the protocol for simulating π with noisy communication channels, which is detailed in Procedure 11.

Note that we can fix the total number of rounds at $5T$: if the simulation has been completed before time $5T$, we let the agents send bits 0 until $5T$; if the simulation has not been completed, we still stop at time $5T$ and claim that the procedure of simulation fails.

Note that, for the potent tree code proposed in Ref. [175], the interactive protocol is still the same.

PROCEDURE 11 ITERATIVE COMMUNICATION FOR SIMULATING PROTOCOL π

1: Input: π: the protocol to be simulated; T: the number of rounds in the protocol π; (z_A, z_B): local inputs; $\mathcal{T}$: the protocol tree of π: $\mathcal{Y}$: the tree of system state.
2: **for** $t = 1 : 5T$ **do**
3: Agent A checks its state s_A in $\mathcal{Y}_A$.
4: Agent A sends out bit $w(s_A)$.
5: Agent A estimates the current state g of agent B such that the Hamming distance $\Delta(W(g), Z)$ is minimized, where Z is the received bits and $W(g)$ is the bit sequence sent by B if its current state is g.
6: Compute the corresponding location of the pebble of B, namely $pebble(g)$.
7: Agent A compares its own pebble location v_A and $pebble(g)$.
8: **if** v_A is a strict ancestor of $pebble(g)$ **then**
9: Agent A does not move its pebble in $\mathcal{T}$ (i.e., the action is $\mathcal{H}$). Reset its own state.
10: **end if**
11: **if** v_A and $pebble(g)$ have a strict common ancestor **then**
12: Move the pebble upstream in $\mathcal{T}$ (i.e., the action is $\mathcal{B}$). Reset its own state.
13: **end if**
14: **if** $v_A = pebble(g)$ **then**
15: Move the pebble downstream in $\mathcal{T}$.
16: **end if**
17: Agent B carries out the same steps as those of A.
18: **end for**

8.7.4 PERFORMANCE ANALYSIS

We first focus on the analysis in Ref. [174], where it is shown that the procedure described in Procedure 11 can simulate the original protocol π with an exponentially small failure probability. This is summarized in the following theorem:

Theorem 59. *We assume a binary symmetric communication channel, and the existence of a tree code with relative distance 1/2. The simulation procedure described in Procedure 11, which is run 5T times (T is the number of rounds in the original protocol π), can simulate π with an error probability less than or equal to 2^{-5T}.*

The rigorous proof of this important theorem is given in Ref. [174]. In this book, we provide an intuitive explanation. The key step is to quantify the level of success in the simulation procedure by defining a concept called mark:

Definition 28. Consider the two pebbles at v_A and v_B in $\mathcal{T}$. The least common ancestor is denoted by $\bar{v}$. The mark of the simulation procedure is defined as

$$\text{mark}(v_A, v_B, \bar{v}) = \text{depth}(\bar{v}) - \max(d(\bar{v}, v_A), d(\bar{v}, v_B)), \tag{8.174}$$

where d is the distance in the tree.

It is important to realize that π is successfully simulated if the mark at the termination (time $5T$) is at least T: if so, the least common ancestor has a depth of at least T, which means that the two pebbles have reached a node with depth T simultaneously and thus implies the success of the simulation. This is rigorously proved in Lemma 4 of Ref. [174]. Hence the estimation of the error probability is adjusted to the problem of evaluating the probability that the mark is less than T.

The second important point is to realize the impact of a good move (when both agents estimate the system states correctly) or a bad move (when a state estimation error occurs):

- A good move increases the mark by 1.
- A bad move decreases the mark by no more than 3.

Then a sufficient condition for a successful simulation of π is that the proportion of good moves is at least 4/5, since at the termination at time $5T$, we have

$$\text{mark} \geq \frac{4}{5} \times 5T - 3 \times \frac{1}{5} \times 5T = T. \tag{8.175}$$

Therefore the problem becomes one of bounding the proportion of bad moves given the tree code with relative distance 1/2. We denote by $l(t)$ the larger one of the pebble location errors. If there is a bad move at time t, we define the error interval as $t - l(t) + 1, \ldots, t$, since there must be transmission errors within this error interval. Each bad move must be within an error interval; hence we can use the number of error intervals to bound the number of bad moves. A more detailed analysis leads to the conclusion in Theorem 59.

In Ref. [175], the requirement of the existence of tree code is relaxed to that of the potent tree code, while the interactive protocol is still the same as that in Ref. [174]. The corresponding performance is summarized in the following theorem:

Theorem 60. *Suppose that the communication channel is binary symmetric. Consider a $(1/20, \alpha)$-potent tree code. Then a protocol π of length T without communication error can be successfully simulated by Procedure 11 with probability $1 - 2^{-\Omega(T)}$, if it is carried out for $N = O(T)$ rounds.*

8.8 CONCLUSIONS

In this chapter, we have discussed the physical dynamics-aware design of the physical layer in the communication network of a CPS. Various approaches have been discussed, ranging from modulation to channel coding. These are still the subject of research in academia. Almost all existing communication protocols of the physical layer in a CPS have not included the awareness of physical dynamics. The main challenge lies in the following aspects:

- The computational complexity of these algorithms is still too high. For example, in the adaptive modulation, the scheduler needs to solve an LMI. Although efficient algorithms from convex optimization can be used, it is still challenging for real-time control in CPSs.

- These algorithms require detailed models and parameters of the physical dynamics, which may be unavailable in many situations. Hence lack of availability of information on physical dynamics could disable these algorithms. Even if the models and parameters can be obtained from system identification or machine learning, it is still not clear whether these algorithms are robust to errors in the models and parameters.

Hence there is still a long way to go before the physical dynamics-aware design of the physical layer becomes useful in practice.

Bibliography

[1] A.G. Phadke, J.S. Thorp, Synchronized Phasor Measurements and Their Applications, Springer, New York, 2008.

[2] H. Li, F. Li, Y. Xu, D.T. Rizy, J.D. Kueck, Adaptive voltage control with distributed energy resources: algorithm, theoretical analysis, simulation, and field test verification, IEEE Trans. Power Syst. 25 (3) (2010).

[3] E.A. Lee, Cyber physical systems: design challenges, in: Proceedings of the IEEE International Symposium on Object Oriented Real-Time Distributed Computing (ISORC), 2008.

[4] P. Kundur, Power System Stability and Control, McGraw-Hill, 1994.

[5] F. Bullo, J. Cortés, S. Martínez, Distributed Control of Robotic Networks, Princeton University Press, 2009.

[6] V.I. Arnold, A. Weinstein, Mathematical Methods of Classical Mechanics, Springer, New York, 2007.

[7] C.G. Cassandras, S. Lafortune, Introduction to Discrete Event Systems, Cambridge University Press, Cambridge, UK, 2009.

[8] P. Kunkel, V. Mehrmann, Differential-Algebraic Equations: Analysis and Numerical Solutions, European Mathematical Society, 2006.

[9] D. Lind, B. Marcus, An Introduction to Symbolic Dynamics and Coding, Cambridge University Press, Cambridge, UK, 1995.

[10] A. Katok, B. Hasselblatt, Introduction to Modern Theory of Dynamical Systems, Cambridge University Press, Cambridge, UK, 1995.

[11] A.S. Matveev, A.V. Savkin, Estimation and Control Over Communication Networks, Birkhäuser, 2008.

[12] T. Downarowicz, Entropy in Dynamical Systems, Cambridge University Press, Cambridge, UK, 2011.

[13] A. Sahai, S. Mitter, The necessity and sufficiency of anytime capacity for control over a noisy communication channel, IEEE Trans. Inform. Theory 52 (8) (2006) 3369–3395.

[14] S. Tatikonda, A. Sahai, S. Mitter, Stochastic linear control over a communication channel, IEEE Trans. Inform. Theory 49 (9) (2004) 1549–1561.

[15] H. Li, Entropy reduction via communications in cyber physical systems: how to feed Maxwell's Demon? in: Proceedings of the IEEE International Symposium of Information Theory, 2015.

[16] M.S. Branicky, S.M. Phillips, W. Zhang, Stability of networked control systems: explicit analysis of delay, in: Proceedings of the American Control Conference (ACC), 2000.

[17] P. Seiler, R. Sengupta, Analysis of communication losses in vehicle control problems, in: Proceedings of the American Control Conference (ACC), 2001.

[18] L.A. Montestruque, P.J. Antsakilis, Quantization in model based networked control systems, in: Proceedings of the 16th IFAC World Congress, 2005.

[19] J.P. Hespanha, P. Naghshtabrizi, Y. Xu, A survey of recent results in NCS, Proc. IEEE 95 (2007) 138–162.

[20] X. Liu, A. Goldsmith, Wireless medium access control in distributed control systems, in: Proceedings of the Annual Allerton Conference on Communications, Control and Computing, Monticello, IL, 2003.

[21] X. Liu, A. Goldsmith, Wireless network design for distributed control, in: Proceedings of the IEEE Conference on Decision and Control (CDC), Paradise Island, Bahamas, 2004.

[22] L. Xiao, M. Johansson, H. Hindi, S. Boyd, A. Goldsmith, Joint optimization of communication rates and linear systems, IEEE Trans. Automat. Control 48 (2003) 148–153.

[23] J. Bai, E.P. Eyisi, Y. Xue, X.D. Koutsoukos, Dynamic tuning retransmission limit of IEEE 802.11 MAC protocol for networked control systems, in: Proceedings of the First International Workshop on Cyber-Physical Networking Systems (CPNS), 2011.

[24] V. Gupta, On an estimation oriented routing protocol, in: Proceedings of the American Control Conference (ACC), Baltimore, 2010, pp. 580–585.

[25] C. Langbort, V. Gupta, Minimal interconnection topology in distributed control design, in: Proceedings of the American Control Conference (ACC), Minneapolis, 2006.

[26] X. Liu, H. Zhang, Q. Xiang, X. Che, X. Ju, Taming uncertainties in real-time routing for wireless networked sensing and control, in: Proceedings of the 13th ACM International Symposium on Mobiel Ad Hoc Networking and Computing (MobiHoc), Head Island, USA, 2012.

[27] L.M. Thompson, Industrial Data Communications, fourth ed., Instrumentation Systems, 2002.

[28] B. Galloway, G. Hancke, Introduction to industrial control networks, IEEE Commun. Surv. Tut. 99 (2012) 1–21.

[29] National Instruments, Fieldbus: Foundation Fieldbus Overview, 2003.

[30] S.A. Boyer, SCADA: Supervisory Control and Data Acquisition, fourth ed., The Instrumentation, Systems and Automation Society, 2009.

[31] M.V. Mynam, A. Harikrishna, V. Singh, Synchrophasors Redefining SCADA Systems, Schweitzer Engineering Laboratories, Inc., 2013.

[32] C.E. Shannon, A mathematical theory of communications, Bell Syst. Tech. J. 27 (1948) 379–423, 623-656.

[33] T.M. Cover, J.A. Thomas, Elements of Information Theory, second ed., Wiley, 2006.

[34] S.P. Lloyd, Least squares quantization in PCM, IEEE Trans. Inform. Theory 28 (2) (1982) 129–137.

[35] C. Berrou, A. Glavieux, P. Thitimajshima, Near Shannon limit error-correcting coding and decoding: tubro-codes, in: IEEE International Conference on Communications, 1993.

[36] A.S. Tanenbaum, D.J. Wetherall, Computer Networks, fifth ed. 2010.

[37] D.P. Palomar, M. Chiang, A tutorial on decomposition methods for network utility maximization, IEEE J. Sel. Areas Commun. 24 (8) (2006) 1439–1451.

[38] M. Chiang, S.H. Low, A.R. Calderbank, J.C. Doyle, Layering as optimization decomposition: a mathematical theory of network architectures, Proc. IEEE 95 (1) (2007) 255–312.

[39] R. Knopp, P. Humblet, Information capacity and power control in single-cell multiuser communications, in: Proceeidngs of IEEE Internatinoal Conference on Communications (ICC), 1995.

[40] L. Tassiulas, A. Emphremides, Stability properties of constrained queuing systems and scheduling policies for maximum throughput in multihop radio networks, IEEE Trans. Automat. Control 37 (1992) 1936–1948.

[41] A. Goldsmith, Wireless Communications, Cambridge University Press, Cambridge, UK, 2005.

[42] D. Tse, P. Viswanash, Fundamentals of Wireless Communications, Cambridge University Press, Cambridge, UK, 2005.
[43] A. Nucci, K. Papagiannaki, Design, Measurement and Management of Large-Scale IP Networks: Bridging the Gap Between Theory and Practice, Cambridge University Press, Cambridge, UK, 2009.
[44] T.S. Rappaport, Wireless Communications: Principles and Practice, second ed., Prentice Hall, 2002.
[45] A. Gersho, R.M. Gray, Vector Quantization and Signal Compression, Springer, New York, 1991.
[46] T. Berger, Rate Distortion Theory: Mathematical Basis for Data Compression, Prentice Hall, 1971.
[47] S. Lin, D.J. Costello, Error Control Coding, Prentice Hall, 2004.
[48] R.E. Blahut, Algebraic Codes for Data Transmission, Cambridge University Press, Cambridge, UK, 2003.
[49] J. Proakis, M. Salehi, Digital Communications, fifth ed., McGraw-Hill Education, 2007.
[50] R. Srikant, L. Ying, Communication Networks: An Optimization, Control and Stochastic Networks, Cambridge University Press, Cambridge, UK, 2014.
[51] J. Li, X. Wu, R. Larioia, OFDMA Mobile Broadband Communications: A Systems Approach, Cambridge University Press, Cambridge, UK, 2013.
[52] J.P. Hespanha, Linear Systems Theory, Princeton University Press, 2009.
[53] J.C. Doyle, B.A. Francis, A.R. Tannebaum, Feedback Control Theory, Dover, 1992.
[54] P.E. Caines, Linear Stochastic Systems, Wiley, 1988.
[55] S. Asmussen, Applied Probability and Queues, second ed., Springer, New York, 2000.
[56] J.S. Thorp, C.E. Seyler, A.G. Phadke, Electromechanical wave propagation in large electric power systems, IEEE Trans. Circuits Syst. I, Fundam. Theory Appl. 45 (1998) 614–622.
[57] H. Lin, S. Sambamoorthy, S. Shukla, J. Thorp, L. Mili, Ad hoc vs. supervisory wide area backup relay protection validated on power/network co-simulation platform, in: Proceedings of the 17th Power Systems Computation Conference, Stockholm, Sweden, 2011.
[58] Z. Zhang, H. Li, Wireless communication aided differential protection in smart grids: a concerted Blizkrieg, in: IEEE Symposium on Smart Grid Communications (SmartGridComm), Vancouver, Canada, 2013.
[59] J. Jiang, J. Yang, Y. Lin, C. Liu, J. Ma, An adaptive PMU based fault detection/location technique for transmission lines—Part I: theory and algorithms, IEEE Trans. Power Deliv. 15 (2) (2000) 486–493.
[60] C. Bennett, S. Wicker, Decreased time delay and security enhancement recommendations for AMI smart meter networks, in: Proceedings of the Innovative Smart Grid Technologies Conference, Gothenburg, Sweden, 2010.
[61] ISO New England, Inc., Overview of the Smart Grid: Policies, Initiatives and Needs, 2009.
[62] S. Borenstein, M. Jaske, A. Rosenfeld, Dynamic Pricing, Advanced Metering, and Demand Response in Electricity Markets, Center for the Study of Energy Markets, University of California, Berkeley, CA, 2010.
[63] M. Balijepalli, K. Pradhan, Review of demand response under smart grid paradigm, in: Proceedings of the IEEE PES Innovative Smart Grid Technologies, Anaheim, CA, USA, 2011.

[64] G. Sharma, L. Xie, P. Kumar, Large population optimal demand response for thermostatically controlled inertial loads, in: Proceedings of the IEEE International Conference on Smart Grid Communications (SmartGridComm), Vancouver, Canada, 2013.

[65] W. Tang, S. Bi, Y.J. Zhang, Online speeding optimal charging algorithm for electric vehicles without future information, in: Proceedings of the IEEE International Conference on Smart Grid Communications (SmartGridComm), Vancouver, Canada, 2013.

[66] M. Roozbehani, M. Dahleh, S. Mitter, Dynamic pricing and stabilization of supply and demand in modern electric power grids, in: Proceedings of the First IEEE Conference on Smart Grid Communications, DC, USA, 2010.

[67] S. Stoft, Power System Economics—Designing Markets for Electricity, IEEE/Wiley, 2002.

[68] H. Li, S. Gong, L. Lai, Z. Han, R.C. Qui, D. Yang, Efficient and secure wireless communications for advanced metering infrastructure in smart grids, IEEE Trans. Smart Grid 3 (3) (2012) 1540–1551.

[69] S. Thrun, W. Burgard, D. Fox, Probabilistic Robotics, MIT Press, 2005.

[70] W.S. Wong, Control communication complexity of distributed control systems, SIAM J. Control Optim. 48 (3) (2009) 1722–1742.

[71] T. Sagawa, M. Ueda, Nonequilibrium thermodynamics of feedback control, Phys. Rev. E 95 (2012) 2.

[72] B.D.O. Anderson, J.B. Moore, Optimal Control: Linear Quadratic Methods, Dover Books, 2007.

[73] S.L. Baldwin, E.E. Slaminka, Calculating topological entropy, J. Stat. Phys. 89 (5/6) (1997) 1017–1033.

[74] G. Froyland, O. Junge, G. Ochs, Rigorous computation of topological entropy with respect to a finite partition, Physica D 154 (2001) 68–84.

[75] R.G. Gallager, Information Theory and Reliable Communication, Wiley, 1968.

[76] M. Gastpar, B. Rimoldi, M. Vetterli, To code, or not to code: lossy source-channel communication revisited, IEEE Trans. Inform. Theory 49 (5) (2003) 1147–1158.

[77] W.R. Ashby, An Introduction to Cybernetics, Filiquarian Legacy Publishing, 2012.

[78] N. Wiener, Cybernetics: The Control and Communication in the Animal and the Machine, second ed., MIT Press, 1965.

[79] R.C. Connant, Information transfer required in regulatory processes, IEEE Trans. Syst. Sci. Cybernet. 5 (4) (1969) 334–338.

[80] H.L. Weidemann, Entropy analysis of feedback control systems, in: Advances in Control Systems: Theory and Applications, Academic Press, 1969, pp. 225–255.

[81] T. Kailath, A.H. Sayed, B. Hassibi, Linear Estimation, Prentice Hall, 2000.

[82] H. Wang, Minimum entropy control of non-Gaussian dynamic stochastic systems, IEEE Trans. Automat. Control 47 (2) (2002) 398–403.

[83] B.M. Brown, C.J. Harris, Neurofuzzy Adaptive Modeling and Control, Prentice Hall, 1994.

[84] W.J.M. Kickert, J.M. Bertrand, J. Praagaman, Some comments on cybernetics and control, IEEE Trans. Syst. Man Cybernet. 8 (11) (1978) 805–809.

[85] H. Li, J.B. Song, Does feedback control reduce entropy/communications in smart grids? in: Proceedings of the IEEE International Conference on Smart Grid Communications (SmartGridComm), 2015.

[86] E. Fermi, Thermodynamcis, Dover, 1956.

[87] C.A. Truesdell, S. Bharatha, The Concepts and Logic of Classical Thermodynamics as a Theory of Heat Engines: Rigorously Constructed Upon the Foundation Laid by S. Carnot and F. Reech, Springer, New York, 1977.

[88] M. Feder, N. Merhav, Relations between entropy and error probability, IEEE Trans. Inform. Theory 40 (1) (1994) 259–266.

[89] N.C. Martins, M.A. Dahleh, Feedback control in the presence of noisy channels: 'bode-like' fundamental limitations of performance, IEEE Trans. Automat. Control 53 (7) (2008) 1604–1615.

[90] I. Koch, Analysis of Multivariate and High-Dimensional Data, Cambridge University Press, Cambridge, UK, 2013.

[91] J.A. Lee, M. Verleysen, Nonlinear Dimensionality Reduction, Springer, New York, 2007.

[92] L.C. Evans, Partial Differential Equations, second ed., AMS, 2010.

[93] D. He, D. Shi, R. Sharma, Consensus based distributed cooperative control for microgrid voltage regulation and reactive power sharing, in: Proceedings of the IEEE Innovative Smart Grid Technologies (ISGT Europe), 2014.

[94] A.C. Yao, Some complexity questions related to distributed computing, in: Proceedings of the 11th ACM Symposium on Theory of Computing (STOC), 1979.

[95] E. Kushlevitz, N. Nisan, Communication Complexity, Cambridge University Press, Cambridge, UK, 2006.

[96] H.S. Leff, A.F. Rex, Maxwell's Demon 2: Entropy, Classical and Quantum Information, Computing, Institute of Physics Publishing, 2003.

[97] L. Brillouin, Negentropy principle of information, J. Appl. Phys. 24 (9) (1953) 1152–1163.

[98] O. Penrose, Foundations of Statistical Mechanics: A Deductive Treatment, Dover Publications, 2005.

[99] S. Toyabe, T. Sagawa, A. Ueda, E. Muneyuki, M. Sano, Experimental demonstration of information-to-energy conversion and validation of the generalized Jarzynski equality, Nat. Phys. 6 (2010) 988–992.

[100] H. Li, S. Djouadi, K. Tomsvic, Flocking generators: a PdE framework for stability of smart grids with communications, in: Proceedings of the IEEE Conference on Smart Grid Communications, 2012.

[101] F.R.K. Chuang, Spectral Graph Theory, American Mathematical Society, 2007.

[102] G. Ferrari-Trecate, A. Buffa, M. Gati, Analysis of coordination in multi-agent systems through partial difference equations, IEEE Trans. Automat. Control 51 (2006) 1058–1063.

[103] M. Jilg, O. Stursberg, Optimized distributed control and topology design for hierarchically interconnected systems, in: Proceedings of European Control Conference (ECC), 2013.

[104] M. Fardad, F. Lin, M.R. Jovanovic, Design of optimal sparse interconnection graphs for synchronization of oscillator networks, IEEE Trans. Automat. Control 59 (9) (2014) 2457–2462.

[105] A. Ghosh, S.K. Das, Coverage and connectivity issues in wireless sensor networks, in: Mobile, Wireless and Sensor Networks: Technology, Applications and Future Directions, John Wiley & Sons, Inc., 2006.

[106] Y. Kim, M. Mesbahi, On maximizing the second smallest eigenvalue of a state-dependent graph Laplacian, IEEE Trans. Automat. Control 51 (1) (2006) 116–120.

[107] J. Marschak, R. Radner, Economic Theory of Teams, Yale University Press, 1962.
[108] Y.C. Ho, K.C. Chu, Team decision theory and information structures in optimal control problems—Part I, IEEE Trans. Automat. Control 17 (1972) 15–22.
[109] R. Radner, Team decision problems, Ann. Math. Stat. 33 (3) (1962) 857–881.
[110] V. Terzija, G. Valverde, D. Cai, P. Regulski, V. Madani, J. Fitch, S. Skok, M.M. Begovic, A. Phadke, Wide-area monitoring, protection, and control of future electric power networks, Proc. IEEE 99 (1) (2011) 80–93.
[111] Y. Zhang, P. Markham, X. Tao, Wide-area frequency monitoring network (FNET) architecture and applications, IEEE Trans. Smart Grid 1 (2) (2010) 159–167.
[112] J.D.L. Ree, V. Centeno, J.S. Thorp, A.G. Phadke, Synchronized phasor measurement applications in power systems, IEEE Trans. Smart Grid 1 (1) (2010) 20–27.
[113] IEC 61850 Technical Issues Overview, tech. rep., URL http://tissues.iec61850.com/parts.mspx.
[114] Gidelines for proper wiring of an RS-485 network, tech. rep., URL http://www.maxim-ic.com/app-notes/index.mvp/id/763.
[115] J. Baillieul, P.J. Antsaklis, Control and communication challenges in networked real-time systems, Proc. IEEE 95 (2007) 9–28.
[116] G.N. Nair, F. Fagnani, S. Zampieri, R.J. Evans, Feedback control under data rate constraints: an overview, Proc. IEEE 95 (2007) 108–137.
[117] D. Niyato, P. Wang, Z. Han, E. Hossain, Impact of packet loss on power demand estimation and power supply cost in smart grid, in: Proceedings of the IEEE Wireless Communications and Networking Conference, 2011.
[118] J. Nutaro, V. Protopopescu, The impact of market clearing time and price signal delay on the stability of electric power markets, IEEE Trans. Power Syst. 24 (2009) 1337–1345.
[119] X. Yu, K. Tomsovic, Application of linear matrix inequalities for frequency control with communication delays, IEEE Trans. Power Syst. 19 (2004) 1508–1515.
[120] P. Cholley, P. Crossley, V. Van Acker, T. Van Cutsem, W. Fu, J. Soto Idia Òez, F. Ilar, D. Karlsson, Y. Kojima, J. McCalley, et al., System Protection Schemes in Power Networks, tech. rep., CIGRE Technical Brochure 2001.
[121] K. Narendra, T. Weekes, Phasor measurement unit (PMU) communication experience in a utility environment, in: Proceedings of the CIGRE Canada Conference on Power Systems, 2008.
[122] S. Kagami, M. Ishikawa, A sensor selection method considering communication delays, in: Proceedings of the IEEE International Conference on Robotics and Automation, 2004.
[123] J. Lunze, et al., Handbook of Hybrid Systems Control, Cambridge University Press, Cambridge, UK, 2009.
[124] H. Li, A. Dimitrovski, J.B. Song, Z. Han, L. Qian, Communication infrastructure design in cyber physical systems with applications in smart grids: a hybrid system framework, IEEE Commun. Surv. Tut. 16 (3) (2014) 1689–1708.
[125] H.S. Witsenhausen, A class of hybrid-state continuous-time dynamic systems, IEEE Trans. Automat. Control 11 (2) (1966) 161–167.
[126] R.D. Middlebrook, S. Cuk, A general unified approach to modeling switching-converter power stages, in: Proceedings of the IEEE Power Electronics Specialists Conference, 1976.
[127] C. Sonntag, O. Stursberg, Optimally Controlled Start-up of a Multi-stage Evaporation System, Technical Report, Technische Universitata, Dortmund, 2005.

[128] A. Balluchi, M.D. Di Benedetto, C. Pinello, C. Rossi, A.L. Sangiovanni-Vincentelli, Hybrid control in automotive applications: the cut-off control, Automatica 35 (1999) 519–535.
[129] W. Zhang, J. Hu, A. Abate, Infinite horizon switched LQR problems in discrete time: a suboptimal algorithm with performance analysis, IEEE Trans. Automat. Control 57 (7) (2012) 1815–1821.
[130] H. Li, Z. Han, A.D. Dimitrovski, Z. Zhang, Data traffic scheduling for cyber physical system with application in voltage control of microgrids: a hybrid system framework, IEEE Syst. J. 8 (2) (2014) 542–552.
[131] H. Li, L. Lai, H.V. Poor, Multicast routing for cyber physical systems with application in smart grid, IEEE J. Sel. Areas Commun. 30 (6) (2012) 1097–1107.
[132] J.C. Geromel, P. Colaneri, Stability and stabilization of discrete time switched systems, Int. J. Control 79 (2006) 719–729.
[133] S. Boyd, L. El Ghaoui, E. Feron, V. Balakrishnan, Linear Matrix Inequalities in System and Control Theory, Society of Industrial and Applied Mathematics, Philadelphia, PA, 1994.
[134] Z. Sun, S.S. Ge, T.H. Lee, Controllability and reachability criteria for switched linear systems, Automatica 38 (2002) 775–786.
[135] C. Seatzu, D. Corona, A. Giua, A. Bemporad, Optimal control of continuous-time switched affine systems, IEEE Trans. Automat. Control 51 (2006) 726–741.
[136] A.V. Savkin, R.J. Evans, Hybrid Dynamical Systems: Controller and Sensor Switching Problems, Birkhäuser, 2002.
[137] L. Meier, J. Peschon, R.M. Dressler, Optimal control of measurement subsystems, IEEE Trans. Automat. Control 12 (5) (1967) 528–536.
[138] Y. Oshman, Optimal sensor selection strategy for discrete-time state estimators, IEEE Trans. Aerosp. Electron. Syst. 30 (2) (1994) 307–314.
[139] D.P. Bertsekas, Dynamic Programming: Deterministic and Stochastic Models, Prentice-Hall, Englewood Cliffs, NJ, 1987.
[140] W.B. Powell, Approximate Dynamic Programming: Solving the Curses of Dimensionality, Wiley, 2007.
[141] A. Rantzer, Relaxed dynamic programming in switching systems, IEE Proc. 153 (2006) 567–574.
[142] N.A. Lynch, Distributed Algorithms, Morgan Kaufmann, 1996.
[143] H. Li, Virtual queue based distributed data traffic scheduling for cyber physical systems with application in smart grid, in: Proceedings of the Second International Workshop on Cyber Physical Networking Systems (CPNS), 2012.
[144] B.D.O. Anderson, J.B. Moore, Optimal Filtering, Prentice Hall, 1979.
[145] A.I. Zečević, D.D. Šiljak, Control of Complex Systems: Structural Constraints and Uncertainty, Springer, Berlin, 2010.
[146] H. Li, J. Song, Q. Zeng, Adaptive modulation in networked control systems with application in smart grids, IEEE Commun. Lett. 17 (7) (2013) 1305–1308.
[147] O.L.V. Costa, M.D. Fragoso, M.G. Todorov, Continuous-Time Markov Jump Linear Systems, Springer, New York, 2013.
[148] O.L.V. Costa, M.D. Fragoso, Stability results for discrete-time linear systems with Markovian jumping parameters, J. Math. Anal. Appl. 179 (1993) 154–178.
[149] I. Matei, N.C. Martines, J.S. Baras, Optimal linear quadratic regulator for Markovian jump linear systems in the presence of one time-step delayed mode observations,

in: Proceedings of the 17th Wolrd Congress of the International Federation of Automatic Control, 2008.
[150] D.L. Neuhoff, R. Gilbert, Causal source codes, IEEE Trans. Inform. Theory 28 (5) (1982) 701–713.
[151] N.T. Gaardar, D. Slepian, On optimal finite-state-digital transmission systems, IEEE Trans. Inform. Theory 28 (2) (1982) 167–186.
[152] J.C. Walrand, P. Varaiya, Optimal causal coding-decoding problems, IEEE Trans. Inform. Theory 29 (6) (1983) 814–820.
[153] V.S. Borkar, S.K. Mitter, S. Tatikonda, Sequential vector quantization of Markov sources, SIAM J. Control Optim. 40 (1) (2001) 135–148.
[154] H. Li, Z. Han, Distributed source coding for controlling physical systems with application in smart grid, in: IEEE Conference on Global Communications (Globecom), 2014.
[155] D. Slepian, J.K. Wolf, A coding theorem for multiple access channels with correlated sources, Bell Syst. Tech. J. 52 (1973) 1037–1076.
[156] J. Garcia-Frias, F. Cabarcas, Approaching the Slepian-Wolf boundary using practical channel codes, Signal Proces. 86 (2006) 3096–3101.
[157] T. Matsuta, T. Ueymatsu, R. Matsumoto, Universal Slepian-Wolf source codes using low-density parity-check matrices, in: IEEE International Symposium of Information Theory, 2010.
[158] S.S. Pradhan, K. Ramchandran, Distributed source coding using syndromes (DISCUS): design and construction, in: Proceedings of the IEEE International Data Compression (DCC), 1999.
[159] S.S. Pradhan, K. Ramchandran, Generalized coset codes for distributed binning, IEEE Trans. Inform. Theory 51 (10) (2005) 3457–3474.
[160] V. Stankovic, A.D. Liveris, Z. Xiong, C.N. Georghiades, Code design for the Slepian–Wolf problem and lossless multiterminal networks, IEEE Trans. Inform. Theory 52 (4) (2006) 1495–1507.
[161] Z. Liu, S. Cheng, A.D. Liveris, Z. Xiong, Splepian-Wolf coded nested lattice quantization for Wyner-Ziv coding: high-rate performance analysis and code design, IEEE Trans. Inform. Theory 52 (10) (2006) 4358–4379.
[162] R. Zamir, S. Shamai, U. Erez, Nested linear/lattice codes for structured multiterminal binning, IEEE Trans. Inform. Theory 48 (6) (2002) 1250–1276.
[163] P.L. Dragotti, M. Gastpar, Distributed Source Coding: Theory, Algorithms and Applications, Academic Press, 2008.
[164] I. Conway, J. Sloane, Sphere Packing, Lattices and Groups, Springer, New York, 1998.
[165] S. Gong, H. Li, Decoding the nature encoded messages for distributed energy generation in microgrid, in: Proceedings of the IEEE International Conference on Communications (ICC), 2011.
[166] Open Smart Grid (OpenSG), SG Network System Requirements Specification, tech. rep. 2010.
[167] D. Simon, Optimal State Estimation: Kalman, H Infinity, and Nonlinear Approaches, Wiley, 2006.
[168] J. Pearl, Probabilistic Reasoning in Intelligent Systems: Networks of Plausible Inference, Morgan Kaufmann, San Mateo, CA, 1988.
[169] R. Ostrovsky, Y. Rabani, L. Schulman, Error-correcting codes for automatic control, in: Proceedings of the 46th Annual IEEE Symposium on the Foundations of Computer Science (FOCS), 2005.

[170] R.T. Sukhavasi, B. Hassibi, Anytime reliable codes for stabilizing plants over erasure channels, in: Proceedings of the IEEE Conference on Decision and Control (CDC), 2011.
[171] I. Sason, S. Shamai, Performance analysis of linear codes under maximum-likelihood decoding: a tutorial, Future Trend Commun. Inform. Theory 3 (2006).
[172] L. Dossel, L.K. Rasmussen, R. Thobaben, Anytime reliability of systematic LDPC convolutional codes, in: Proceedings of IEEE International Conference on Communications (ICC), 2012.
[173] A. Tarable, A. Nordio, R. Tempo, Anytime reliable LDPC convolutional codes for networked control over wireless channel, in: IEEE International Symposium on Information Theory (ISIT), 2013.
[174] L.J. Schulman, Coding for interactive communication, IEEE Trans. Inform. Theory 42 (6) (1996) 1745–1756.
[175] R. Gelles, A. Moitra, A. Sahai, Efficient coding for interactive communication, IEEE Trans. Inform. Theory 60 (2014) 3.
[176] M. Braverman, Towards deterministic tree code constructions, Tech. Rep. 64, Electronic Colloquium on Computational Complexity, 2011.

Index

Note: Page numbers followed by *f* indicate figures and *t* indicate tables.

Printed in the United States
By Bookmasters